JAN 2019

D1591677

The Man Who Saw Tomorrow

The Man Who Saw Tomorrow

The Life and Inventions of Stanford R. Ovshinsky

Lillian Hoddeson and Peter Garrett

The MIT Press
Cambridge, Massachusetts
London, England

© 2018 Massachusetts Institute of Technology

All rights reserved. No part of this book may be reproduced in any form by any electronic or mechanical means (including photocopying, recording, or information storage and retrieval) without permission in writing from the publisher.

This book was set in ITC Stone Sans Std and ITC Stone Serif Std by Toppan Best-set Premedia Limited. Printed and bound in the United States of America.

Library of Congress Cataloging-in-Publication Data

Names: Hoddeson, Lillian, author. | Garrett, Peter K., author.
Title: The man who saw tomorrow : the life and inventions of Stanford R. Ovshinsky / Lillian Hoddeson and Peter Garrett.
Description: Cambridge, MA : The MIT Press, [2018] | Includes bibliographical references and index.
Identifiers: LCCN 2017032336 | ISBN 9780262037532 (hardcover : alk. paper)
Subjects: LCSH: Ovshinsky, Stanford R. | Electrical engineers--United States--Biography. | Inventors--United States--Biography.
Classification: LCC TK140.O97 H63 2018 | DDC 621.3092 [B] --dc23 LC record available at https://lccn.loc.gov/2017032336

10 9 8 7 6 5 4 3 2 1

Contents

Preface (by Lillian Hoddeson)   vii
Acknowledgments   xiii
Image Credits   xvii

Introduction   1

I   Becoming an Inventor   13

1   Young Years (1920s–1930s)   15
2   Passion for Machines (1940–1944)   29
3   Smarter Machines (1944–1952)   45
4   Love Story (1950s)   65
5   New Beginnings in the Storefront (1960–1964)   93

II   Inventing with Others   121

6   The Birth of ECD: An Invention Factory (1965–1979)   123
7   The ECD Community: A Social Invention (1965–2007)   147
8   Solar Energy: Working at the Edge of Feasibility (1979–2007)   171
9   Hydrogen and Batteries: The Genie and the Bottle (1980–2007)   187
10   Information: Displays and Memory Devices (1981–2007)   209
Interlude: Science, Art, and Creativity   225

III   Later Years   239

11   Losing Iris, Losing ECD   241
12   New Love, New Company   253

13   Last Days   267
Epilogue: Deaths, Survivals, and Revivals   285
Conclusion   291

Appendix I: Interviews   297
Appendix II: Ovshinsky's Major Honors and Distinctions   301
Notes   303
Index   357

Preface (by Lillian Hoddeson)

It took me a long time working as a historian of science to come to the subject of this book, though in retrospect I realize that I might have come to it much sooner. In the mid-1980s, I was surprised by a postcard from the eminent physicist Sir Nevill Francis Mott, who had shared the 1977 Nobel Prize for research on magnetic and disordered systems. Having recently heard me speak on a topic in the history of solid-state physics at a small meeting he had organized in London, Mott suggested that I consider writing about the new area of amorphous and disordered solids.[1]

I simply did not know what to do with this suggestion. Amorphous and disordered materials—materials lacking the rigidly ordered structure of crystals—were already receiving increasing attention from physicists but not from historians. I did not have the background I thought necessary to open up this new historical research area on my own, and in any case I was already over my head with co-editing a massive history of the entire field of solid-state physics.[2] Had I taken Mott's suggestion I would have encountered the work of Stanford R. Ovshinsky two decades earlier than I did, but any history I might have written then would have been more narrowly focused and less engaging than this biography.

When I later came to write about Ovshinsky, it was at the suggestion of a different senior physicist. The adventure started in December 2004, when Peter Fritzsche, my colleague in the History Department at the University of Illinois, gave a copy of my recently published biography of the physicist John Bardeen to his father as a Christmas present.[3] Hellmut Fritzsche, a condensed matter physicist who had been working with Ovshinsky for decades (as, indeed, had Mott), got in touch with me a few weeks later. He said he had enjoyed the Bardeen biography but had an even better subject to suggest for my next book. I had never heard of Ovshinsky, nor (again) did I have time to spare from other commitments.

But Fritzsche did arouse my curiosity: he told me that Ovshinsky, whose formal education ended with high school, had made crucial discoveries leading to the growth of a new branch of materials science, and that he was using his new materials to create alternative energy technologies aimed at reducing global warming. Fritzsche's anecdotes—such as ones about how travelling with Ovshinsky could get complicated because, true to his union roots, he would never cross a picket line—provided intriguing glimpses into Ovshinsky's personality. "This is a fascinating story waiting to be told," Fritzsche wrote to me a few days later. Eventually, after almost a year, I gave in to my curiosity and visited Ovshinsky's laboratory, Energy Conversion Devices (ECD), in a northern suburb of Detroit.

On that visit I met Stanford Ovshinsky as well as Iris Ovshinsky, his wife and partner of some fifty years, who was constantly at his side. Both were intensely serious about promoting clean energy, and they proudly showed me examples of ECD's recent achievements. I was impressed by the company's huge 25-megawatt solar panel machine (Iris whispered to me, "You're supposed to say *wow!*").[4] And it was a thrill to drive ECD's converted Toyota Prius, which ran on hydrogen and was certainly the quietest and cleanest car I had ever handled. Hearing Ovshinsky describe his vision of an energy economy based on solar and hydrogen power, I felt as if that future had already arrived. I also realized that I was in the presence of a genius quite different from any of the outstanding scientists I had known or written about previously. For one thing, his aim was not simply explaining the world. He hoped to change it.

I was soon convinced that Ovshinsky's life was worth examining. His many inventions (he had been awarded over four hundred patents) and the broader applications of his materials in the areas of energy and information made him a significant figure in the history of technology. Working long after the era of great independent inventors like Edison and Bell, he showed what an exceptional individual could still accomplish in a time when so much innovation and discovery came from industrial and university laboratories.[5] My interest in Ovshinsky also involved more general questions about scientific and technological creativity. Writing about Bardeen had led me to approach scientific thinking from the perspective of cognitive psychology, and I was just then examining the scientific use of analogy, which seemed to play an even larger role in Ovshinsky's work than in Bardeen's. I became eager to understand how Ovshinsky's extraordinary mind worked and how his life experience had contributed to his inventions.

I still hesitated to take on the task of writing about Ovshinsky because of my other commitments. At the same time, knowing all too well how impermanent living sources are, I didn't want to lose the opportunity to record Ovshinsky telling his own story. I decided to conduct a few in-depth oral history interviews and turn them over to one of my former collaborators, who had expressed interest in using the materials to write a book. The interviewing began in January 2006.

Half a year later, my relationship to Ovshinsky suddenly shifted. My presence at Iris's tragic death (described in chapter 11) involved me more closely in his life, and I decided to write his story myself.

That was ten years ago. I believed then that the work would not take very long, because there was plenty of documentation accessible in the files at ECD, and more in the Ovshinsky home. Most of the characters were alive and seemed eager to help. Soon I was visiting ECD every few months, sifting through papers, conducting interviews with Ovshinsky and his colleagues, drafting preliminary chapters, and occasionally sharing my thoughts about his work with colleagues in seminars, conference papers, or colloquia.

But my expectations turned out to be much too optimistic. The largest cache of documents, those at ECD, would be destroyed by the time I needed to use them (for reasons explained in chapters 11 and 12). Moreover, Ovshinsky turned out to be the most difficult interview subject I had encountered in thirty-some years of conducting oral history interviews. His memory was still vivid despite his age (he was already over eighty). But not only was his memory highly selective (as all human memory is), his stories had been repeated so many times that they had hardened into a mask of the image he wanted to project.[6] My efforts to penetrate this mask and get a more complex account typically just made him angry.

Ovshinsky also seemed to lack a sense of his own intellectual development. When I would ask him to describe how some line of thought had evolved, or when he had come to some realization, his usual answer was "I always knew." I suspect that his prodigious early, independent learning, combined with his ability to quickly recognize possibilities, made him believe that he had indeed always known what he later understood.

Even more frustrating was his inability to explain anything—not even how one of his own inventions worked—in terms I could understand. He would good-naturedly go on at length in his attempted explanations, but the more he spoke the more confused I became. For some years I thought this confusion was owing to my lack of background, but in time I learned that nearly everyone he worked with had trouble understanding him. His mind raced so quickly that he would leave out crucial connections and often skip entire sentences as he flitted from topic to topic. Iris, in contrast, had been easy to interview: clear, consistent, and straightforward. Sitting in on my early sessions with her husband, she was a considerable asset, for she could push him to be clearer and stay focused. It is impossible to overstate the immense setback her death dealt to the research for this book, though it was also the very fact of her death that brought me to commit to writing it.

Another difficulty that caused considerable delay came from Ovshinsky's opposition to my interviewing many former ECD consultants or staff members he feared might give unflattering accounts. I ended up postponing most of those interviews until after his death in October 2012. In addition to much useful information, they did indeed include critical comments, but those also helped me get a fuller sense of Ovshinsky's character. This book is thus by no means the completely laudatory biography that Ovshinsky hoped I would write, but in recognizing many of his human failings in addition to his impressive achievements it better represents the range of Ovshinsky's brilliance and humanity. I hope that had he lived to see it, he would have come to appreciate that.

Another obstacle, which turned into an advantage, came when I fell ill in December 2012. For a short time I didn't think I would be able to complete the book, and I asked my husband Peter Garrett, a retired English professor, whether if the worst happened he could bring the book to press. He offered instead to help complete it if I managed to recover. I did, and Peter became my co-author. Bringing his forty years of writing and interpretative experience to bear on the unfinished manuscript has made this biography a deeper and more intricate representation of Ovshinsky's life and work.

Completing the book required dealing with several problems, and some did not allow simple solutions. One unexpectedly complicated issue was what to call our protagonist. Because everyone he knew or worked with, even his critics, addressed him as Stan (at his insistence!), we initially did that throughout the manuscript. Then two anonymous reviewers objected to our calling him Stan, which seemed too informal and also made us seem too closely aligned with his perspective. Yet using "Ovshinsky" to tell about his private life also felt awkward. We finally opted for the compromise of referring to him as Stan when dealing with his personal experience (family life, marriages, and the like) but using Ovshinsky when narrating his career as an inventor. The organization of the book involves similar compromises, such as alternating (even within some chapters) between Ovshinsky's personal and his professional experience, and sometimes shifting the focus away from him to include others' experiences and contributions.

To mark the different phases of Ovshinsky's story, we have divided the narrative into three parts. In part I we tell the story of how Ovshinsky became an inventor. Beginning with an account of his family background and early years, we trace his transition from machinist to independent inventor (chapters 1 and 2). We then focus on his first major invention, the Benjamin Lathe; his efforts to automate this massive machine tool led him to study cybernetics and neurophysiology, and ultimately to invent his first amorphous switch, the Ovitron (chapters 3 and 4). After his life became linked with Iris Dibner's, the two founded their company, Energy Conversion Laboratories (ECL, later renamed ECD), dedicated to using science and technology to address social problems. With that began the most creative period of Ovshinsky's inventive career, resulting in his first chalcogenide switches and phase-change memory devices (chapters 4 and 5).

In part II we trace the major collaborative work of ECD, an unusual institution that was both a fruitful research and development laboratory and a social invention based on Stan and Iris's progressive values (chapters 6 and 7). Three short technical chapters cover simultaneous teamwork at ECD devoted to solar energy (chapter 8), energy storage (chapter 9), and information technology (chapter 10).

After a brief interlude sampling Ovshinsky's art and ideas about creativity, we present his later years in part III. Devastating losses came with Iris's death and Ovshinsky's forced departure from ECD (chapter 11). He achieved personal and professional recovery in his autumnal love story and marriage to the physicist Rosa Young, and in his last bold, ambitious, and unfinished solar energy project (chapter 12). Finally, we trace the events of his last year (chapter 13).

In the brief epilogue we summarize the later development of Ovshinsky's major projects through 2016, and then offer some concluding reflections.

Acknowledgments

Among the institutions that supported this work, I particularly want to thank the Department of History of the University of Illinois at Urbana-Champaign. Besides providing this book's primary institutional home, with ample encouragement, criticism, and countless forms of administrative help, the department offered me repeated periods of released time from teaching to work on it. The Department of Physics provided a supportive secondary home. Most important were interactions with colleagues and students who helped me think about the meaning of the project. Thanks especially to the university's Center for Advanced Study (CAS) and its Illinois Program for Research in the Humanities (IPRH) for supporting released time and student assistants. I am grateful to Tom Siebel for endowing the Siebel Chair in the History of Science at the University of Illinois, which I was honored to hold during much of the time I spent working on this book, and which provided support for travel and research assistance. Special thanks go to the Department of History's business manager Tom Bedwell for coordinating the book's complex budgets.

For extensive help with finding and copying documents and photos, we are most indebted to: Rosa Young Ovshinsky, Ovshinsky's widow; Freya Saito, Ovshinsky's dedicated assistant at ECD and later at Ovshinsky Innovation; Kimi Williams, who continued Freya's efforts for a time after Ovshinsky's death; Irina Youdina, the Ovshinskys' devoted caregiver, housekeeper, and friend in his last five years; and the physicist Genie Mytilineou, Rosa's close friend. Thanks to the Bentley Historical Library at the University of Michigan, where the Ovshinsky papers are now preserved and available to researchers, and especially to Shae Rafferty for her help making the collections accessible to us. Thanks to my ever helpful friend Tracy McAllister for expertly processing photos and helping to prepare and circulate interim drafts to many readers during the later stages of this work.

For the many opportunities I've had to share parts of this work while it was in progress, I would especially like to thank the following additional institutions: the

Lemelson Center for the Study of Invention and Innovation at the Smithsonian Institution, especially Art Molella and Joyce Bedi, whose conversations about invention and whose enthusiasm for this project were most helpful in the early stages of conceiving this book; the Department of the Philosophy and History of Science at the University of Athens, especially Professor Stella Vosniadou; the International History of Science Society; the Society for the History of Technology; members of the Laboratory History Conference; the Physics Department of the University of Minnesota; the History and Philosophy of Science Department of Notre Dame University; the Physics Department of the University of California at San Diego; the Cohn Institute for the History and Philosophy of Science and Ideas in Tel Aviv; the Van Leer Institute in Jerusalem; the History of Science Society; the Center for History of Physics of the American Institute of Physics; and the Consortium for the History of Science, Technology, and Medicine.

Given the enormous gaps in the existing physical documentation for this book, interviews (despite the hazards of using them) proved as important as documents in writing this manuscript. We are deeply indebted to the hundreds (most listed in appendix I) who offered their time and precious memories in interviews for this book. Without these contributions, our story would have been greatly impoverished. We are especially grateful to those who agreed to be interviewed repeatedly. Thanks also to those who worked to transcribe the interviews: Janet Abrahamson, Kathryn Dorsch, Kristen Ehrenberger, Ryan Jones, Tracy McAllister, Melissa Rohde, Erin Sullivan, and in the last three years, the anonymous and efficient transcription staff employed by Rev.com.

Thanks to Mike Lehman and Kristen Ehrenberger, then history graduate students and now PhD historians, who in the course of their research assistantships helped me to learn basic concepts (e.g., of climate change, renewable energy, and amorphous solids) that lay outside my training and professional experience. Thanks also to my daughter Carol Baym for clarifying certain neurological concepts.

Many of our interviewees helped greatly with our struggle to grasp new technical concepts essential to Ovshinsky's work, but three worked especially hard to educate us: Hellmut Fritzsche, Rosa Young Ovshinsky, and Guy Wicker. We cannot thank them enough. To understand Stan's personal life, we are indebted to many of his friends and especially to members of the Ovshinsky clan (Rosa, Herb, Ben, Harvey, and Dale Ovshinsky, and Robin and Steven Dibner). Special thanks go to Rosa, Herb, Ben, and Harvey for taking the time to read multiple drafts. Thanks also to Harvey for his extensive and energetic help with shaping the narrative, drawing on his own memoir, *Detroit and the Art of the Impossible*, as well as his personal archives from his decades of helping

Acknowledgments

Stan tell ECD's story in videos, brochures, annual reports, and the like. A list of Harvey's interviews used in writing this book is included in appendix I. Harvey also provided our title, *The Man Who Saw Tomorrow*.

Many others also read our drafts and offered comments and corrections. They include Nancy Bacon, Dennis Corrigan, Ben Chao, Ruth Cowan, Subhash Dhar, Elif Ertekin, Subhendu Guha, Steve Hudgens, Chet Kamin, Alex Kolobov, Boil Pashmakov, Benny Reichman, Rita Smith, Srinivasan Venkatesan, Zvi Yaniv, and the anonymous reviewers who improved the manuscript by pointing out problems in an earlier draft. Many thanks also to the excellent staff at the MIT Press, including our copy editor Mary Bagg, our production editor Marcy Ross, and especially our acquisition editor Katie Helke, who worked patiently with us and the reviewers to bring this book to press. To all those who provided feedback, we want to say that while we have not taken all their suggestions, we are grateful for their help and hope that we have made the best use of their information to improve the book.

Last but not least, thanks to my extraordinary "extended family" that supported my work on this book. It is impossible to adequately thank my loving husband Peter for stepping in to help me complete the work at a point when doing so alone seemed impossible. As it happened, his efforts proved to be just what was needed to turn my crammed and awkwardly written manuscript into a much more readable representation of a fascinating life. And it was really fun!

Lillian Hoddeson
September 2016
Urbana, Illinois

Image Credits

Figure 3.1: Used by permission from David G. Kanzeg.

Figure 5.6: Adapted from Richard Zallen, *The Physics of Amorphous Solids* (1983), 12. Copyright Wiley-VCH Verlag GmbH&Co. KGaA. Reproduced by permission.

Figure 6.1: From the *New York Times*, November 11, 1968, 1. All rights reserved. Used by permission and protected by the Copyright Laws of the United States. The printing, copying, redistribution, or retransmission of this Content without express written permission is prohibited.

Figures 7.7 and 13.11–13.14: Used by permission from Lillian Hoddeson.

Figures 2.2, 3.6a, 4.5, 9.1, 9.4, and 12.3: Public domain.

All other figures: Used by permission from the Ovshinsky Family and the Bentley Historical Library, University of Michigan.

Introduction

Stanford Robert Ovshinsky not only saw tomorrow, he helped make it happen.[1] His inventions and discoveries led to many developments in later twentieth- and early twenty-first-century technology that we now take for granted but that seemed like science fiction when he foresaw them decades earlier. To take just one emblematic example, in 1968 he said we would one day have flat television screens that could be hung on the wall like a picture, a prediction that most electronics experts of the time scornfully dismissed. But Ovshinsky could confidently make it because he foresaw how the flat screens would be made from the new kind of materials whose possibilities he had discovered.

That prediction, along with another anticipating the creation of affordable "small, general purpose desktop computers for use in homes, schools and offices," was made almost incidentally in a press conference announcing Ovshinsky's creation of a new kind of electronic switch, resulting in a front-page *New York Times* article, "Glassy Electronic Device May Surpass Transistor," published on November 11, 1968.[2] The story sent shock waves through not only the world of commercial electronics but also the world of academic science. Ovshinsky's announcement was met by outraged denunciations, both from big corporations that were economically invested in transistor technology and from established scientists who were intellectually invested in theories that could not account for this new glassy device, or for any semiconductor device that wasn't made from crystalline materials. Yet on the same day as the *Times* story, Ovshinsky's discovery was published in one of the most prestigious physics journals, *Physical Review Letters,* giving scientific credibility to his unforeseen findings.[3]

Ovshinsky also played a key role in creating the rechargeable nickel metal hydride (NiMH) batteries that have powered everything from portable electronics to hybrid cars. He invented a system for mass-producing affordable thin-film solar panels, and he is furthermore responsible for rewritable CDs and DVDs and the electrical phase-change memory that is poised to enable the next advance in computer architecture.

All these inventions and many more followed from his use of amorphous and disordered materials, whose possibilities he continued to explore and exploit throughout the rest of his career.[4] As "the master chef of the periodic table," he experimented with combining as many as eleven elements to tailor the properties of his new materials, using them in his own inventions and, by showing their possibilities, also contributing importantly to the growing field of materials science.[5]

No wonder *The Economist* called Ovshinsky "the Edison of our age."[6] Like Edison, he moved from being a solitary inventor to creating a large research and development laboratory, Energy Conversion Devices (ECD), and to manufacturing several of his inventions. Also like Edison, Ovshinsky not only produced important new individual technologies but also linked them in technological systems.[7] Just as Edison's light bulb was part of a system that included power generation and transmission, the basis for mass electrification, so Ovshinsky conceived his energy technologies as part of a continuous system he called the "hydrogen loop."

Yet even the comparison with Edison, which several others have also made, doesn't capture all of Ovshinsky's achievement. As Robert R. Wilson, the Manhattan Project physicist and founding director of Fermi National Laboratory, pointed out, "Edison was primarily an inventor. There is a larger component of pure science in Stan than there was in Edison."[8] For unlike Edison, Ovshinsky made his discoveries on the frontiers of late twentieth-century physics, manipulating molecular structures and envisioning the new materials they could yield, learning "to put together something that nature hasn't done."[9] And perhaps most remarkably, he did all this without any advanced training: his formal education ended with high school. He drew instead on his lifelong voracious reading, his hands-on experience, and his penetrating intuitions.

Ovshinsky was distinguished not only by his inventive genius but also by the purpose that came to guide it. Unlike most successful innovators and entrepreneurs, he was not concerned with empire building or getting rich. Instead, while always intent on commercial success, he believed that technology should tackle important social problems, and he focused his inventive efforts on finding solutions to those he considered the most urgent. When in 1960 he and his partner Iris Miroy Dibner began a research company, they called it Energy Conversion Laboratories because they considered energy a crucial social problem, and as the company grew it developed ways to replace fossil fuels with alternative energy sources—with solar power (and batteries to store it), with hydrogen in fuel cells or internal combustion engines. To Ovshinsky, the information technologies the company also developed were another way of addressing social problems; he envisioned a world where greater access to information would empower citizens and promote democracy.

Driven by this vision of a better future, Ovshinsky also had the charismatic power to inspire others and gain their support, though at every stage he also met strong resistance. Ovshinsky was always a controversial figure, an outsider who many dismissed or mistrusted, though many others, including the most gifted, recognized his genius. But even most who admired him did not know the whole story of this remarkable man.

Becoming an Inventor

Ovshinsky is best known for inventions that built on his pioneering work in materials science, but that work came after he had already been an inventor for nearly twenty years, following a twisting path. He began as a machinist and toolmaker in the machine shops and factories of Akron, Ohio, where he was born in 1922, and his first invention, made in the mid-1940s, was an advanced machine tool, a high-speed automatic lathe (chapters 2 and 3). His development of other kinds of automation and control devices later took him to Detroit in 1951, where he created new automotive components such as an electrical automatic transmission and power steering (chapter 4).

These devices all used sensors and feedback mechanisms. To gain a deeper understanding of such processes, Ovshinsky plunged into the scientific literature on neurophysiology, following the lead of Norbert Wiener's cybernetics, which considered "control and communication in the animal and the machine" in the same terms.[10] Ovshinsky not only studied but also contributed to the field, and on the basis of his writings on nerve impulses and intelligence, he was invited by Wayne State University Medical School to join in pioneering brain research (chapter 4).

Ovshinsky's scientific investigation of how the brain sends and receives signals was both a departure from his work as an inventor and a way to advance it. He considered the nerve cell as a switch that will fire when incoming signals accumulate and reach a threshold, releasing energy through the cell's semipermeable membrane. To model this mechanism, he created a device he called his "nerve cell analogy," in which the analogue for the cell membrane was the thin film of oxide on two strips of tantalum immersed in an electrolytic solution. The result was a new kind of electrochemical switch he named the Ovitron (chapter 4).

Another twist came when, prevented from developing the Ovitron by the settlement of a lawsuit, Ovshinsky had to find new materials for his switches. (His career was punctuated by many legal disputes, mostly resolved in his favor.) His search led him to study the electronic properties of amorphous and disordered (non-crystalline) materials, particularly chalcogenide glasses. (These are compounds of chalcogen elements such as selenium and tellurium. See the fuller discussion in chapter 5.) Making this choice showed Ovshinsky's willingness to follow his independent intuitions, for in the late 1950s nearly all researchers and manufacturers focused on crystalline materials for microelectronics. Ovshinsky, however, sensed that amorphous materials offered more possibilities. Working in the newly established Energy Conversion Laboratories, housed in a modest Detroit storefront, he experimented with combining various elements and compounds, grinding powders like a modern alchemist and pressing them into thin layers.

This solitary work led to Ovshinsky's crucial breakthrough, his discovery in 1961 of the new switching mechanism that is now called the Ovshinsky effect (chapter 5).[11] Thin films of variously composed disordered materials formed the basis for the reversible action of his threshold switch and phase-change memory, inventions whose importance for information technologies is still growing. These were the culminating achievements of Ovshinsky's work as an independent inventor, and they were also the pivot on which his career turned toward more collaborative creation.

Inventing with Others

With the growing recognition of his discoveries, as well as the growing revenues from licensing them, Ovshinsky expanded his operations. In 1964 the company was renamed Energy Conversion Devices to signal its increased commercial orientation, and in early 1965 it moved to a much larger building in the northern Detroit suburbs (chapter 6). Recruiting an increasing cohort of scientific consultants and hiring many highly trained researchers, Ovshinsky developed ECD into a productive research and development company, supported by both public and private funding.

With this support, and with a staff that eventually numbered over a thousand, Ovshinsky pursued several concurrent R&D programs in energy and information technologies. Some of these grew into significant manufacturing operations. The program in thin-film amorphous silicon solar cells, and Ovshinsky's revolutionary technology for making them "by the mile," grew into the United Solar subsidiary that became for a time the largest US producer (chapter 8). The program in hydrogen storage unexpectedly yielded the NiMH battery, still in wide use (chapter 9). Both these energy technologies were the result of Ovshinsky's collaboration with the many scientists at ECD who worked over decades to realize his ideas and continually improve the devices. The inspiration and direction always came from him, but the work of innovation was now teamwork.

In ECD's information programs, the capacity of amorphous semiconductors to cover large areas that had been exploited in the thin-film solar panels also enabled the construction of flat panel displays, just as Ovshinsky had predicted in 1968. Though

ECD was a pioneer in developing this technology, the Japanese and Korean electronics giants, who in fact had help from ECD in getting started, ended up dominating the multibillion-dollar industry (chapter 10).

Phase-change memory, an offshoot of Ovshinsky's original threshold switch, became the basis for optical devices in the late 1960s after he proposed using a laser to set and reset his amorphous switching material. The resulting rewritable CDs and DVDs were the first successful commercialization of the Ovshinsky effect, though the revenue from licensing the technology was never large. More important but much longer in gestation were electrical memories based on Ovshinsky's chalcogenide alloys. Although these had several advantages over silicon-based memories, the computer industry saw no compelling reason to make the large investments required to adopt them. Recently, however, announcements of new phase-change devices by several major manufacturers make it seem likely that this invention will play an increasingly important role (epilogue).

The Intuitive Mind

Ovshinsky's early independent work and later collaborative inventions both arose from the qualities of his exceptional mind. In some ways, his lack of formal scientific training beyond high school was a disadvantage, and he had to rely on others, first Iris and then his scientific consultants, to help him communicate his ideas. But in more important ways it was a great advantage. He was from the beginning self-educated, and his life-long, wide-ranging reading gave him an enormous and diverse store of knowledge to draw on. Many who knew him marveled at how quickly he could read and later recall everything; others were struck by his ability to deal with several issues at the same time—multiple phone calls, simultaneous meetings, or interrupted conversations—without losing track of any. Chester Kamin, Ovshinsky's long-time attorney and adviser, observed, "His mind was capable of processing on all these different tracks, and he would actually be working on all the issues at the exact same time. It really is an astounding ability." As Ovshinsky explained, this ability also fueled his creative process. "At any one time I have four or five deep things I'm thinking about simultaneously, and they feed upon each other. I'll read a book or paper or journal and see something that has no obvious connection to what I'm looking for. That will spring another idea into my head. Then I start putting things together, and then I come up with something."[12]

Instead of proceeding logically step by step, the process of "putting things together" often depended on seeing unexpected analogies. "When I do a new problem," he noted, "I have much more to draw on in my mind in terms of analogies or things that other people are not associating at all, things that nobody else would think there are any connections to." And "that's where new invention, new discoveries, new science comes from," he believed.[13] A crucial instance of such creativity is Ovshinsky's "nerve cell analogy," the Ovitron. Disregarding the obvious differences between the organic and inorganic to focus on the structural similarities between the nerve cell membrane and the thin film of oxide enabled him to invent a new kind of switch, an essential step in his inventive career.

Ovshinsky's innovations also drew on his highly developed visual imagination, which gave him a way to grasp the structure of atoms and molecules and sense their possible combinations. "I *see* electrons," he would say. "I *feel* atoms. I know what they want to do." This sounds at first like a grandiose claim to unique special powers, but like his use of analogies it was just a heightened form of a common cognitive strategy that served as an alternative to the more formal and abstract approaches of trained scientists. Richard Flasck, a physicist who worked for several years as an ECD researcher, recognized that "Stan thought in pictures, not in numbers and principles, and sometimes that gives insight that you can't get from standard mathematics."[14]

One example of the insight that Ovshinsky's idiosyncratic visual equivalents for standard scientific formulations could yield comes from Arthur Bienenstock, a professor of physics at Stanford and one of Ovshinsky's early scientific consultants. He recalled a time when they had talked about the structure of germanium telluride. "And Stan drew these pictures on the blackboard, little drawings, squiggles of chains of telluriums and germaniums. If Stan had gotten up and given a talk on that at a scientific meeting, no one would have paid attention to him. But I come along and I take those pictures and I put some mathematics to them and I do some x-ray diffraction and they're well received. But those pictures of Stan's were central and key to the paper. It was critical what he contributed to the thing. And they turned out to be right."

Besides illustrating Ovshinsky's scientific insights, this anecdote also shows why he often had trouble conveying them. An unsympathetic audience at a scientific meeting might well have dismissed his "squiggles," but even those who wanted or needed to understand what he was saying could become exasperated by his frequently opaque and tangled efforts to express his thoughts. He needed his scientific consultants to communicate his insights effectively. As another Stanford scientist, John Ross, said, "Stan is a genius, but he's not a scientist. Stan knows that certain things are correct, but he can't possibly tell you why. He can't write an equation. He feels science; he can't explain it to you."

Ovshinsky may not have been able to write equations, but sometimes his way of feeling science could reveal possibilities that escaped highly trained and accomplished scientists. Here is another example from Arthur Bienenstock, who told of a discussion with a group of physicists.

> The rest of them were arguing with Stan as to whether quantum mechanical tunneling could be an important mechanism in some of the materials of interest. And Stan was saying yes, and they were saying no, and they said no because they said that the relevant interatomic distances are too long for tunneling to be a factor. And Stan listened to this for a second, and he said, "You're thinking statically. Remember that the tunneling probability is very strong. It has a very, very strong dependence on interatomic distance. And when the atoms vibrate, in the brief period when they are closer together, the tunneling probability would go way up, and therefore you could have tunneling." And I remember that I thought it was remarkable that a man with only a high school education could invent that on the spot. Stan saw it intuitively and I just thought it was evidence of his enormous, educated intuition.

Stan the Man

Behind Ovshinsky's achievements as an inventor and the qualities of his brilliant mind were the formative experiences that gave him the values and character traits that shaped his career. Both his vocation and his social values owed much to the influence of his father, Ben, who was a scrap-metal collector and took the young Stan with him to the machine shops and factories where he worked. It was there that Stan began to sense what he called the "glamour" of manufacturing and to feel the passion for machines that lasted all his life. Ben was also a highly cultured radical activist and took Stan to meetings of the Workmen's Circle, a fraternal organization dedicated to promoting social justice and creating "a better and more beautiful world."[15] There Stan was exposed to the progressive political culture of working-class Eastern European Jews, which fed his commitment to causes like labor and civil rights. While still in school, he was a leader in the Young Peoples' Socialist League, and when he began working he quickly became involved in union struggles (chapters 1 and 2).

Not until Ovshinsky joined his life with Iris Miroy Dibner's, however, did these values begin to direct his work as an inventor. When they met in the early 1950s, they were both already married with children, and from the beginning of 1955, when they realized they had fallen in love, until late 1959, when they could at last be together, they talked and wrote to each other constantly about their beliefs and goals (chapter 4). Iris had her own progressive values, influenced by the philosophical anarchism of her parents, and when in 1960 she and Stan started ECL they aimed to use technology to address social problems. For over forty-five years they made the company not only a center for innovative research but also a social experiment based on their beliefs about how a just society should be organized (chapter 7).

Ovshinsky's inventive ingenuity and idealism were hardly all he needed to succeed; he also had to be tough. He enjoyed boxing when he was young and in school preferred contact sports like football. Later, when he went to work in the Akron machine shops and rubber factories, he was not only learning to be a machinist and discovering his vocation as an inventor, but he was also learning to deal with violence. As he said, "I was brought up in a class war situation in a Midwestern town where the CIO had to face tear gas and bullets and clubs and blacklists," and he was ready to fight when necessary. With his social democratic values, he soon became involved in union activities, and at age eighteen he led a work stoppage and picketing in protest over B. F. Goodrich's violence against organizers at another plant. He was recognized as a leader not only by his fellow workers but also by the management, whose thugs tried to kill him (chapter 2).

Ovshinsky's toughness in the face of this physical intimidation carried over into his later resilience in dealing with the intense and sometimes vindictive opposition that met his scientific claims, and as an executive he never shrank from confrontations in patent litigation with large, powerful adversaries like Toyota and Matsushita.[16] As one of his patent attorneys, Larry Norris, observed, "Stan liked a good fight, particularly when he was in the right," and Chester Kamin, his attorney in many of these battles, added, "Stan was never afraid. That's not his personality."

This combativeness, however, could also be a liability. Ovshinsky was a tough and effective negotiator, but there were times when he was too aggressive or intransigent, derailing talks rather than reaching agreement. His success in negotiations with the Japanese, whose social codes were quite different from his blunt American manners, often depended on the tact of a trusted translator (chapter 7), and Iris, usually by his side in meetings, sometimes had to intervene to restrain his temper.

On the other hand, when confronting technical challenges, Ovshinsky's courage and determination could be decisive, leading to large financial commitments and ambitious technological advances. In the late 1970s, when ECD researchers were making experimental thin-film solar cells of just one square centimeter, Ovshinsky announced his plan to make them "by the mile" in a continuous process rather than one batch at a time. To those who understood the plan's extreme technical challenges, it seemed impossibly bold, but with the major new funding Ovshinsky secured and the long, hard work of ECD's scientists and engineers, it succeeded (chapters 6 and 8). Similarly, in 1982 when researchers on hydrogen storage found that one of the disordered materials they were testing could be used to make a battery, Ovshinsky seized on the idea. At a point when there was only a laboratory demonstration in a small beaker, he announced that ECD would develop and manufacture the new NiMH batteries and predicted that they would not only replace all existing rechargeable batteries but also one day power an electric car—all of which came true (chapter 9). And near the end of his life, when he had lost both Iris and ECD, he boldly committed his own savings to launching a new company, Ovshinsky Innovation, to develop his idea for a new process to vastly increase the rate of producing solar cells and so make solar energy cheaper than coal (chapter 12).

The courage to trust his insights and his belief in their potential to make the world better also made Ovshinsky a powerfully persuasive advocate for his programs. As the physicist Brian Schwartz said, "There was not a better negotiator in terms of being convincing, in terms of getting resources than Stan. It was his passion, conviction." The success, and at times the survival, of ECD depended heavily on his charismatic advocacy. The unconventionality of Ovshinsky's intuitive mind could make his convoluted efforts to explain his inventions exasperating for both his listeners and himself, yet his passionate conviction in expounding his vision of the future could make him eloquent and inspiring.

Finally, besides the formative experiences we have surveyed, there is a story that shows something deep in Ovshinsky's character that preceded all influences, an incident from his childhood that became an enduring part of his legend. One of his aunts was handing out cookies to a group of children, prompting each in turn with "What do you say?" Each dutifully responded with the expected "Thank you," until it was his turn. "What do you say?" his aunt asked. "I want more," he answered. Like other legends, this story exists in several different versions, but they all end with "I want more." Ovshinsky told it himself, and many others told it about him because they recognized how it captured an essential trait: his insatiable hunger not just for personal gratification but also for greater achievement and the fulfillment of his vision.

A Long Trajectory

Beyond the interest of what Ovshinsky accomplished and how he did it, beyond the interest of his story as an individual, there is the interest of how his story is related to its larger historical context. Spanning the period from his first inventions in the mid-1940s to his death in 2012, Ovshinsky's work was part of the economic and cultural transformation that led from the industrial to the information age. Beginning as a machinist, toolmaker, and machine builder, he had his roots in the rubber, machine tool, and automotive industries, as well as the social and political values of industrial unionism. But even with his earliest inventions he was intent on using the power of information to advance manufacturing (and, he believed, to liberate workers) by building smarter machines and automating controls. He was already beginning to see tomorrow, sensing the direction the whole economy would take.

Ovshinsky's work on control devices for automation led to his creation of switches based on amorphous materials, linking him to the exponential growth of microelectronic devices that became the basis of the information economy. Through the growth of ECD, his discoveries enabled the creation of other information devices and to the development of alternative energy technologies that have also become important parts of the postindustrial age.

Ovshinsky's career thus tracks the profound socioeconomic changes since World War II, changes to which his inventions significantly contributed, but he also stood somewhat apart from those developments. For all his advanced technologies and visionary aims, he remained loyal to his roots, to the social democratic politics of his youth and to the working-class culture of the shop floor. More important, and of more than personal interest, Ovshinsky's path from the shop floor to the research laboratory offers an alternative perspective on the birth of the information age itself.

Those origins are usually located in a very different world from Ovshinsky's, in the efforts of highly trained physicists working in cutting-edge research facilities like Bell Laboratories, where the transistor was invented.[17] But Ovshinsky developed his amorphous materials in a setting that was much closer to the dirty environment of the shop floor than to the purified atmosphere of the modern cleanroom. It was, as we shall see, the powdered materials contaminating the air in his storefront lab that led him to discover the Ovshinsky effect, produced with the micrometer he carried in his machinist's apron (chapter 5). Here we can see the new age as not simply opposed to the old industrial world but arising out of it, and indeed it was partly the messy impurity of Ovshinsky's new physics, not to mention its outsider origins, that antagonized some established researchers.[18]

In time, Ovshinsky would become recognized and honored with a long string of honorary degrees and other awards (appendix II), and in time ECD would build its own cleanrooms for developing its cutting-edge information devices (chapter 10). But just as his pivotal discoveries were marked by connections with his early industrial experience, so he never stopped trying to re-create the world of well-paid manufacturing work that the postindustrial age had eclipsed, and he promoted his energy technologies as the basis for "new industries."[19] Instead of seeing only a disruptive break between the two ages to which he successfully contributed, we can see Ovshinsky's career as a bridge between them.

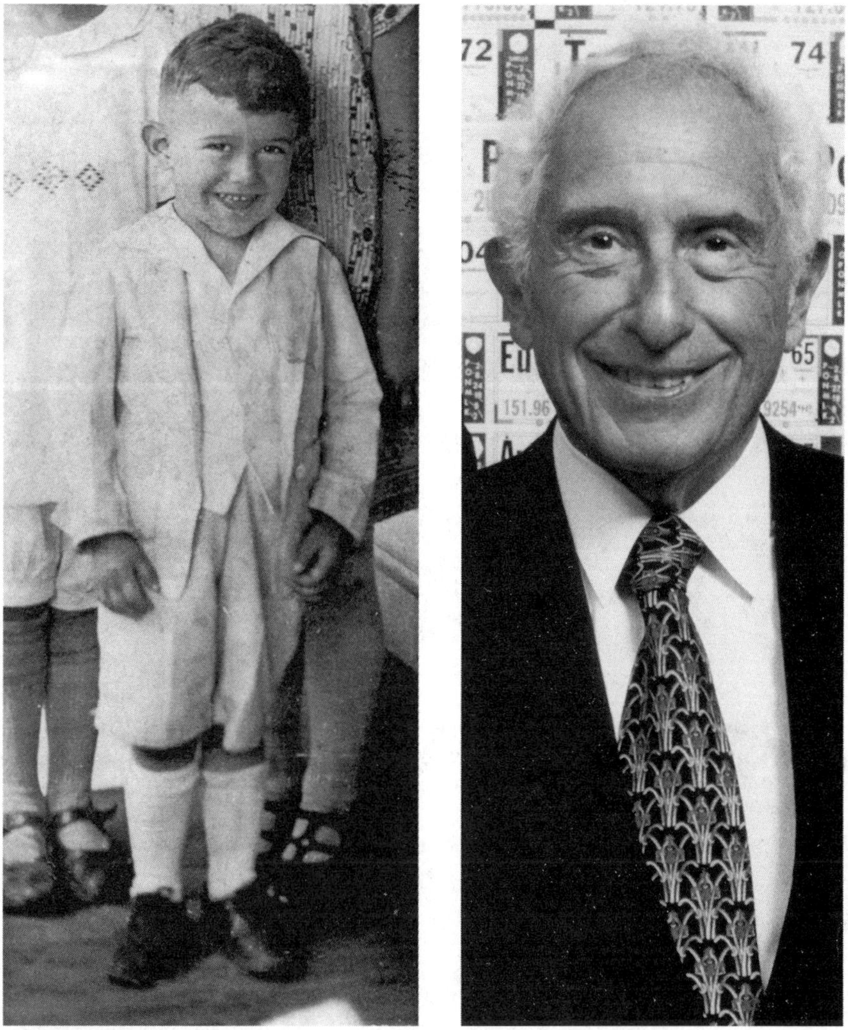

Figure 0.1a–b
Stan Ovshinsky, young and old, with the same mischievous grin.

I Becoming an Inventor

1 Young Years (1920s–1930s)

Akron's location on the Erie-Ohio canal and its railroad connections had already made it an important commercial center at the time Benjamin Franklin Goodrich visited in 1869. He had planned to move his rubber factory from New York to Chicago, but the enthusiastic reception he received from local businessmen, and the development funds they offered, made him decide to locate in Akron instead.[1] B. F. Goodrich was followed by Goodyear, Firestone, and other rubber companies, whose growth paralleled the rising automobile industry. Between 1910 and 1920, Akron, touted as the "Rubber Capital of the World," was the fastest growing city in the country. By November 1922, when Stan Ovshinsky was born, fully a third of its population were recent arrivals, mostly from Appalachia and Eastern Europe.[2] Among the latter group were Stan's parents, Ben Ovshinsky and Bertha Munitz, who each worked for a time in the rubber factories.

Ben Ovshinsky

Ben came to America at age fourteen from the shtetl of Calgory (Kalvarija), on the East Prussian border of Lithuania. At that time, Calgory was part of the Russian Empire, where in the later nineteenth century the czar had granted Jews the opportunity to serve in the army. Such service was not always voluntary. At about age ten, Ben's father, who was already carrying goods and people in his horse-drawn wagon, was impressed into the Russian cavalry, joining the ranks of the Nicolaischen Soldaten. After his thirty years of service, he was given a pair of boots, a sword, and a plot of land; he then raised horses and used horse-drawn wagons for deliveries. As Stan said, "Horses were in our family."

The youngest of six or seven children, Ben was born in 1892 when his father was sixty-two. As a child he was put to work driving a wagon transporting Russian and German officers across the border. Under the wagon seat, the boy also carried contraband literature for the Jewish Labor Bund (part of the democratic socialist movement). A few years later, his political involvement brought him to St. Petersburg at the start of the 1905 revolution. Indeed, he seems to have been in Palace Square in front of the Winter Palace in January on the fateful "Bloody Sunday," when czarist troops attacked peaceful demonstrators. Before he could escape, the charging cavalry rode him down: a horse stepped on his head, knocking him unconscious and leaving a permanent dent in his skull. In later years, Ben would comb his hair down over the hoof print. "And as a kid I used to try to push it up to look at it," Stan recalled.

About a year later, on hearing rumors that the authorities were planning to send Ben to Siberia, the family quickly gathered money to help him escape. After being smuggled into Germany, he found his way to America, entering the country in 1906 without any

Figure 1.1
Ben Ovshinsky's family. Ben, about ten, stands behind his father and mother. On the left is Ben's oldest brother "Alter." The woman on the right may be Ben's older sister. The young girls, Sarah (front) and Rachel, are the daughters of Stan's aunt, Bashe Garlovsky (not in photo). Thanks to Herb Ovshinsky for identifications.

money in his pocket. He never saw his parents again and later received news that his mother had died of cold and starvation.

After landing in New York, Ben stayed for a time in Bayonne, New Jersey, with an older brother. He worked in the needle trade, the mainstay of many Jewish immigrants, but "couldn't stand it," according to Stan's brother Herb.[3] Learning about the union activity in Chicago, the politically motivated youth joined his twenty-years-older sister and her family there. Her husband, Stan's uncle Lou, was a large, heavy-set man "with the appearance of a tough guy," Stan recalled. He had worked in the Mesabi Range iron mines, and "his children all became plumbers." In Chicago, Ben initially found work in a horsehair factory but soon grew restless working indoors.[4] He was very strong and preferred more physical work outdoors, eventually finding a job he loved as a Chicago teamster driving four large horses.

While teamstering, Ben started organizing for the Industrial Workers of the World, also known as the Wobblies (or the One Big Union). Ben would later tell his children memorable stories about how the Wobblies would chain themselves to a lamppost while giving their talks, or how he and his friend Bill Haywood, a leading radical labor leader, would get together and eat Mulligan stew. "Big Bill" advocated uniting workers in large, industrial unions rather than the separate craft unions of the AFL and wanted to give workers control of the means of production. However much Ben may have been attracted by such ideas, he didn't stay long with the Wobblies because, according to Stan, he was annoyed by the way many of them would just "sit around the office talking" instead of organizing.

Ben next moved to Duluth, Minnesota, where he found work erecting telegraph poles along the railroad, and where for recreation he helped show wild mustangs for auctions. Stan remembered Ben telling how much he disliked the way the cowboys broke the horses: "he thought it was way too rough." Ben's gentler approach was far more effective and earned Ben the reputation of being "a fellow who spoke to horses." Stan remembered how people in Akron would bring their "horses that had problems" to see Ben.

Ben's railroad work eventually took him to the Pacific Northwest and California. He would sometimes jump trains and ride the rails like a hobo. (He often mentioned that he had met Jack London in a California hobo camp.) If stopped by a cop, he would use a pass he had obtained from one of his buddies whose father worked for the railroad.

After a decade of wandering, Ben settled in Akron, which offered not only jobs but also a sizeable community of Jewish workers. He worked in the rubber factories for a time, then purchased a horse and wagon and set up a one-man business

Figure 1.2
Drawing of Ben Ovshinsky made during the Depression by an itinerant artist.

in which he could work outdoors, gathering scrap metal from machine shops and foundries and selling it to dealers. Because Ben was effective at networking with the local industries, his business prospered, and he survived the Depression. Harvey Leff, a childhood acquaintance of Stan's, reported that, "most machine shops in Akron exclusively did business with Ben because of his reputation as a man of integrity and reliability."[5]

As for Ben's horses, he would buy ones that no one else wanted but that he saw were smart and easy to work with. Stan remembered a particular blind white horse who knew his way home when Ben let the reins go. Ben continued to work alone with a horse and wagon until 1934, when he replaced them with a truck, but he never fully made the transition from driving a horse. When he wanted to stop the truck he'd say "whoa."

Young Years (1920s–1930s)

Bertha Munitz

The daughter of a farmer in White Russia (now Belarus), Stan's mother Bertha Munitz came to America in 1914 at age sixteen. In the 1890s, her widowed father had married her mother, Rebecca Daitch, "the town beauty" in the shtetl of Dauschitz (Dokshytsy), about 65 miles north of Minsk, where the family lived on their farm. When the Russians evicted all Jewish farmers, she and her parents and two younger sisters left on one of the last boats to sail before the outbreak of World War I.

Arriving in New York, the family spent some time in Brooklyn, where an older brother from her father's first marriage lived. Unfortunately, Bertha's father was totally disoriented and couldn't find work, so Bertha helped support the family by working on a punch press in a primitive machine shop.

They eventually settled in Akron, where one of Bertha's sisters from her father's first marriage lived. Soon Akron became the destination for many more relatives from Dauschitz—Munitzes, Mermans, and Kobatzniks. Growing up, the Ovshinsky children had hundreds of relatives in Akron on their mother's side. But the support of relatives was not enough to ease the transition for Bertha's parents. Two years after coming to America and still in his fifties, Bertha's father died of pneumonia. Unable to adjust to life in America, he had, according to family lore, "spent most of his time sitting in shul with a book." Nor did Bertha's mother Rebecca, then about forty, ever assimilate. Out of her element from the very first day, she always demanded special treatment. Her grandchildren resented the way she would order them around and "carried herself like the Empress of Russia," as Stan and Herb recalled.

As the oldest child, Bertha had to work, while her younger sisters were sent to grade school. She later learned to read and write English through the Workmen's Circle, where she joined a women's book club. Although she resented being denied the opportunity to attend school, Bertha liked working, which gave her a sense of dignity. One of the places where she worked was the old Goodrich Miller Plant 2, where years later her son Stan would work as a toolmaker (see chapter 2).

Ben and Bertha Ovshinsky

Ben Ovshinsky needed a room when he came to Akron in the winter of 1917–18. One of his friends, a kosher butcher who was Bertha's half-brother, put him in touch with the Munitz-Merman-Kobatznik clan, which took Ben in. On meeting, Ben and Bertha were immediately taken with each other. Ben hesitated for a while to give up his bachelor freedom, but before long they married on May 2, 1918. He was twenty-six and she was twenty.

Figure 1.3
Ben and Bertha Ovshinsky, c. 1918.

The young couple lived at first in a small apartment in Warner Court, the poorest section of Akron. Bertha, who often used her Yiddish name, Teibel (dove) developed a reputation for being "a very kind and hospitable person who often invited neighbors and friends to stop by for conversation and food."[6] Always charitable, she took care of anyone who needed help. Bertha was religious; she kept a kosher household, attended holiday services, lit candles on the Sabbath, and wished she could go to temple more often than she did.

Ben was not religious like his wife, but he too was generous, agreeing to Bertha's condition that her mother live with them and later to taking in other members of her family during the Depression.[7] Ben's sympathies were broader, though, as expressed in his social and political efforts to make the world better for others, while Bertha resented Ben's political activities for taking time away from their immediate family. Recognizing the substantial differences between his parents' values, Stan once asked Ben, "How could you marry her?" "Well, she was very pretty," Ben replied. Stan concluded, "It was a case where enough physical attraction sort of conquered all."[8]

The Birth of Mashie, Stan, and Herb

On March 5, 1919, about ten months after Ben and Bertha were married, their first child, a girl they named Myrtle, was born. Her nickname was Mashie, but to her friends she was Sandra.[9] Like Bertha, she was religious, but she did not accept strict Jewish orthodoxy and became a reformed Jew who subscribed to Socialist Zionism.

Stanford Robert Ovshinsky, Ben and Bertha's second child, came into the world on November 24, 1922. He was delivered in the upstairs part of the Wooster Avenue office of a German doctor, who, Stan said, "took care of all these immigrants." The doctor nicknamed the boy "Schnuckelfritz" (cuddly boy), and "he was Schnucky to the whole family," Stan's brother Herb recalled.

By the time Herbert Ovshinsky, Bertha's and Ben's third and last child, arrived on July 7, 1928, almost six years after Stan, the family had moved from Warner Court to Moon Street, where most in the neighborhood were Jews. Stan proudly took on the role of protective older brother, and bragged that as Herb grew up and attended school, the bullies would say, "Leave him alone. He's Shinsky's brother." (For his part, while acknowledging the deep kinship he shared with Stan, Herb could not remember ever needing his brother's protection from bullies.)

The extended family clan was large, warm, and diverse and included many half-brothers and half-sisters because women often died in childbirth. They ate their big weekend meal on Friday night, and on Sundays typically had more visitors. As Ben's

Figure 1.4
Ovshinsky family. Back, left to right: Rebecca Daitch Munitz, Herb (about a year old), Bertha, Ben. Front: Mashie, Stan (not yet seven).

business prospered, even during the Depression, the Ovshinsky home, with the only telephone on the street, became the gathering place for the family.[10]

Weekend shopping on Wooster Avenue, the bustling commercial street not far from the Ovshinsky home, was memorable family time for the children. Most stores were closed for the Sabbath on Saturdays, so Jewish families usually shopped on Sundays, or on Saturday nights after dark. The Ovshinskys did most of their shopping in a string of small shops, including a kosher butcher, a Hungarian delicatessen, a Jewish bakery, a chicken store, a hardware store, a drug store, a barber shop, a hat shop, and an all-purpose grocery store whose cans and boxes on high shelves were pulled down with the help of a retractable fork on a long pole. The smelly delicacies of the nearby fish

Figure 1.5
Herb and Stan (at about ages six and twelve).

market could be sensed a block away. Herb fondly recalled shopping on Wooster with his dad, going from one end to the other in his truck and stopping off at Roseman's delicatessen bar on the corner for grilled hotdog sandwiches. Some of the shop owners were family. Uncle Morris, the kosher butcher, had years earlier introduced Ben to the family, and Uncle Abe, a baker who "was my mother's older sister's husband, made the best jelly rolls in the world."

While many of Herb's childhood memories of time spent with Ben dwelt on food, Stan's were largely about work. Sometimes, especially when Ben was sick, Stan helped with metal collecting. He treasured this shared time, despite the fact that it was "the hardest work I have ever seen." Stan remembered getting up at 4 a.m. to clean the stable, which was half a block from home, and having coffee with his father afterward. "He of course had coffee," Stan said. "He gave me milk with coffee." At night, when Ben came home, Stan would run out in the street to meet him; Ben would let him drive the horses home and help feed them. It was while going with Ben on his scrap collecting rounds that young Stan "fell in love with factories, machine shops, foundries."

Ben was also an important influence on Stan's social values. Both Stan and Herb remembered their father as a knowledgeable, self-taught socialist and intellectual who was always aware of events elsewhere in the world. He didn't read English well, but he read widely in the other languages he knew—Yiddish, German, Polish, Lithuanian, and Russian—despite having poor eyesight. He could speak intelligently on many subjects, and wrote columns for Yiddish newspapers. As he had in Lithuania and Russia, Ben continued to be politically and socially active in Akron, both in the labor movement and in the Jewish segment of the Socialist Party. Drawing on his experience with the Bund, he helped to form the Akron branch of the Workmen's Circle, where Ben's children and grandchildren received much of their early education. Ben was also involved in starting a union that helped Akron peddlers resist mistreatment by the police and others.[11]

The aim of the Workmen's Circle—to create "a better and more beautiful world"—was a goal Stan and Ben shared, and they would talk at length about how to achieve it. Stan considered Ben "my best friend," and Stan was a best friend for Ben as well. Each found in the other the approval they could not get from Bertha. "He was just as much of an outlaw in the family as I was," Stan reflected. Ben also set an example of tolerance; his progressive politics didn't make him hostile to those with very different views. He even befriended some men in the plants who were in the Ku Klux Klan, recognizing that though some were dangerous, others had simply been misled. "He was a very physical sort of a guy too," Stan explained, "not easily intimidated."

Ben also loved the theater and music, sometimes acting in plays put on by the Akron Workmen's Circle, practicing his lines at home with Bertha. In his earlier Chicago period, Ben had performed in the Yiddish Theater alongside the well-known actor Paul Muni (Frederich Weisenfreund).[12] Ben had a good singing voice and loved to sing Jewish cantorial songs. Although he was an agnostic, he would sing in the choir of an Orthodox shul wherever he happened to be. When Ben sang in the Workmen's Circle chorus, Stan liked to sing with him, "folk songs, songs of struggle, songs of the labor movement, in Yiddish."[13]

These family and cultural experiences were strong influences, but for young Stan Ovshinsky, the most wondrous and influential place was the small public library on

Wooster. He started to read at an early age, and would bring home books on many subjects—history, archeology, astronomy, chemistry, physics, politics, poetry, theater, cosmology, biology, and art. He loved reading plays by Clifford Odets or Chekhov as well as the novels of Jules Verne and H. G. Wells, and not only for their science fiction visions of the future. "I liked their politics," he said. "When I was five or six years old," Stan recalled, "I would read anything."

For this insatiably curious boy, the librarians on Wooster Avenue made an exception to their rule that children could borrow only two books at a time from titles appropriate to their age group or grade in school. They allowed Stan to borrow all the books he wanted, on any subject and on any age level. In later years Stan could not remember how he had managed to gain this privilege, but he did recall that one day, when he was "tottering out with all these books," one librarian remarked, "Stanford, what will happen to you? When you grow up you'll have read all the books."

In addition to reading all the books, Stan began reading the Sunday *New York Times* regularly, and he also profited from the Workmen's Circle Yiddish school, whose teachers were well informed about current events. But "I learned nothing in grade school," he said. "And they screwed me up in mathematics." He blamed this on the school's adoption of the progressive Winnetka plan, but it seems more likely that his difficulties in math came from the ways his mind worked differently from others'.[14]

Outside of school, Stan could follow his own interests, including science. "I always wanted a microscope set and a chemistry set." Once there was enough money, "I got both when I was maybe eleven or twelve. I also wanted to get a book by H. G. Wells, *The Outline of Science*, that was advertised in the newspaper. But by the time I talked my parents into giving me the money they were out of it, so I bought *The Outline of History* instead."

Stan's scientific curiosity early in life had led to some ill-judged experiments. At the age of three or four, expecting that the air would hold him up, he jumped from the family's second story porch with an open umbrella and nearly killed himself. (After that, Ben put a wire fence on top of the porch.) Stan also studied all the appliances in his home, taking most of them apart. Watching one of his aunts do laundry in the cellar at the age of four or five, he became interested in the rollers of the wringer. When his aunt stepped away for a moment, he almost lost an arm trying to see how the wringer worked. A bit later, he nearly electrocuted himself by poking his finger in a light socket. Bertha didn't know what to make of Stan's fascination with machines. Once, when he was playing with her sewing machine, eager to solve the mystery of how it worked, she asked him whether he wanted to sew. "No, no," he replied.

Like many scientifically inclined boys of that period, Stan read popular magazines about science and invention, sending away for kits to make various gadgets. Among those magazines were some edited by the well-known science fiction writer Hugo Gernsback, who had come to the United States about the same time as Ben. His magazines like *Popular Invention* helped many other budding inventors of the period form their identity.[15]

Growing up in the Depression was an education of its own. Stan was almost seven when the stock market crashed in October 1929. Akron was particularly hard hit, and Stan was shocked when he saw evictions. Though his own family was not economically affected, seeing so much misery around him fed his desire to make life better for others and helped bring him into politics at an early age.

Stan became deeply committed to democratic socialism and strongly opposed to Communism and Stalinism. Adults took the boy seriously when he expressed his views. He recalled that when he was about eight, "I went to a barber at the end of my street, Moon Street, who was a member of the Socialist Labor Party, and so I'd end up arguing politics with this guy while he was cutting my hair. But nobody thought it was unusual that I was a kid taking on this grown man." Stan was also invited to give political talks at the Workmen's Circle and elsewhere when he was nine or ten. The Russian immigrants didn't think twice about this. "Back in Russia many of them had started their political and their cultural activities at a very early age." He was active in the Young People's Socialist League, but for Stan, the point of socialism was practical, "to make a better life for working people, with education and so on. The people who believed that were called sewer socialists and so I became known as a sewer socialist."[16]

Stan especially liked attending the Friday night meetings of the Workmen's Circle, to which speakers were brought in from New York and elsewhere, many having left Europe to escape the Nazis or the Communists. These meetings were typical expressions of the secular, radical culture of Eastern Europe that Stan called Bund culture, and he looked back at it with fondness and regret. "We had a very rich life that won't be duplicated again. Yiddish culture is almost dead. It was tremendously cooperative." Its members "stuck together, helped each other. They were all bright and intelligent, even though they were carpenters, toolmakers, painters, rubber workers, shopkeepers, shoemakers, tailors."

Stan's intense early interest in politics did not rob him of a normal childhood. In many ways Stan was just a regular kid. He played with friends on the streets year-round and had many girlfriends. Through his relatives in Chicago, Stan got interested in boxing and would sometimes go to the park near their house and use the rings to box "just for the joy and the hell of it." By the time he was in high school, he recognized the dangers and decided to stop.

One of Stan's friends in this period was Branko J. Widick, who was always called BJ. Though he was twelve years older, Widick was attracted to Stan: "He was totally different than anybody I knew. He obviously had a brilliant brain." They met at the Akron Workmen's Circle. Widick was not Jewish, but he felt that the Workmen's Circle was "the only place that had any culture in Akron. If you wanted to hear a good lecture, you'd go there."[17] Widick remembered being stunned when at the age of twelve Stan challenged the famous Marxist theorist Max Shachtman while he was lecturing at the Workmen's Circle on the 1905 revolution. "Who is that brilliant young bastard?" Shachtman asked Widick. An orthodox Marxist in that period, Widick recalled that while he and Stan were both in the Socialist Party, they had considerable political differences, for Widick was a Trotskyite, and Stan was not. Their disagreements intensified to the point where they stopped seeing each other, but the break in their friendship proved temporary (see chapter 4).

Stan's political work also put severe strains on his relationship with Bertha, who could not understand why her son preferred to spend his time in meetings rather than play with friends. She would say no when Stan asked to attend a political meeting, but if Stan was determined to go he would visit a friend and slip out from there to the meeting. At one point, perhaps recalling the Red Scare of 1919–20 and fearing that Stan might be targeted, Bertha tried to protect him. Stan came home one day to find that his mother had burned all of his books.

Stan did not share Bertha's religious values, which caused further strain between mother and son.[18] When Stan reached the bar mitzvah age of thirteen, marking a Jewish boy's transition to manhood, he aligned with Ben (as Herb did later), who saw the ceremony as just a ritual in which children received presents like fountain pens. Stan's refusal to have a bar mitzvah was extremely upsetting to Bertha and her family, who could not understand his reasons and wanted him to see a psychologist. He refused to do that too. Still, both boys were sent to a Yiddish afternoon school for many years, typically four days a week plus Sunday, and both developed a love of Yiddish language and culture. It was, in fact, through Yiddish that Stan found he could write for others, producing articles for a Yiddish Sunday paper published by the Workmen's Circle for young people when he was eight or nine.

Well before this, Stan had also discovered that he could draw very well. To amuse himself he would spend time drawing on any sheet of available paper, but his mother did not approve of this means of expression. One of his kindergarten drawings was included in an art show in a downtown department store, but Bertha never told him.

Stan continued drawing for the rest of his life, producing a huge collection of quick caricatures as well as a few paintings (see the interlude.)

Early High School Years

Stan found new interests and new frustrations at John R. Buchtel High School, which he entered in the fall of 1937. Still interested in many subjects, including writing, astronomy and other sciences, anthropology, and art, he continued to read passionately and extensively. As a result, he paid less than full attention to his teachers, whose knowledge often lagged behind his. Stan remembered one occasion when a high school English teacher asked the class to write a book review. Stan liked this particular teacher, from whom he had heard poems by Blake for the first time. But when he turned in a review of a book by André Gide he received an F. The teacher told Stan he failed him because he had made up both the name of the book and the author. "There's no such person named Gide," he said. From then on, whenever Stan realized that he knew more than his teacher, "I just didn't argue about it."[19]

In Yiddish school, on the other hand, some teachers appreciated Stan's extensive knowledge. One allowed him to teach the literature class occasionally, and he was invited to give a class on current events and the history of the labor movement. The librarian recognized Stan's contributions by letting him choose two books to keep. He selected Whitman's *Leaves of Grass* and *The Machinist's Handbook*.

More important than anything Stan learned or did in school was his dawning sense of vocation. By the time he reached high school, he had already been going along with Ben on his rounds for some years, experiencing the machine shops and foundries and at times helping with the hard work of using a pitchfork and shovel to load his truck with heavy steel scrap. He also enjoyed "looking at the machines, watching what they did, learning about things, talking to people in the various factories, shops, foundries." This was when Stan "really fell in love with machines" and "learned to love industry." "To me," he said, "manufacturing has always had glamour to it. The glamour was being in the foundry, the flames, the sand, the noise, the machine shop, the smell of the oil and the chips coming off." That growing romance offered Stan a ready answer when Ben asked him one day, "'Simcha,' which was my Yiddish name and means joy and pleasure, 'what kind of trade are you going to learn?'"[20] Without thinking, Stan said, "I think I'll be a machinist and toolmaker. I just like that kind of thing."

2   Passion for Machines (1940–1944)

Stan's first step toward becoming a machinist was a job for the summer of 1940, before his senior year of high school. Through Ben's connections, he was hired at a company that made molds for automobile tires. Not yet eighteen, he lied about his age to obtain his work permit.[1]

Akron Standard Mold

The machine shop at Akron Standard Mold was primitive compared with others where Stan would work later, but its belt-driven machines included a drill press and other basic tools that he was eager to learn to use. Stan's pay was minimal, and came to even less because he needed to buy his own tools. But he was nevertheless very proud that at seventeen he was already being paid for serious shop work. And he loved buying his own tools, which were his to use for whatever he wanted. Paying Stan his two dollars at the end of the week, the foreman would caution him: "Don't spend it all in one place." Stan recalled, "I'd take the two dollars, find a bookstore with the prettiest girl as a clerk and go in and buy a book. So I had two pleasures from the money."

He had to begin with the shop's most humble tasks. "You had to file, you had to make sure the belts ran, all the things an apprentice does." Stan enjoyed the work, but he found that learning new skills required help that was not easy to get. "Nobody wanted to help you," Stan said, "because your accomplishments threatened their job security." When Stan asked a Russian friend of Ben's who was then working at Akron Standard Mold to show him how to grind drills, the fellow at first refused to share his expertise, explaining, "we don't get paid much. And everybody's job here is very precious." As it turned out, the Russian made an exception for Ben's son, and he agreed to show Stan just once how to grind drills. Stan recalled breaking into a cold sweat because he felt sure he'd lose his job if he didn't learn the skill. But he practiced repeatedly until at last, "I could improve on what he showed me, and I got a much better drill."

Stan's growing experience fed his intuitions about materials and machines, which in turn enhanced his experience. "I had these hunches," he explained, and guided by them "I could go and learn." For example, in working with cutting tools for lathes, Stan figured out an improved design involving a change of angle that made the chip roll off before it could heat up. This was a significant improvement, the first of several ways Stan found to machine metal more efficiently that eventually led to the invention of his novel center drive lathe (see chapter 3).

Stan also learned, however, that such resourcefulness would not enhance his popularity in the shop, where "busting ass" was frowned upon. The older workers resisted changes that increased productivity because they had learned from hard experience that they would not be rewarded. The higher rate would become the new norm, effectively reducing their wages.[2] Stan recognized that this was an abuse, and he started to get involved with issues of justice in the workplace.

Following in Ben's footsteps, Stan agreed to represent his fellow workers in the Akron Central Industrial Union Council, which represented the CIO industrial unions in town. Their choice recognized Stan's earlier experience giving lectures on labor issues and helping to organize picket lines when he was leading the Young People's Socialist League. But while Stan often supported the work of the unions, he never wore an official union pin, nor did he ever run for union office, because he resisted its organizational constraints. "I just wanted to be a worker, doing my job and being able to believe in what I believed in. I was really inherently a Wobbly who hated bureaucracy anywhere, including in the unions, and they did things I didn't like."[3]

Late that summer, Stan came home from work and shocked his parents. "I'm quitting school," he said. He had had a conversation that day with Fitzpatrick, the shop superintendent at Akron Standard Mold. This demanding man, known only by his last name, was famous throughout Akron for his toughness. Stan later compared him to J. R. Williams's well-known 1940s cartoon character "Bull of the Woods."[4] The term, Stan explained, was used in that period to describe a "tough son of a bitch foreman."

Fitzpatrick had gained considerable respect for Stan and his work over the course of the summer, and that day he had asked Stan to consider not going back to school. "You're going to be a very good machinist," he told Stan. "I like what you're doing and your attitude. And I think you know I never went beyond the third grade and look at me. You've got the stuff. You don't need school." Stan's goal at that time was to become a great machinist, so his immediate response to Fitzpatrick was "sounds good." Ben and Bertha did not, however, allow Stan to quit high school. As a compromise they let him work part-time at Akron Standard Mold during the fall semester.

Senior Year

When high school began again in the fall, Stan planned to study art and develop his gift for drawing. But when he found that the art class mainly consisted of rote exercises, he switched to shop class, where he was able to practice some of the skills he had recently learned at Akron Standard Mold, such as cutting threads and making tapers. He also learned new skills. "I loved forge work," Stan recalled, learning to tell the heat of metal by its color and how to temper it, "whether you quench it or anneal it to get different properties." Stan's understanding of materials deepened in this period. He learned about the properties of different steels, and how and why they differed from other metals like cast iron. This fueled his growing interest in metallurgy, and, he recalled, "one of the first things I did back in those days was I would make powdered materials solid." Years later, he would draw on this metallurgical experience in creating the new glassy materials that became the basis of his groundbreaking inventions (see chapter 5).

Stan enjoyed the high school shop courses so much that he decided to attend a public trade school (Hower Trade School) at night to learn how to operate all the machines then typically found in a tool-making shop. For the trade school course he bought more tools: "micrometers, Johansson blocks, calipers, different scales with steel rollers, and thread gauges." Stan recognized that making and using tools is an art, and he became "very disciplined about studying and working and thinking about tools."

When Stan learned that he was short a course needed for graduation, he approached the mechanical drawing teacher, Mr. Wetzel, and asked what he could do to get the semester's credit he needed. Like several other teachers, Mr. Wetzel considered Stan a trouble-making socialist, but he respected his abilities. He asked Stan whether he knew anything about diesel engines. Stan replied, "I know about gas engines, but I never really got interested in diesels." Mr. Wetzel told Stan that on Friday afternoon he would give him a demanding take-home exam on diesels and on Monday morning would give him the credit he needed if he passed. "Are you ready to take the challenge?" Stan answered, "Sure." He realized from the teacher's mischievous smile that he did not expect him to succeed, but he went home and studied "a little of this and that," putting things together for himself as he wrote the exam. On Monday he had answered the questions, and Mr. Wetzel kept his promise to give Stan the credit.

Figure 2.1
Stan's high school graduation photo, June 1941.

Goodrich Rubber

After his bad experiences in grade and high school, the idea of attending college didn't even enter Stan's mind at the time he graduated in June 1941. (And in any case, his teachers didn't consider him "college material.") Had he wanted to, his parents would probably not have stood in his way, though "I think they would have thought it was peculiar," Stan remarked. "My father would have said something like, 'How can you make a living doing that?'"

Stan began looking for a job. Although the United States was not yet at war, military production was already building up, and the increased demand worked to Stan's advantage. First, however, he had to overcome some predictable anti-Semitism. (Stan's job at Akron Standard Mold had been possible only because Ben was highly respected there.) When he applied for a tool-making post at B. F. Goodrich's Miller Plant 2, the hiring officer told him, "We don't hire Jews." Stan asked, "Under what condition would you ever hire me?" "I'd hire you if you had a recommendation from Fitzpatrick," the officer responded. "That tough son of a bitch would never give anybody a recommendation." As it happened, Stan had a recommendation from Fitzpatrick in his pocket, for when Stan left he had had the presence of mind to ask for the reference. On seeing the letter from Fitzpatrick, the hiring officer said, "You've got the job." When Stan told his mother that he would be working at the Miller plant, she said, "Oh, that's where I worked too." She had been employed there during World War I, when workers were so badly needed that being both Jewish and a woman did not keep her from being hired.

The Miller plant was one of the oldest buildings Stan had seen, and the machines, all belt-run, were also extremely old fashioned. But Stan threw himself into his work, making "anything that required precision machining, out of cast iron, bronze, steel."

Figure 2.2
Nineteenth-century factory with belt-driven machines.

He gained a sense of pride in his trade, especially in doing things the way the old-timers had. In his first assignment, Stan machined special parts using enormous no-frills machines in a large maintenance and repair shop. With a gigantic belt-driven nineteenth-century lathe, he was expected to make or remake huge threaded parts. The machine had no attachment for cutting threads; "all they gave you was big calipers," he said, so Stan had to improvise. "I would take a piece of chalk and put marks on the slide of the machine, and control it by just the chalk marks to cut the threads I wanted. Everybody was very impressed." Such experience, Stan reflected, "gave me the self-confidence I needed to be a real inventor." Improvising control methods would also help Stan to think more generally about intelligence, setting him on a path that led in time to his invention of smart, self-regulating machines (see chapter 3).

Stan's time at Goodrich allowed him to attend the night school offered there for staff, gaining knowledge that later fed his inventions in unpredictable ways. What he learned about polymers from a class on the chemistry of rubber, for instance, later contributed to his breakthrough creation of chalcogenide switches (see chapter 5). More immediately, working at Goodrich gave him powerfully formative experiences of class conflict. Although he took no official role in the union, he eagerly joined in collective efforts to improve conditions for his fellow workers. Becoming known as an activist marked him as a target for violent reprisals that endangered his life several times, but those struggles also helped him gain the toughness he needed in later struggles as an independent inventor and scientist.

A dramatic confrontation came early in October 1941, when Stan helped to organize a one-day strike to protest recent violence at another Goodrich plant in Oaks, Pennsylvania. Thugs hired by the company had attacked Joseph Feineison, an organizer who was passing out leaflets, beating not only him but also his pregnant wife. "We were all of us very upset that the company would be so cruel as to beat up a pregnant lady," Stan recalled. Stan offered to help by disconnecting drive belts and cutting off power to close the plant in protest.[5]

On the day of the strike, Stan rose early, as usual. As he was leaving home, Ben came out and said, uncharacteristically, "We're worried about you." Stan was confused. "Look," he replied, "you've been in more dangerous situations. Did you ever not go?" Ben said no. When Stan saw the tears in his mother's eyes, he knew that she was the one who was worrying. On arriving at Miller, Stan knocked the belts off the pulleys, shut the plant down, and organized the picket line for a nonviolent general strike that lasted about twenty-four hours. But violence arose in the evening as Stan walked out of the plant. He noticed a spotlight from a tower following him around. When a car without headlights drove straight toward him in the dark, he feared for his life. Stan

Passion for Machines (1940–1944)

recalled, "The light was still on me, and I heard a whirr. Then I heard just a roar from our picketers, and they grabbed that car, these coalminer guys from West Virginia. Nobody said anything; they all grabbed a hold of the car spontaneously." The picketers stopped the car, started to shake it and were about to overturn it when Stan stepped in to prevent the men inside from being killed and thanked the workers warmly. "They went back in the picket line. They were used to that kind of thing."

After that incident, Goodrich moved Stan over to its main plant, Plant 1. "There was no reason except the fact that I was an active member of the union, and that I was on the industrial union council." He was there by early December 1941, when "we were having a union meeting on Sunday, and the chairman stopped the meeting and said, 'The Japs are bombing Pearl Harbor. Meeting adjourned.'" In Plant 1 Stan was put in a big machine area and given a number of jobs that proved to be life-threatening. One required inserting a huge key about five inches in diameter into a big machine to lock the shaft. He was sent to help the man who hammered the key into place, who "looked like a poster of a Soviet worker, built very strongly." Stan's job was to kneel and hold the key while the huge man hit it with a sledgehammer. "Now I'm on my knees and he said, 'Stan, I want to talk to you. I know why you're here.'" He had heard through the grapevine that the management wanted to get rid of Stan. His previous helper had been killed. Stan recalled his account: "'I told him, don't look at me. Keep your eyes focused on that shaft, and hold like hell, but he kept looking back, and when I see that, bang. Hit his face. So that's what you're here for, they kill you that way.'" Stan followed instructions. "I never moved; I didn't look at him."

Afterward, Stan spoke more with this powerful man. He was a German and turned out to be a Social Democrat who had been part of the Kiel mutiny at the end of World War I that helped end the German Empire. As fellow socialists they discovered they had much in common. As Stan explained, "The Workmen's Circle, the Yiddish labor movement, copied what the German Social Democrats had," including singing groups and theater. So instead of getting rid of Stan, the company had helped him find a new ally.

There were several more such attempts. The most dramatic came when a group of managers in suits came and told Stan they had a special job for him. "They knew I liked special jobs." They took him to the powerhouse and told him to climb up to a higher level, where he would find a piece to machine. When he got there, it was very dark, and, as Stan soon discovered, the floor was just wire mesh. "When I got out on the floor, my foot went through a huge hole. It'd have been the end of me when I hit the bottom. I reached over and grabbed the wires and pulled myself with great difficulty out of the hole." When Stan went back down, the men were still there. "'You know, fellas,' he said, 'I could have been killed up there.' And they said, 'Well, you're getting the idea, Stan.'"

Stan realized that he would not last long at Goodrich. A union representative told him, "They're out to get you. It's not worth your life." Early in 1942, the union helped Stan find a job in another union shop, working in the tool room at Imperial Electric, a company that made motors.

Imperial Electric

Imperial Electric was located in a very poor "red light" section of Akron. Stan remembered the scene on Saturday nights when he worked alone in the tool room on special jobs. "The cops would go up and down the street, calling out the names of these quite unattractive women carrying screaming children." The choice for these poor women was between paying off the cops and going to jail. The building at Imperial Electric was shabby, too. Stan described it as "a terrible place, with rats." When there were blackouts the "rats would come walking across our feet."

But despite the dilapidated setting, Stan enjoyed his tool-making work there, learning from the challenging assignments. "They'd give me things that I didn't know I could do, because nobody else could do it." One major project involved machining a new part for a gigantic generator that the company was about to ship. Something was wrong with it, and Stan "had to figure out a way to get inside that huge thing" and machine the part. Working for an hour or two, he succeeded. Stan got great pleasure from such problem solving, but as at Akron Standard Mold, Stan's ideas for improvements were not welcomed at Imperial Electric. "It was put to me very squarely" by Frenchy, the shop foreman, "a very nice guy. He came over and said, 'I got to give you my advice for friends. You got to remember you're paid to work, not think.' That's when I first started thinking that I would end up going on my own. I love my work, but if I can't be creative it's not what I want." By discouraging such creativity on the job, Frenchy "was the one who caused me to first think that I ought to set up my own shop."

But if Frenchy discouraged thought, the workers at Imperial Electric were receptive to Stan's efforts to get them to think. In the Jewish labor tradition of the Workmen's Circle, Stan organized cultural activities, including a reading group for which workers were to write book reviews. Being more widely read than the others, he agreed to choose readings, "books that made you think," and to help the others learn how to write reviews. At the union meeting where this idea was discussed, Stan suggested that the workers draw lots to go first. As it happened, the first person chosen to present a book review was "this big brute of a guy" who had what was probably the worst job in the plant, dipping pieces in strong acids. Nobody in the plant paid any attention or even talked to the man who drew the lot, and the others were surprised when he said he would try. Stan then worked with him, explaining, "You read this and then you say to yourself, 'What did I learn from this?' And you do the best you can." Stan had tears in his eyes when this man managed to present his review. It was not the best possible book review, but "for the first time he was thinking about something, reading something. It opened his mind." In addition, "this fellow was so proud, and he got respect." Stan felt that his most important work in the shop was to help raise consciousness in such ways. He wanted his fellow workers to "understand why you would want to change the world, not just be mad at it."

The owner at Imperial Electric was "a one-armed guy who was a follower of Mussolini" and "treated people very badly." At one point, he replaced the congenial Frenchy with a new, hostile foreman, probably to combat the union. To agitate Stan, a known union activist, this new foreman used a Jewish slur. Stan's instinct was to hit him, but the other workers shouted, "Don't hit him! That's what they want you to do! They'll fire you immediately." So instead, Stan pointed his finger at him and said, "I'll see you outside tonight, and we'll see whether you can repeat that to me." But when the time came, the other workers kept Stan late. When he finally went outside, he didn't see anyone, and after waiting for a time went home. About forty-five minutes later he got a phone call and learned what had happened. When the new foreman had come out, some of the other workers had grabbed him. They had planned to simply scare him so that he would not try it again, but when a large ball-peen hammer dropped from his jacket, clearly meant for attacking Stan, they beat him up. "How bad off is he?" Stan asked. "Pretty bad off," he was told. "Call the ambulance, but give a few minutes for Tony or whoever it was to get out of there, and tell him he can't go home tonight, that we'll sort this thing out tomorrow." The next day, Stan and a union representative went to see the owner and threatened him with a strike if he tried to prosecute the attackers. The owner fired the new foreman, and Frenchy returned. Stan was pleased, he said, that at both Imperial Electric and Goodrich "I was protected by workers. I saw that real solidarity."

Attempts to Enlist

Meanwhile, the country was mobilizing for the war effort, and Stan was eager to help in the fight against fascism. After the Japanese attack on Pearl Harbor, when Stan was still at Goodrich but preparing to leave, he tried to enlist in the navy, thinking that would be a better place for a machinist than the army. Not until he had moved to Imperial Electric did he get his physical exam; to prepare, "I got me a military haircut." But both the navy and the army rejected Stan because of his asthma, which had by then become serious.

Stan's asthma had started to trouble him when he was in his teens and would bring home stray dogs and cats. When the family kept one of the cats to deal with mice in their house, Stan's asthma worsened, flaring up when he was exposed not only to animals but also to dust. No consistent steps were taken to treat this problem, and he became so sick at times that he "couldn't even walk." During high school and during his first year of working he was unable to breathe through his nose. "That was a tough time in my teens," he recalled. At Goodrich and Imperial Electric he had to work in settings filled with fumes, sawdust, and metal chips, which aggravated his asthma. He was also smoking and chewing tobacco, but strangely, cigarette or pipe smoke didn't begin to bother him until much later, after he gave up smoking. "I didn't like smoking cigarettes, actually. But to me it was a bonding thing with my father. He'd smoke and hand it to me."

Stan had had several nasal surgeries "just to breathe." The military examiners could see the scar tissue and rejected him. In retrospect, Stan recognized that he would have been rejected in any case because he was "a well-known agitator." He also realized that his severe allergies would have made it unbearable for him to wear the rough woolen uniforms. Throughout the rest of his life he had to avoid wearing wool or other coarse fabrics, which were extremely irritating to his skin.

Stan next thought he would go west to Sacramento, where positions for machinists were being advertised. When he was rejected without explanation, he realized that he had been blacklisted in Akron for his organizing work. He decided instead to try Arizona, which for Stan was an exotic place that he had read about as a child, with good weather, Indians, and mesas. Its dry climate also promised to help his asthma. This time he didn't try to arrange a job in advance. He was nineteen, and adventurous.

Marriage to Norma Rifkin

Some months before leaving for Arizona, Stan married his beautiful high school sweetheart Norma Rifkin. Born on January 18, 1924, she was Jewish, and her parents came from the working class. Her father, Abe, from Ukraine, began by repairing umbrellas and rose to owning a furniture store. Her mother, Ida Moon, was the daughter of a German-Jewish immigrant. When Ida later left Abe, Norma, at the age of twelve, took responsibility for her younger brother Jerry. Norma was not religious, but she socialized as a teenager at the Akron Jewish Center and probably met Stan there.[6]

Stan recalled that Norma's family had cautioned her against marrying him because he was "a troublemaker" who carried a lunch pail. They had hoped she would rise in status and marry an accountant or a doctor. But as soldiers left for war, "everybody was getting married," and so did they. He was nineteen, she eighteen. At that age, Stan was too young to realize that while the two were physically attracted they were poorly matched in their values and interests. Norma was artistic, creative, and levelheaded, but she cared little about Stan's political and intellectual interests, which she would eventually come to resent as a diversion from his family life.

Several dozen friends and relatives witnessed their Jewish wedding ritual on August 9, 1942, at the Akron Jewish Center.[7] Stan wore a yarmulke and performed the symbolic act of stepping on a glass. A small party followed. After the wedding, Stan was dismayed when Bertha presented him with all the money he had brought home from his job at Akron Standard Mold, which she had lovingly saved for him. "She had a sweet heart," Stan said.

Figure 2.3
Stan and Norma, c. 1942.

Arizona

In the cold of winter, probably in December 1942, Stan and Norma boarded the train to Phoenix, taking only several changes of clothes. Phoenix was a small city at that time, and its climate was indeed beneficial: "I actually did feel better with my asthma," Stan recalled. After staying for a few days with a cousin, they rented a room at 733 West Portland and found jobs.[8] Stan took the first job he found, in a small machine shop housed in an open-ended garage, a very hot place to work. The job was but a temporary stopgap, however. Not only was the pay inadequate, but the owner also wanted him to run the machine that Stan had been hired to set up. Stan knew that he could not do repetitive work. When the owner told him he didn't want union guys in his place, it was even clearer to Stan that he had to leave.[9]

He then found an excellent job in the tool room of the Goodyear Aircraft bomber plant in Litchfield Park, which was in the desert about 30 miles outside Phoenix. Much bigger than the shops Stan had worked in earlier, and far more modern, its "really good all around tool room didn't have belts. It had real lathes, real milling machines, big planers, and some very good toolmakers." Stan enjoyed the variety of the job's demands. "Unlike a machinist who becomes a specialist on one machine, I had to run every machine in the shop." He also gained valuable experience working with materials that were not in common use before then, such as titanium, used in making airframes. Stan was also pleased that in making tools for manufacturing airplanes "I was doing my patriotic duty."

Although he enjoyed his work, Stan found many practices at the Litchfield plant upsetting. He hated the waste: the workers would routinely bury both their scrap and imperfectly made parts. More important, because it was virtually impossible to get good inspectors, many airplanes had defects and sometimes crashed. Stan felt the underlying problem was that Goodyear received a fixed profit of 10 percent no matter what the actual production was. Stan criticized this policy at meetings where workers were asked for suggestions, but the management ignored his views.

The wartime need for workers, heightened by the diversion of many young men into the military, meant that those hired in the tool room were often unskilled and had to be trained on the job. Back in Ohio the men who worked with Stan came off the farm, but they had also done machine work. In Arizona, however, many were itinerant workers who showed up without a toolbox. They had "just come back from building mine equipment in South America or a railroad shop somewhere," said Stan, who was glad to help train them.

Meanwhile, Stan learned much about prejudice and racism. Since he did not then own a car, he usually took the bus from Phoenix to the plant. During his forty-five-minute daily commute he regularly encountered badly treated American Indians, African Americans, and Mexicans. He also found prejudice between groups that were themselves the targets of discrimination, as in a hot dog place run by a Mexican that refused to serve a nicely dressed black soldier. Even though Stan had lived in a mostly black neighborhood of Akron, he had not before then encountered this degree of prejudice. He advised a black sweeper at the plant to go north where he might have a better chance but later learned that even the union didn't accept black members.

Stan did not have to deal with anti-Semitism toward himself in Arizona: "Nobody could believe I was a Jew." Many thought he was an American Indian because of his dark tan, black hair, high cheekbones, and aquiline nose, so he sometimes experienced racism on that count. He also found it awkward when he would have to tell Mexicans that he didn't speak Spanish. "It was sometimes quite dicey," he recalled, when members of zoot suit gangs would stop him, thinking he was one of them. "When I couldn't answer," he remembered, "they would say, 'You are ashamed to be a Mexican.'"

Stan admired the resourcefulness of one American Indian in the shop who was denied the use of a surface gauge. "He just went up to this big piece of work, took out a rule, marked it off and did a perfect job of machining with a piece of chalk. That's a thing I used to do." But the racism in the shop was so severe that Stan's own position was threatened when he tried to teach skills to American Indians who were interested in getting into the tool room. He was told, "You can't work with those people. They're animals." One racist came to see Stan to tell him, "You've got to stop working with Indians and showing them. They're not any good." Stan asked him how he knew that, and was shocked when told, "Before I came here I was a teacher at an Indian school." Stan found racism and prejudice even between different tribes. An Apache man spat when Stan spoke with him about a friend of his who was a Pima. "The Pima are dog eaters," muttered the Apache.

Herb's Visit to Stan and Norma

During the fall of 1943, Stan's brother Herb, accompanied by Bertha, took the train to Phoenix. Sleeping two nights on the train and eating the foods that Bertha had packed for the trip remain among Herb's fond memories. The fifteen-year-old really enjoyed "seeing the United States through the window. I just could not get over the mountains and desert."

Figure 2.4
Norma, Bertha, and Stan in Arizona.

When they arrived in Phoenix, Bertha and Herb slept on the pullout couch in Stan and Norma's living room. After Bertha went on to California to visit other family members, Herb stayed for two more months, the entire fall of his sophomore year of high school. During the visit Herb and Norma spent much time together and became very close. Herb's relationship to Stan also developed, changing from kid brother to younger brother: "I finally got his attention." The two talked "about everything from life, death, sex, to science, technology, work, and certainly politics." On the weekends, they played chess. At the beginning, Stan was so much better than Herb that he could read the *New York Times* while he played. But Herb recalled that after some time, Stan "finally had to put down the paper to beat me." In Herb's view, Stan's transition from craftsman to inventor and entrepreneur also "happened in Phoenix during the time that we spent on weekends sunbathing." Among the many topics of their conversations was Stan's evolving idea for a center-drive lathe.

Stan's Return to Akron

Soon after Herb left Arizona, Stan received a letter from home reporting that his father was very ill. After a large meal with much to drink at a Labor Zionists party at the Workmen's Circle, and then vigorously dancing the *kazatzka* (which calls for kicking out the legs alternately while in a low squat), Ben had had a heart attack. By this time Stan had already been investigating other possible jobs, including one in a machine shop with Mormon friends in Salt Lake City. "They called themselves Black Mormons because they drank and did other things you're not supposed to do as Mormons." But when he heard that Ben was sick, he decided to return to Akron as soon as he could.

It was spring 1944 by the time he and Norma again boarded the train and traveled back to Akron via Chicago. When they arrived, he immediately went to see Ben.

Figure 2.5
Bertha and Ben, c. 1944. Photo by Herb Ovshinsky.

3 Smarter Machines (1944–1952)

Figure 3.1
The Goodyear Airdock. The low building in front housed the tool room where Stan worked.

Returning to Akron in the spring of 1944 with excellent references from the Goodyear plant in Arizona, Stan was soon hired to work in the tool room of Goodyear's Airdock. Then the world's largest building without interior support, the Airdock was nearly four times the length of an American football field. The impressive hangar-like structure, built by the Goodyear Zeppelin Corporation in 1929 for constructing huge lighter-than-air dirigibles, was used during World War II to build US Navy blimps. It was so large it had its own weather conditions; as Stan recalled, "it rained inside" from condensation.

Working at the Airdock

While the Airdock's architecture was remarkable, Stan found the culture of its tool room alienating. The contrast with his Arizona experience was dramatic. The Litchfield plant in Phoenix had "a real tool room with real toolmakers" who took pride in their craft; even those who just trained there worked hard. But in the Airdock most workers did not exert themselves. Because neither the machinists nor the toolmakers belonged to the working-class culture of skilled craftsmen, Stan felt little kinship with them. It especially irked him when workers who were building airplanes for the war tried to learn about their machines by reading books while lying in the grass. None of this seemed to concern the company, which was guaranteed its 10% wartime profit. To make matters worse, Stan's bosses would complain when he showed imagination. Like Frenchy at Imperial Electric, they would tell him that he was being paid to work, not think. But a job was still a job, and he needed to pay the bills. Despite his dissatisfaction, Stan worked at the Airdock for almost a year.

Life in this period was a struggle for Stan in other ways, too. Since housing in Akron was extremely tight in the early years after his return, he and Norma initially tried living in the Ovshinsky home, but Norma and Bertha did not get along.[1] Even worse for Stan, Bertha rarely allowed him to see Ben, who remained in serious condition after his heart attack. Stan was eager to tell Ben about his frustrations at the Airdock and about his plan to open his own shop and build his ambitious center drive. But when Stan, whose work schedule allowed him little spare time, arrived at his parents' home in his work clothes to see Ben, Bertha typically said, "No, you don't want to disturb him." She did, however, let Stan tell Ben early in March 1946 that Norma was pregnant.

Exchange Auto Parts

In summer 1945, shortly before the war ended, Stan found an attractive opportunity to work with a socialist friend in a small automotive shop. Barney Baranoff, an auto mechanic then rebuilding carburetors and generators in Cleveland, had married one of Stan's close childhood friends from the Workmen's Circle, Frances Wolinsky.[2] Barney

Smarter Machines (1944–1952)

wanted to open an auto parts business in Akron, and as auto parts were in short supply, he needed a skilled machinist like Stan to make them.

Stan and Barney named their new shop Exchange Auto Parts for its location on Exchange Street. For their workspace, they had cleared out and closed in the portico of an old-fashioned 1920s-era gas station that Frances's father bought and let them use. Barney focused on repairs, while Stan ran the machine shop and machined "anything you couldn't buy"—airplane steps, a Buick transmission shaft, standard molding, a parasol, even a urinal. Meanwhile, Akron Standard Mold, the company that had employed Stan when he was in high school, offered them job shop work. Stan gratefully recalled how the jobs for Akron Standard Mold then and later kept him alive. "I would go to them in my work shirt and shop pants and ask for the money ahead of time. They were wonderful people."

Stan and Barney handled most of the work, but Herb, now a seventeen-year-old high school senior, would occasionally help out as an unpaid apprentice. Stan also brought in a talented one-armed machinist named Ernie, who had worked with him at Imperial Electric. (Stan later learned he could be violent and "get into fights," e.g., when he refused to pay for his service at a whorehouse because he was unsatisfied.) Ernie was the one who showed Stan the headlines when the United States dropped the atomic bomb on Japan on August 6, 1945. Stan remembered that he was running the lathe at the time. Having read about the atom and nuclear fission, he sensed that "something terrible was happening."

Stan took pride in making things better and more cheaply than was possible for the big well-equipped machine shops. By beating everyone else's price, he could bring in large contracts, and once again he took pleasure in applying his ingenuity to adapt the "older than hell" lathe to perform a variety of functions. "I could take a little lathe, for example, and make you think it was a drill press, a shaper, this or that," Stan said. But the business remained "a struggling, hand-to-mouth operation," Herb recalled. Also, as the shop had no fan, it was usually filled with dangerous fumes that made Stan's asthma and allergies flare up. Worst of all, the shop was too small for Stan to build his lathe. By late summer, Stan was getting ready to leave Exchange Auto Parts.[3]

During his last months at the shop, Stan looked for a suitable place to build his new lathe. While he searched, whenever he could find some free time he designed and built small parts for the lathe in the tiny back room of a garage owned by a Hungarian friend. A master mechanic at Babcock & Wilcox, Mr. Ricetti (as Stan knew him) had been in the Hungarian army during World War I. After being captured by the Russians, he came back "antiwar and a radical," Stan recalled. Ricetti and his wife had once hoped that Stan would marry their daughter. In Mr. Ricetti's garage, Stan said, "I became a machine builder."

Benjamin Ovshinsky's Death

The time when Stan was beginning to build his new lathe was exciting and hopeful, but it was also a sad time because Ben remained seriously ill. When Stan left Arizona, he had hoped to nurse Ben back to health, but since Bertha rarely allowed him even to see his father, he felt helpless. For his part, Ben had been trying to extend his life. He tried to stop smoking, but it was an uphill battle because he had smoked cigarettes since the age of ten. Losing weight was difficult too. (Ben had been extremely thin as a young man but had gained pounds with the years.) In spite of these efforts, Ben suffered another heart attack, and on March 30, 1946, he died at the age of fifty-four.[4] The hospital nurses let Stan see Ben one last time a few minutes after he died. Stan recalled, "I went in and kissed him on his forehead, held his hand then left to call my mother, and the nurses wouldn't let me use the phone. A most despicable thing. 'No, go downstairs and pay for a phone call!'"

Stan was overcome with sadness. He had not realized just how ill his closest friend and "comrade in arms" had become toward the end of his life. He understood that in keeping Ben's condition from him Bertha was simply trying to protect both father and son, but he never got over regretting that Ben "wanted me and I couldn't come. It was very hard on me." Stan also regretted that Ben had not been to see a specialist, as some of his Workmen's Circle friends had urged. Ben, however, had trusted his favorite doctor and told Bertha he didn't need help from anyone else.[5]

More than a hundred mourners attended Ben's funeral. Large numbers of factory heads, whose respect Ben had earned for his honesty and good values, as well as many workers filled a long hall. Afterward, Bertha sold their house and moved into the second floor of another. Although he often couldn't afford it, Stan continued to support Bertha until her death on January 12, 1972.[6]

The Barn on Chaffin

Still grieving for his father, Ovshinsky (as we should now call him) proceeded to set up his company with support from Akron and Cleveland investors.[7] He named it after himself, the Stanford Roberts Machine Company, and to honor his father he would name his first invention the Benjamin Center Drive Lathe. For a workplace, he had rented an old barn on the outskirts of town on Chaffin Road, off Waterloo. Complete with steel stanchions for holding cattle, its narrow space was heated only by a potbelly stove, and in winter the icy wind blew through the open slats. It did offer better ventilation than Exchange Auto Parts, but Ovshinsky and his staff would have to work in gloves.

To pay everyday bills, he continued to work in the barn on contract for Akron Standard Mold, machining parts and building small machines. His initial costs were high, because he needed to purchase equipment, including a lathe and a drill press. Fortunately, as wartime production wound down, it was possible to find good surplus machinery and good staff. Bertha loaned him money for a used lathe, and by late summer 1946, Ovshinsky had started building his new machine tool. The half dozen people he hired to help in the barn included several toolmakers from shops where he had worked earlier, and he taught some others just out of the army how to do machine work. Ruth Heinnig, who had worked at Goodyear, made the drawings for the first Benjamin Lathe. She was, Ovshinsky recalled, "one of the first female draftsmen that I had ever known." One of Norma's relatives, Sam Schankler, played an important role. A trained engineer, Schankler was, as Herb recalled, the one who helped the most to translate Ovshinsky's design ideas into a working prototype.[8]

Figure 3.2
Ovshinsky in his barn on Chaffin, behind the first Benjamin Center Drive Lathe. Ovshinsky is third from the left, next to Ruth Heinnig.

In the early months, Herb Ovshinsky also helped. He remembered hanging the last fluorescent lights in the barn shortly before going off for his freshman year at the Case Institute of Technology in Cleveland. A month after Herb started at Case, his girlfriend Selma, whom he had been dating since high school, called to tell him that her forty-six-year-old father had been killed in a factory accident. Herb came back to do what he could to help, and they now became a committed couple. Selma was seventeen and Herb was eighteen.

The First Invention

In building his new lathe, Ovshinsky did not simply make partial improvements on existing machines; here at the outset of his inventive career, he created the design by working from basic principles. He understood that unnecessary movement—whether mechanical play or chatter of the cutting tool—limited a lathe's speed and efficiency, wasting energy and damaging both the machine and the piece being worked through destructive vibration and heat. He knew the standard tricks machinists used for quieting chatter such as resting your hand on the machine bed, treating it like "a woman you wanted to woo." But Ovshinsky aimed to prevent chatter before it could ever begin by eliminating all mechanical looseness. He began by replacing the traditional cast iron bed, typically on legs, with a massive block of welded steel. The slide, which held the cutting tool, was another piece of hardened and ground steel, and there were no gears, belts, or screws, just a chain drive from the powerful motor (initially 25 horsepower, later increased to 50) to the chuck.

Most important was the design of the chuck, the radial clamp whose jaws held the piece to be machined. It was also massive, and instead of being tightened by screws it was held closed by compressed air (later replaced by hydraulic pressure) with such force that, as Ovshinsky proudly said, "you couldn't make the part slip." With this design, nothing could move except the simple wedge pushing in the slides and moving them forward. As a result, the lathe could make much faster and deeper cuts than anyone had thought possible. There was no chatter and no heat buildup, so there was no need to use coolant. While the heat of other machines produced blue chips, the chips from Ovshinsky's lathe were white, and, as later tests showed, their microscopic structure was undisturbed. Other features further increased the lathe's efficiency. The powerful chuck gripped the piece in the middle (hence "Center Drive") so that it could be machined at both ends without taking it out, and as Ovshinsky had learned to do with older machines, he could use his lathe for operations that usually involved several machines, such as boring, milling, or threading. (In later versions there would be different stations corresponding to the different operations.)[9]

In later years, Ovshinsky liked to compare his contribution to the tool-making field with John Wilkinson's invention of a machine that could accurately bore a hole large enough for the cylinders of Watt's steam engine. The steam engine was, arguably, the key technology of the Industrial Revolution, and Ovshinsky envisioned comparable revolutionary consequences from his Benjamin Lathe. Although the lathe's actual impact was more evolutionary than revolutionary, it was clearly an important advance in machine tools, and there is no question about how important it was to Ovshinsky's development and sense of himself as an inventor.[10] In his work as a machinist, he had devised several ingenious improvements that contributed to the lathe, such as grinding his tool-bits to minimize heat, but this was the first time he conceived and created a whole new device. Long afterward, he would retell the story of inventing the lathe, and he chose it as a "case history" for his autobiographical account of his creative process.[11] Building the lathe confirmed his belief in his intuitions, contributing to the profound self-confidence that he drew on throughout his life.

On November 18, 1946, while the new lathe was being created, Stan and Norma's first child was born. They named him Benjamin too, after his grandfather. Herb, by now studying at Case, prepared a clever birth announcement in the form of a blueprint for a machine tool called Baby Benjamin. The drawing specified the new machine's features, including feed rates, inputs and outputs, and performances (including "self-lubricating"). Two more sons would follow before long: Harvey was born on April 9, 1948, and Dale on August 27, 1949.

Stanford Roberts in Dover

The first Benjamin Lathe, the one Ovshinsky built in his barn, was a prototype to establish what he would later typically call "proof of principle."[12] He planned to make the next lathe much bigger and to make it completely automatic, reducing to seconds the time for jobs that often took minutes or hours. For that project, he needed a larger space.

After raising money with the help of an accountant friend, Ovshinsky found a suitable shop in Dover, Ohio, about 50 miles south of Akron, where he moved during the summer of 1947. The shop was in a small modern building at 619 East Iron Avenue. The landlord, who lived next door in an old farmhouse, was "a machine shop guy himself," who "had built this plant for himself before realizing that he was too old to run it." He was building an automobile in his garage, because, as he told Ovshinsky, "I don't think these new-fangled cars are what they should be." Ovshinsky's assessment of the heavy car he was building was that it would never work. "He would probably build it until he died. That's the sort of thing that went on in parts of rural America."

Ovshinsky rented a small house for the family next to the plant. He found that he loved living in Dover both for "the wildness" of the surrounding Amish countryside and the novel experience of a small town nestled in the Appalachian foothills. There were only a few restaurants, he recalled, mostly small Italian ones.

In Dover, Ovshinsky could focus on building a larger lathe, which incorporated several improvements. He replaced the compressed air controlling the chuck and slides with hydraulics driven by a second motor. By this time he was also taking advantage of the new carbide cutting tools; as Herb observed, they were available in the late 1940s but not widely used because machine tool companies were slow to adopt them. Such tools allowed machining at much higher speeds. Also, because of the need to have hardened surfaces in a part like a crankshaft, Ovshinsky drew on his high school forging experience to build the process of heat treating right into the lathe, applying a flame to the part as it revolved and adding a water tank to rapidly quench the heated surface. Like the different machining operations that could be performed on the lathe, this was another way of integrating and expediting the manufacturing process.

As Stanford Roberts grew, Ovshinsky made many business contacts. The most important was with Ralph Geddes, a public accountant who had become wealthy buying distressed or bankrupt companies, then rebuilding and selling them. Earlier, he had been the head of the Peerless Motor Company, a Cleveland manufacturer of luxury automobiles, and though Geddes worked with a variety of firms, like Chicago Gear and Gibson Refrigerator, he particularly favored the automotive business. In that connection he would for a while become a towering figure in Ovshinsky's world.

Geddes recognized Ovshinsky's talents as an inventor because his company in Cleveland had one of the Benjamin Lathes. He was particularly impressed by how quickly it could machine heavy shafts, including crankshafts, whose eccentric lobes and weight made them extremely tricky and labor-intensive. So in 1947 Geddes invited Ovshinsky to be part of an ambitious plan "to make the best cars that could be made," Ovshinsky recalled. The plan included Danish-born William Signius Knudsen (1879–1948), who had formerly been the president of General Motors (May 1937 to September 1940, succeeding Alfred P. Sloan). When Knudsen finished his service directing US wartime production, he was available to work with Geddes, having had a falling out with General

Figure 3.3
The Dover shop. Ovshinsky's improved double-end centering lathe is in front and his larger 50-horsepower version in back.

Motors. Geddes created a plan in which Knudsen was to be the builder of the new cars, Ovshinsky was to be the inventor and developer, and Geddes would serve as the financial man and run the plant.

Ovshinsky's lathe was a crucial component of this plan, so Geddes hired M. Kronenberg, a leading expert in machine design. The Cincinnati-based mechanical engineer came to Ovshinsky's Dover shop to evaluate the Benjamin Lathe in April 1948. Some weeks later, he tested it "for the purpose of determining its metal cutting capacity in comparison with present American practice." Using sintered carbide lathe tools to run a series of tests, he found "unusual cutting performance in spite of the fact that the machine was not in first-class condition but needed general overhaul, even before tests had begun." But despite the defects of this "hand-made machine," Kronenberg reported that it offered "excellent service," finding that its cutting capacity averaged about 200% (i.e., a factor of two) "above present practices."[13]

But when Knudsen died on April 27, 1948, just as Kronenberg was drafting his highly favorable report, Geddes had to give up the idea of starting a new automobile company. He had by then become the majority owner of the Hupp Motorcar Corporation in Detroit, which had formerly made the well-respected Hupmobile. Geddes turned Hupp into an auto supply company, producing gears and other components, such as the first electric windows (a device that Ovshinsky later worked on; see chapter 4). Geddes still wanted to work with Ovshinsky and said he was going to invest in Stanford Roberts Machine Company. The Sunday night before the signing was scheduled, Geddes called to tell Ovshinsky, "Everything's all set. You're going to get your money tomorrow morning. I want to congratulate you Stan. We'll always remain friends." But by morning Geddes had changed his mind. He told Ovshinsky he was backing out, noting only that it was nothing personal. It was not the last time that Ovshinsky would be subjected to Geddes's caprices.

Meanwhile, Herb Ovshinsky had been getting more involved with Stanford Roberts. He had learned a lot in his freshman year courses at Case, but his grades weren't good enough for him to continue. He dropped out of the engineering program in June 1947 and spent the summer helping to set up the Dover shop. Herb planned to continue working there, but in September he contracted polio and could not work, or even walk. He was in the hospital until November and was still on crutches in March 1948, when he and Bertha drove to California. Herb remembers feeling "empowered" because the car had Hydramatic transmission, so he could drive it using only his right leg. A month later, he was walking without a cane when Stan called to tell him about Geddes's proposed investment: "Mr. Geddes wants to do this. You better get back here." Even though that plan fell through, Herb returned and took charge of the second shift. He was now nineteen and rented a room with a family in Dover's twin town, New Philadelphia.[14] Herb enjoyed coming to work later in the morning: "Norma would make my favorite sandwich, salami and sliced bananas and peanut butter and lettuce." Often Herb brought along his "pet," young Ben, by now a toddler.[15]

As he now traveled regularly between Akron, Toledo, Detroit, and Cleveland, Ovshinsky's reputation as an inventor grew, and he recalled meeting "all the big shots in the auto industry." Among the many excellent mechanics and managers of automobile firms Ovshinsky met was the independent and outspoken master mechanic John Dykstra, later president of Ford. Ovshinsky said he "knew Ford was changing when they hired John Dykstra," for he always spoke his mind. Ovshinsky recalled eating with him at the country club when a man tugged at Dykstra's sleeve and said "You can't say that, Mr. Dykstra." Dykstra replied, "I'll say any damn thing I want."

With Stan often traveling on business while Norma was isolated in Dover, their marriage became strained, although it would be years before anyone outside the family realized it. To Herb, who had observed the couple in Arizona when they were still happy newlyweds, the tensions were very clear. As the relationship unraveled, Stan often confided in Herb, who "knew many years before the breakup that Stan was very, very unhappy." Stan continued to find Norma physically attractive, but he felt that she rejected everything that mattered to him, including his politics and his science. She thought of his work as just a job, while to him it was the most important thing in his life. He came to realize that in marrying Norma he had made a mistake. As he later reflected, "I should have rectified it sooner, but I was worried about the kids."

Norma had her own quite valid grievances. She had not wanted to move to Dover and hated living there. She also hated the fact that she and Stan were constantly struggling financially, and that Stan traveled so much to see clients. She wanted him to spend more time with her, but it seems clear that he didn't want to. Norma "didn't want to get divorced," said Herb, and "did everything she could" to save the marriage. But Stan did not reciprocate. When Norma urged him to see a counselor with her, he said he was too busy. "I'm not the one who needs help," he insisted.[16]

Moving Back to Akron

By the late summer of 1948, Ovshinsky had decided to change the business plan for Stanford Roberts. Instead of manufacturing the lathes, the company would focus on engineering and sales, contracting with others to build the machines. The family moved back to Akron and lived for two years on Madison Avenue, while the company moved to offices on East Exchange Street.[17] Ovshinsky set up the manufacturing operation with Heidrich Tool and Die Corporation in Detroit, where he commissioned an even larger center drive machine and also maintained an office. Within a year, Stanford Roberts had to change both locations. In Akron, they were evicted from the Exchange Street offices and found new quarters on Main Street.[18] In Detroit, Ovshinsky realized that Heidrich was cheating him, so he found another manufacturing partner, Baker Brothers in Toledo.[19] For their Toledo offices, Stan and Herb rented space in the vacant Harbauer pickle company building.

About six months later, however, Ovshinsky ran into more serious problems with his investors.[20] As Herb recalled, "Our investors realized it would take a lot more money to be a player in the machine tool business. They didn't have the faith that this was the way to go. They weren't visionaries." Ovshinsky recognized that since they were unwilling to make further commitments, he would have to sell the business and return their money.

Ovshinsky returned to Geddes. "What do I do now?" he asked. Geddes offered to organize a meeting at the Cleveland City Club to help Ovshinsky sell Stanford Roberts to a machine tool company. He told Ovshinsky he planned to invite "an old style machine company," cautioning him that he wouldn't "get any money, but you can see that your machine gets made," Ovshinsky recalled. At the meeting in the spring of 1950, a Cleveland company that was part of the New Britain Machine Company Group of Connecticut showed strong interest. New Britain had established itself as the "Hardware City of the World," and the New Britain Machine Tool Company was one of the largest and best in Connecticut. The resulting deal gave the company rights to the Benjamin Lathe and included hiring the Ovshinskys to work in New Britain.

Machine Intelligence

In the interval between making the deal and moving to Connecticut, Ovshinsky worked in Akron on a longstanding interest, the further automation of his lathe. As the patent (2,619,710) announced, it was "particularly ... adapted for automatic operation," and some of its functions had already been programmed with relays using ladder logic.[21] But Ovshinsky envisioned a completely automatic machine, where "you could dial in whatever sizes you wanted to machine," and "all you had to do was push the button; the machine did the rest." To accomplish that, he felt he needed to think more deeply about the whole question of control mechanisms. He also needed a place to work. As he recalled, "There was an old bootlegger who took pity on me." For a very low price he let Ovshinsky use a high floor of his ancient building. The building was "like out of a horror movie; all the floors were empty," recalled Ovshinsky, who was afraid to ride in its creaky elevator.

Instead of concentrating only on making machines smarter, however, Ovshinsky also tried to understand human and animal intelligence. This expansion of focus was a crucial move. It opened up a line of investigation that would eventually lead to his revolutionary work with amorphous materials about ten years later (see chapter 4).[22]

Ovshinsky was led in this direction by his reading in the new field of cybernetics, which systematically crossed the boundary between animate and inanimate intelligence: the full title of Norbert Wiener's influential 1948 book is *Cybernetics; or Control and Communication in the Animal and the Machine*. There, as well as in his less technical 1950 popularization, *The Human Use of Human Beings*, Wiener showed how both mechanical and animate control functions can be understood in terms of the same processes, such as feedback, "the property of being able to adjust future conduct by past performance."[23] Ovshinsky corresponded with Wiener and even wrote a review of *The Human Use of Human Beings*.[24] He also studied the work of the British psychiatrist and neurophysiologist William Grey Walter, the Russian physiologist Ivan Pavlov, and Hans Berger, the German neurologist who invented the electroencephalograph for measuring electrical activity in the brain.[25]

The particular problem that Ovshinsky addressed in 1950 was a special case, controlling a paralyzed or prosthetic limb. He wrote a paper titled "The Use of Electro-Mechanical Motion to Replace the Loss of Human Movement," in which he proposed using electricity to replace nerve impulses. It was a typically cybernetic problem and solution, as were those considered in his later, more general "Nerve Impulse" paper, which devoted much attention to epilepsy.[26] Wiener also discussed such disruptions of normal motor control processes, and Ovshinsky followed his lead in considering them as communication failures, or disordered circuitry. ("Circuits are circuits," he would say.) But Ovshinsky would go much further in trying to understand precisely how neural impulses were propagated. In the mid-1950s, he not only read deeply in the neurophysiological literature but also contributed to it himself and conducted original research (see chapter 4). That work might seem like a wandering detour for a master machinist and machine builder, but it proved to be a crucial turn in his development as an inventor and a confirmation of his deep intuition about the nature of intelligence in humans and machines.

Building Lathes in New Britain

Stan arrived in New Britain alone in the fall of 1950. Norma and the boys planned to join him after he had found a place for the family to live. Stan remembered that the weather was brisk and that football season had just begun. He tried to go to work, but the factory was closed. The watchman told him no one would be there on Saturday, "and besides, there's a Yale game on."

New Britain's machine shop culture was again different from those he had known in Akron and Phoenix. In Akron, the machinists and toolmakers Stan had worked with were mainly Scots-Irish, while in Arizona many had been itinerant laborers from all over. Here, while the Yankees owned the factories, the shop workers and foremen were immigrants from Europe or Canada. Most were Swedes, French-Canadians, Portuguese, Poles, or Italians. Each group had its own culture, and Stan found it all interesting. He also found Connecticut "a beautiful state," where he learned to like seafood. "I had this strange thing called scallops. Boy, did they taste good. And I had Italian food and all kinds of herring. The Swedes made their own herring. And we'd drink a lot and go to their house and we'd have herring." Polish refugees, many of whom had arrived in New Britain after World War II, were the city's largest ethnic group, and

Herb recalled that they called Stan "Stashu." "In Polish, Stashu is a common nickname for a young man."[27]

Norma and the boys joined Stan in early December, and Herb and Selma arrived about the same time, having been married in late September. By this time, Stan had found a small place for the family on Westover Road. "It was still after the war and there was very little housing," he said. Ben was four, Harvey not yet three, and Dale about eighteen months old. It was becoming clear that Dale was having developmental problems, especially with language skills. When he was diagnosed with aphasia, Norma and Stan took him to Yale to see the aphasia experts there, and also to an independent speech therapist, who encouraged Stan to work with Dale using flashcards, an activity that Stan took very seriously. But since Stan continued to travel a great deal, it was Norma who bore the brunt of raising the boys and creating a supportive learning environment for Dale. "It took its toll on my mother and certainly their relationship," Harvey said. "Later I read letters she wrote to Dad, imploring him to spend more time with us."[28]

At the New Britain Machine Tool Company, the Ovshinsky brothers formed their own separate engineering department. Stan, who was head, "acted as a very senior employee," said Herb, who served as his brother's assistant. The company initially had no place to put them, so they were moved into the office of George Gridley, the

Figure 3.4
Ben, Harvey, Dale, and Stan with a 1950 Studebaker.

designer of the multiple screw machine that the company was known for. "Gridley was mostly in Florida," Herb recalled.

The Ovshinskys had expected to start building Benjamin Lathes in New Britain. "When we walked in there we had orders for about ten machines, making transmission cases, making crankshaft lathes, and all the choice stuff," Herb recalled. They had already begun working on their crankshaft machine for Chrysler when suddenly they were told to stop. After the Korean War, which had begun in June 1950, escalated with the Chinese intervention that fall, the machine tool companies could not build anything without a government rating. All orders not on the Pentagon's priority list, including the Benjamin Lathe, were put on hold. As salaried employees, the Ovshinsky brothers were asked to help with the company's other work.

But that was not why Ovshinsky had come to New Britain. When he asked the management of the company what their highest priority was now, he learned it was making artillery shells, because with the Chinese attacking, there was a serious shell shortage. Shell cases had gone from brass or bronze, which are expensive, to a special steel that was extremely slow and cumbersome to machine. "What the hell," Ovshinsky said to himself. "I'll tell them I can do it on this machine."

To show the New Britain managers what the Benjamin Lathe could do, Ovshinsky staged a formal demonstration of the machine he had built in Dover. Herb recalled showing the chief project manager that their machine operated at a thousand surface feet per minute, while the New Britain machines operated at a hundred. In the demonstration Ovshinsky showed how his lathe could machine a shaft on both ends and do it much faster, taking minutes to do a job that typically took hours. That convinced the higher-ups at New Britain, including the president, the chairman of the board, and the chief of marketing, that with his Benjamin Lathe they could meet the shell case demand. Asked for quotes, Stan and Herb quickly drafted a proposal for machining the 105-millimeter shells, including drawings, time charts, tooling, and estimated costs.

In building more of his Benjamin Lathes at New Britain, Ovshinsky also had to deal with resistance from the older workers, who shared many beliefs about how machines must be built. When they learned that the bed would be made of welded steel, they exclaimed, "That's impossible! You take castings!" Not only should it be made of cast iron, they said, but it also had to be properly seasoned, left outside to ripen and mellow from being exposed to rain, snow, and sunshine for six months. Then each machinist would scrape in his own favored pattern of "curlicues," grooves that channeled the flow of coolant, and only then was it ready to be made into a machine tool. Ovshinsky liked these old-timers and felt they were "good men" who represented a bygone era

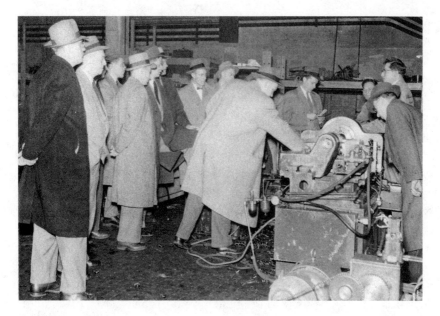

Figure 3.5
Demonstration of the Benjamin Lathe in New Britain. All but Ovshinsky and his assistant (back right) are wearing hats.

"when America was at its peak in machine tools," but like a true revolutionary he dismissed what he considered their superstitions.[29]

Around Christmastime 1950, Bob Frisbee, the company's sales manager, traveled to California to present the Ovshinskys' proposal to Norris Thermador, a company that ordinarily made stoves and refrigerators but now had a government contract to develop a steel shell case. The lathe was now even more advanced. Unlike others, it made use of carbide tooling and was highly automated, with a hundred relays dedicated to its transmission case alone. As Herb noted, "We had DC variable speed motor drives for the spindle, which was brand new at that time." Frisbee had hoped just to secure an order for building a prototype, but when he came back early in February 1951, he brought huge orders for two versions of the machine, each for about a hundred machines.[30] It was going to be an enormous effort.

To make the deadline required mobilizing the whole plant (plus two or three others) and dedicating a building to assembling the machines. Soon, Herb recalled, "all you could see down the middle of the main assembly area was our machines, because they had to build hundreds of them. We were building and shipping machines like mad. I started traveling all over the country to help install them."[31] The Ovshinsky brothers

Smarter Machines (1944–1952)

did not receive royalties on the lathes, but the company made so much money using the lathe that the government had to renegotiate. "So it was quite a winner. They loved it," Ovshinsky said, "but they didn't understand it."

Even during the feverish effort of building the lathes in New Britain, Stan and Herb found time to talk about other topics, especially automation and forming their own company for making smarter machines, which they eventually did in 1953 (see chapter 4). Ovshinsky tried during the summer of 1951 to enroll in an MIT course on cybernetics and servomechanisms. But even though he was working and actually inventing in this area, the school turned him down because he didn't have the formal prerequisites for the course. Rejected by MIT, Ovshinsky studied on his own. "I never thought that there was something that I couldn't do," he said, an attitude that made many projects possible for him over the course of his career.

The Automatic Tractor and the Industrial Computer

Ovshinsky worked on several inventions on the side in New Britain, projects that contributed to significant later developments. Two of them, an automatic tractor and what he called his "industrial computer," grew directly out of his interest in intelligence and control. For these side inventions, Herb put a lathe in the basement, and later they had the use of a barn. As Ovshinsky recalled, "I had complete freedom at New Britain. I worked day and night; I just made the time to go down to this place and work on it. We'd just get something made in the shop and put it together and see if the principles worked." For developing the automatic tractor, Ovshinsky started with a miniature tractor a few feet long he had bought that summer just before moving to New Britain. He had already been thinking about the invention in Akron, but not until then did he have time to follow through and make it work. With Herb's help, he automated the tractor by adding an electrical control system. "I've got a patent for it."[32]

To run the tractor on autopilot, Ovshinsky "set a program that would tell you how far you went. And the program would tell you to turn right and do a 180 (and of course today with GPS it would be a no-brainer)." The tractor would work a field by running down one side and, on receiving instructions from a drum and a gyroscope, turn around and repeat the process along a parallel line, continuing until the entire field was worked. It was a closed loop system.[33]

For Ovshinsky, the automatic tractor was a direct extension of his work on automating the Benjamin Lathe. The basic idea was to "machine your plot of land like you would machine a piece of steel." Following the war, "there was still a lot of hunger out there. I wanted to make an automatic tractor which would use industrial type thinking

a.

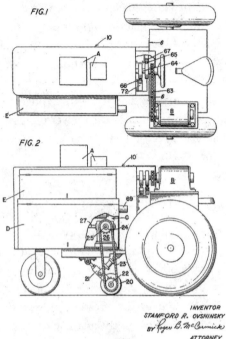

Figure 3.6a
Patent for the automatic tractor.

b.

Figure 3.6b
Automatic tractor.

for agriculture." Ovshinsky had for some time been interested in modern agriculture, reading authors like Louis Bromfield.[34] He thought automation should not be "limited to manufacturing. I'm going to look and see what can I do that would be helpful to the world." The notion that automated tractors could reduce hunger may not seem particularly persuasive, but this appears to be the earliest instance of Ovshinsky's thinking about how science and technology can help solve important social problems, a concern that would increasingly shape his career.

The other smart machine that Ovshinsky began working on in the Connecticut barn was the apparatus he called his "industrial computer." He had long dreamed of automating his lathe with a system that could control any machine. It was conceived as an array of servo and feedback mechanisms like the autopilot that controlled the automatic tractor but much more complex. He had been giving himself a crash course in logic, reading Bertrand Russell and others, and with that knowledge he designed a circuit to be built with relays and vacuum tubes. The instructions for the part to be machined would be on magnetic tape, and a roller in contact with the piece would provide feedback through potentiometers. It is unclear how far Ovshinsky got with this ambitious project in New Britain, but he would return to it in a few years (see chapter 4).

Leaving New Britain

About six months after Ovshinsky arrived in New Britain, Ralph Geddes began trying to bring him back to the Midwest, contacting him in letters and phone calls. He offered substantially more money than Ovshinsky was getting in New Britain to serve as his adviser and as the director of research for Detroit's Hupp Motorcar Corporation. "And the offer got better as time went on," said Ovshinsky. "I was making $8,000 at New Britain Machine Company, and with Geddes I went up to $10,000, and then $12,000, and he would give me bonuses." It was only a matter of time before Ovshinsky would accept the offer. He wanted to devote all his time to developing smarter machines, and he felt that Herb could handle the remaining work on the lathe in New Britain. Once Geddes had increased his offer to one that Ovshinsky couldn't refuse, he went to Herb Pease, the president, to see whether New Britain could match the offer. They could not. When he left late in 1951, Ovshinsky recalled, Pease "came over and gave me a check and said we owe you at least this."

For the next several years after Ovshinsky left New Britain, his Benjamin Lathe continued to break tool-making records. He was so proud that he decided to send for the 1954 performance reviews of his machines. Most were raves.[35] Ovshinsky was disappointed but not surprised, however, to learn that after he left New Britain there was a dramatic drop-off in orders, and the company slowed the speed of the production machines and put coolants back into them. It was clear that while the old-timers had agreed to work with his machine while Ovshinsky was there, they had not accepted its principles. "It bothered me," he said, but this regressive response confirmed his decision to leave New Britain. "I wasn't going to spend my life trying to teach people enslaved in their own minds." Nor did he want to wait, like Moses, for forty years in the desert for a new generation.

4 Love Story (1950s)

Late in 1951, Stan, Norma, and their three young boys left New Britain and moved to Detroit so that Stan could begin his work as the director of research at the Hupp Motorcar Corporation. It was a move to the center of Ovshinsky's world, for Detroit, with its booming automotive companies, was the mecca of the industrial age and was to be the setting for the rest of his career.

Detroit

The city had developed dramatically since the seventeenth century, when Louis XIV authorized building a French settlement on *le détroit* (the strait) of the Detroit River connecting Lake St. Clair and Lake Erie. The fertile region that had attracted French fur traders became a site of territorial dispute, was taken over by the British in 1760 during the French and Indian War and was ceded to the United States at the end of the American Revolutionary War. Shipping, shipbuilding, and manufacturing drove the city's growth through the nineteenth century.[1]

Detroit's proximity to the Great Lakes was also an asset for the motor vehicle industry that grew there in the early twentieth century. Raw materials were shipped by boat and train—coal from Pennsylvania and West Virginia, iron and copper ore from northern Michigan and Minnesota, and steel from Pittsburgh, Youngstown, Cleveland, Gary, and Chicago. In 1908, Henry Ford's motorcar company began producing its legendary Model T, which by 1914 was being manufactured incredibly quickly and cheaply thanks to Ford's invention of the moving assembly line. As was memorably dramatized in the opening segment of Charlie Chaplin's classic film *Modern Times* (1936), the efficiency of producing this historic car often came at the cost of the workers who mindlessly performed repetitive tasks.[2]

General Motors (GM), founded in 1908, arose as Ford's major competitor when the Detroit-based Buick company controlled by the salesman William Durant began to incorporate other car lines, including Oldsmobile, Cadillac, Oakland (later Pontiac), and Chevrolet.³ With the start of Chrysler in the mid-1920s, the third member of Detroit's "Big Three" automobile companies was in place. Unlike other Detroit car companies, Ford, GM, and Chrysler all survived the Great Depression because of their size and their many innovations. A decade later, the Detroit auto industry profited enormously from war production.⁴ It was thus mature and thriving by the time Ovshinsky joined it in late 1951, hoping to modernize it with his innovations.

Arrival in the Motor City

The weather was icy when Ovshinsky arrived in Detroit. "I remember the weather more than anything," he said. But he was warmed by his enthusiasm for moving to this "vibrant city with everything I wanted." Stimulated by the rows of humming machines in the plants, he felt "in the middle of things that were going to change. There was an excitement to Detroit," he recalled, "a dynamism that I liked, and I thought I'd be freer to express my ideas." He looked forward to working with the kinds of talented people he had met earlier at Ford and Chrysler, and he already knew many of the city's industrial leaders. As for family needs, Detroit had an active Jewish community, excellent public schools, and a Workmen's Circle school.

Ovshinsky was also eager to help solve Detroit's seething social and political problems. Like Akron, Detroit had attracted large numbers of immigrants and migrant workers to its thriving industries, producing ethnic and racial tensions that at times exploded in violence.⁵ Ovshinsky initially stayed in touch with the Detroit labor unions, especially the United Automobile Workers (UAW), but he slowly withdrew when they opposed his talks about the promise of automation. In his view, his smarter machines were designed to free workers from mind-numbing repetition. "I was trying to convince the leadership that automation needn't be their enemy and we should start training people now who could understand it and respond." But the unions continued to see automation primarily as a threat to jobs, a view that subsequent history seems to have confirmed.⁶

In December 1952, the Ovshinskys moved into a modest three-bedroom ranch, ample for the family, at 19935 Forrer Avenue, in the 7 Mile and Greenfield section of Detroit.⁷ Its unfinished basement became an office space, which Ovshinsky crammed with papers and books, including medical texts that he later encouraged eight-year-old Ben to read, as he himself had done eagerly at that age. Ben, however, recalled it as "terrifying stuff" and preferred baseball or basketball to such reading.

Not long after moving to Detroit, Ovshinsky reconnected with his old Akron friend B. J. Widick, now also living there. Irving Howe, the eminent left-wing intellectual and critic, with whom Widick had co-authored the important book *The UAW and Walter Reuther* in 1949, was launching the intellectual quarterly *Dissent,* which Howe would edit for the rest of his life. Howe wanted Ovshinsky to help find subscribers, and he invited him and Widick to a meeting at a working-class restaurant near the General Motors building. Reluctant to attend, Ovshinsky explained to Howe that he and Widick "used to have violent quarrels." "For Christ's sake," Howe responded. "We've got to get this journal together and it's got to represent the point of view of non-totalitarian, democratic socialism." He added that Widick had changed. Although dubious, Ovshinsky joined the dinner. "And then BJ comes in," he recalled. "And the first words out of his mouth were, 'Hi Stan! Gee it's great to see you. You know, you were right and I was wrong.'" As their friendship instantly resumed, it seemed to both that only days had passed since their last discussion. To help publicize Howe's magazine, Ovshinsky organized a series of talks, which were well attended.

Still active in Detroit's labor movement, Widick chose Ovshinsky as his campaign manager when he ran as the socialist candidate for mayor. Ovshinsky assembled a large audience, but it was strangely unresponsive. "I just couldn't move that crowd," Widick recalled, "and I was a pretty good speaker in those days." He later learned that everyone in the group Ovshinsky had gathered was Russian-speaking. "I don't think I'll hire you again as campaign manager," said Widick, who failed to get many votes. The two socialists remained close friends for the rest of their lives. Years later, Widick remarked that Ovshinsky "really didn't belong in formal radical politics," but he added that he knew "Stan was going to be somebody" when he "started telling me some of his dreams," particularly about getting affordable power from the sun.

Ralph Geddes Redux

The dominant industrial figure for Ovshinsky in his early Detroit years continued to be Ralph Geddes, always known to the family only as Mr. Geddes. Ben recalled that during the winter when he was seven, the family drove to Cleveland in their luxurious new green 1953 Packard Clipper to visit Geddes at his home in Shaker Heights, a posh section with "block after block of lovely Victorian Gothic, Colonial brick, and Tudor mansions." Norma had dressed the children up in their best clothes, with their long coats. After "knocking and knocking on the door of Geddes's mansion in the snow," the family was ushered into the presence of Mr. Geddes, who to the seven-year-old looked like "the tycoon with a big mustache and a cigar" in the Monopoly game. Even at that age, Ben could sense that Geddes embodied power, for his father treated him with uncharacteristic deference.

Ovshinsky's feelings toward this old-school industrialist were complex. On the one hand, he said, he resented the fact that Geddes "depended on me for everything," not only for business decisions and technical advice, but also for help with raising his son.[8] And Ovshinsky was bothered that the rich Taft Republican "didn't share his wealth as he said he would," not to mention that he was "a bit of a paranoid." At the same time, he considered Geddes "very talented as a manager," especially "when it came to the financial side." He felt indebted to Geddes for hiring him at Hupp after helping him sell the failing Stanford Roberts Machine Company.

Ovshinsky was also grateful for Geddes's protection during the McCarthy era, when he came under government surveillance because of his socialist politics. During his last years in Akron, the FBI had been checking on him regularly. When investigators visited the Workmen's Circle and asked the old-timers whether they considered Ovshinsky a radical, they answered, "He's the best radical we ever produced in Akron." Such testimony confirmed the FBI's worst suspicions, and they pressured Geddes to fire Ovshinsky as a threat to national security. Geddes refused, pointing out that Ovshinsky had made "hundreds of machines working for the military." He added that Ovshinsky's association with the socialist Norman Thomas did not make him a disloyal American, noting that his own banker had voted for Thomas. Ovshinsky had in fact always been part of the anti-Communist left, but the FBI was oblivious to such distinctions. Although he did not lose his job, he was denied clearance even to enter places where his own Benjamin Lathes were serving the government. "It was surreal," he recalled.

Smarter Cars

Among other projects, Ovshinsky worked at Hupp on developing cybernetic components for cars. Most of these used sensors to control basic automotive functions, such as braking, steering, or power transmission. As he later put it, he wanted "to put sensors all over your car," an aim achieved by automakers decades later with the advent of the microchip. One of Ovshinsky's automotive inventions was a new kind of automatic transmission. Unlike the existing hydraulic systems, which wasted power and fuel because of slippage, Ovshinsky's design used electromagnetic clutches that adjusted gear ratios in response to road conditions. "It shifted by itself," he said, anticipating the torque converters in today's systems. Ovshinsky filed for a patent on this invention in January 1955, and it was issued in October 1957.[9]

Another invention responded to reports that children were being injured or even killed by the new power windows that Hupp Motorcar was the first to make. The children would push the button to close the windows "and then they couldn't stop the thing and it was like a choker or a guillotine," Ovshinsky explained. To remedy this tragic situation, he invented a sensor-based device to prevent the window from closing if it sensed a hand or other object in the way. But when Ovshinsky showed this invention to an engineer at Ford (which, unlike Hupp, actually made and sold cars), explaining that it would cost only pennies when produced in volume, the reply was "we don't spend pennies when we don't need to." The incident illustrates the kind of frustration Ovshinsky routinely encountered in his work at Hupp.[10]

The most important automotive invention Ovshinsky produced at Hupp was electric power steering. Herb remembered the night his brother phoned him in Connecticut soon after starting at Hupp to talk excitedly about his idea to adapt the closed loop electromagnetic steering mechanism they had developed in New Britain for the automatic tractor (see chapter 3) into a design for electric power steering in cars. "That will be great," Herb said.[11] The invention was a simple application of cybernetics. A sensor based on a few inches of rubber tubing inserted into the steering column registered turning resistance, and a potentiometer would respond by sending current to two slipping electric clutches to match the resistance with the right amount of power assistance. Such feedback offered the same kind of closed loop intelligence that human brains use in making corrective decisions based on sensing. Even in icy weather, Ovshinsky claimed, his power steering was more sensitive, and much safer, than GM's recently invented, expensive, open loop hydraulic approach, which could generate too much power and lead to a loss of control. Ovshinsky's device also made for a smoother drive by eliminating road vibration because the slipping of the clutches changed with the resistance. Moreover, he said, "It was simple, cheap, and could be put on any car as an add-on."

On Saturdays, Ovshinsky tested his power steering at the General Motors proving ground in Milford, Michigan. Sometimes he brought along Ben, who remembered, "We packed pastrami sandwiches, and salami sandwiches, and we would drive around the track all day in a Class 8 truck tractor-cab, talking about his father, Yiddish theater, Bolshevism, politics, history, mythology." When Ovshinsky felt satisfied with the tests, he and Geddes showed the invention to General Motors, which had recently set up their big hydraulic power steering manufacturing unit in Flint. Engineers from GM's Saginaw Steering group tested Ovshinsky's device and "wanted to make a deal with us."

But as the attorneys who met with Ovshinsky and Geddes on the top floor of the General Motors building in Detroit were writing up the papers, Geddes suddenly walked out of the room. Ovshinsky and Geddes's son-in-law followed him, thinking

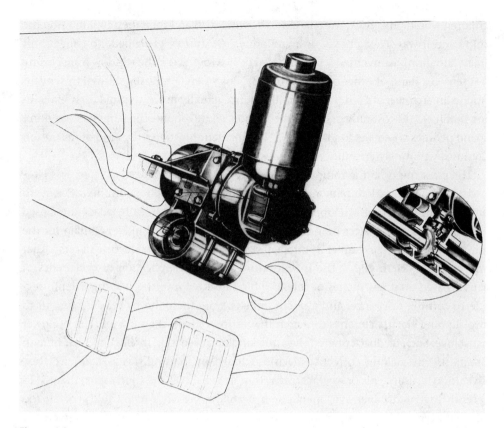

Figure 4.1
Ovshinsky's Electric Power Steering.

that he might want to have a last-minute talk before signing. But Geddes continued on toward the elevator. When they asked him what he was doing, he replied, "Walking out. They're out to screw us." Ovshinsky objected: "There's not one show of any kind except extreme gratitude that we're going to work with them." But Geddes insisted, "You guys just don't know."[12] And when he took the elevator down, it marked the end, said Ovshinsky, "of one of my great inventions that was never used and still today is better than what you have now."[13]

Ovshinsky felt even worse when he later learned that the patent application he had filed for his electric power steering invention had disappeared from the files of the patent office. One of Hupp's vice presidents had secretly asked their patent attorney, Richard (Dick) Dibner, to cancel the application, but Ovshinsky didn't hear about it until after he had left Hupp.[14] "It couldn't have been done without Geddes knowing it," he said. These disappointments and frustrations might well have discouraged someone else, but Ovshinsky displayed the resilience and self-confidence that continued throughout his life. "I learned that from my boxing days," he once said. "It's not just about the punches. How long you survive in the ring also depends on how you take the blows."

Hupp and General Automation

Like several other companies, Hupp had received a government order to machine a large number of 105-millimeter artillery shells. Ovshinsky explained to Geddes, as he had earlier to the New Britain management, that the company could produce them efficiently by commissioning one or more Benjamin Lathes; Geddes approved rebuilding two lathes, one for the shells and one for aluminum rocket tubes. To help with this project, Geddes arranged to bring Herb back from New Britain. Driving back to the Midwest during the summer of 1953, Herb brought along the automatic tractor that he and Stan had worked on (see chapter 3). Playing on Detroit streets with the bright orange tractor remains one of Ben and Harvey Ovshinsky's fond childhood memories from that time.[15]

Meanwhile, Herb's wife Selma and the couple's ten-month-old baby Pam flew from New York and moved in with Selma's mother in Toledo. Herb initially lived with Stan and Norma in Greenfield, visiting Selma and Pam on weekends. But the arrangement "got old for Norma pretty quick," Herb said. She helped him find a townhouse on Wyoming, where Herb and his family moved in later that summer.

For rebuilding the Benjamin Lathes, and for later developing other smart machines, the Ovshinsky brothers formed the General Automation Corporation on March 16, 1954. The new venture was initially funded by a $5,000 loan from the National Bank of Detroit countersigned by Geddes. The Ovshinskys had already conceived it before the move to Connecticut, and now with the help of their lawyer friend, Nate Peterman, they filed the papers.[16] Ed Watkins, a talented advertising man and general adviser, developed announcements.[17] They also contacted two Chrysler master mechanics Ovshinsky had come to know during his Stanford Roberts days, Charles Vanderkirk and Arthur Swigert.[18] Herb remembered meeting with them to talk about plans for General Automation at Cliff Bell's, a Detroit restaurant where automotive and manufacturing people often ate. Vanderkirk then put them in touch with Agnew Machine Company, a small machine tool firm in Milford, roughly 30 miles from northwestern Detroit, which would serve as the base for rebuilding the two lathes.

The Ovshinskys also worked together at the old Hupp plant. On Saturday mornings, Stan often brought along Ben, who recalled an "archetypical old-fashioned" plant in "a cavernous building with hundreds of machine production tools lined up and down hundreds of yards." The mostly empty plant would "reek of machine oil and chips and sawdust and hot metal, which I loved," recalled Ben, who also remembered Uncle Herb "sitting at one of these big drafting tables. And I would look at the blueprints and smell the blueprinting." Saturday work at Hupp often also included a visit to one of the ethnic eating places where automotive executives and engineers would eat lunch in the 1950s in downriver Detroit, not far from Henry Ford's famous River Rouge plant. Besides the food, Ben remembered the hundred-year-old talking parrot at one Italian restaurant, and "the wonderful ambiance downriver. Factories all over." Even a child of seven or eight could sense that Detroit's industry was "very, very vibrant."

Cone Company and the Programmable Automatic Lathe

After completing the two Benjamin Lathes for Hupp, the Ovshinskys turned to their next General Automation project, an ambitious automatic lathe that could be fully programmed. Adapting the drum control from the automatic tractor and the electronic controls of the "industrial computer" he had worked on in Connecticut, Ovshinsky envisioned a machine intelligent enough to make a part automatically from blueprints. The contours of the part would be traced on an oscilloscope and locked in to form an electronic template that the drum-controlled cutting tool would follow. Through a Cleveland friend, Charlie Coffin, then working in the sales office of the New Britain Machine Company, the Ovshinskys now connected with the manufacturer's representative for Cone Automatic Machine Company in Windsor, Vermont. A good-sized, old-fashioned machine tool company that made reliable automatic screw machines, Cone knew Ovshinsky's reputation and agreed to support his plan for the programmable lathe.

In April 1954, while the negotiations with Cone were proceeding, Herb located a suitable space for building the programmable lathe. It was a modest double storefront in a working-class Detroit neighborhood at 14121 West McNichols Road, a section of 6 Mile Road not far from its intersection with Schaefer Highway. It was common in that period to house small engineering companies in storefronts, because they were low-rent properties. The storefront would not only house General Automation but would later become the site of Ovshinsky's most important inventions (see chapter 5).

The Ovshinskys worked for sixteen months in the storefront on the programmable automatic lathe, but once again they were frustrated by the shortsightedness of their patrons when the owners of Cone discontinued the project before it could be completed. "They thought that it could never work," Ovshinsky said, adding regretfully that "it would have been a marvel of its time."[19] Cone went out of business some years later.

Leaving Hupp

One summer day in 1955, Geddes surprised Ovshinsky with the news that he had just sold out his interest in the Hupp Corporation. Ovshinsky was disappointed that Geddes reneged on the promise he had made when he hired him to split his profit with him if he ever sold the company. But Ovshinsky decided to overlook this injustice, realizing there was nothing he could do, and he and Geddes continued to have a limited friendship until Geddes's death.

Geddes's departure, however, gave Ovshinsky the opportunity to leave Hupp too, a move he had been considering since he realized that he was again working in an industry that was resistant to change.[20] But, as Herb explained, Stan "wouldn't just walk out" on Geddes because he felt indebted to him. A day or two after Geddes's departure, the company held a management meeting. Ovshinsky was appalled when "these guys who kissed his ass all the time" now were tearing Geddes apart. When they turned to Ovshinsky and said, "Stan, you're now going to become the vice president," Ovshinsky replied, "No, I'm not. I'm leaving."[21] He informed Herb about his decision to leave Hupp soon afterward, when they had dinner together. "There's no reason for me not to join General Automation full time and make a go of it," Stan told him. Ovshinsky's realization that the machine tool and automotive industries were far behind the times anticipated their later decline and hastened his own growth into the kind of inventor who helped foster the new high-tech economy that succeeded them.

Tann and the Multiple-Ball Switch

The Ovshinskys now turned to new projects. One of the first was a version of the electrical power steering that Ovshinsky had invented at Hupp. He also created a new kind of switch for the steering device that used small (0.235-inch diameter) ball bearings, which a varying magnetic field could align to conduct more or less current. An additional advantage of the design was that the infinitely changing points of contact offered by the surfaces of the balls prevented the switch from wearing out like ordinary mechanical relays. The switch was featured in the February 1958 issue of *Control Engineering*; as the article noted, it could also be used in many other devices like "mechanical rectifiers, potentiometers, function generators, logical elements, and proximity switches."[22]

Seen in retrospect, the multiple-ball device anticipated Ovshinsky's crucial invention of switches based on disordered materials, even though here the disorder was on a macroscopic scale. The experience of working with such switches may even have helped him to think about how the change from a disordered to an ordered state on an atomic level could become the basis of a new kind of switching. It also anticipated some of his later information devices. By varying the applied magnetic field, he said, "I showed logic uses for the balls. I could have them partially on, partially off. They were models of what I wanted to do by then, to copy the brain."[23] His later project of creating what he called his cognitive computer would also depend on such variability (see chapter 10).

When the initial $5,000 bank loan proved insufficient, Ovshinsky realized he needed a financial partner. Early in 1956, he found backing for the work on electromagnetic devices from the Tann Corporation, a family business run by several brothers whose father had started it after World War I. The company began by manufacturing huge dies for automotive companies and had become very successful during World War II. By the mid-1950s it included not only the original die shop and another making die castings but also a division called Congress Controls. Interested in developing the magnetic switches through this division, the Tann Corporation supported the Ovshinskys' work, but after some months, the Tanns insisted that they join their company instead of working independently. Ovshinsky recognized the move as a takeover. "Of course Stan wouldn't stand for that," Herb said, so their association with the Tanns ended abruptly in 1957. It would take until October 1963 to settle the suits that the Tanns and the Ovshinskys brought against each other.[24]

Iris Dibner

Meanwhile, Stan's personal life was changing dramatically. About a year after he began at Hupp, a shy, warm, attractive, and highly intelligent woman in her mid-twenties stepped into his life. Iris Miroy Dibner was deeply concerned about the same social and political issues that engaged Stan. He could hardly believe it. "I didn't think a girl like Iris existed." It took some time for them to realize they were powerfully attracted to each other, but once they did they became inseparably bonded. As one friend later put it, "They were soul mates. Two bodies and one soul."

Iris was born in Manhattan on July 13, 1927, into a radical culture. Her parents subscribed to the philosophy of anarchism, which advocated freedom from the domination of religion, property, and government. In the words of the anarchist and feminist Emma Goldman, it envisioned a social order that would "guarantee to every human being free access to the earth and full enjoyment of the necessities of life, according to individual desires, tastes, and inclinations."[25] Stan admired philosophical anarchism as a "beautiful" movement aimed at a better life for the oppressed, but like his father Ben, he considered it an impractical ideal.

Iris's pacifist father André Herrault had fled conscription in the French army when World War I broke out, and he lived for a time in Canada with a changed name, Miroy. There he worked on the docks loading and unloading ships. When he later moved to New York, he became a teacher and translator of Spanish. He spoke five languages, including Yiddish, "even though he wasn't Jewish," Iris noted, adding that at the time he died, "he was reading *War and Peace* in Russian with the dictionary."

Iris's mother, Anita Spiegel, was born of poor Jewish immigrants in Haverhill, Massachusetts. She and André met at a meeting of the Francisco Ferrer Association, founded in 1910 by Emma Goldman and others in honor of the recently executed Spanish anarchist educator (1859–1909).[26] Like many anarchists, Anita and André did not believe in institutionalized marriage and simply lived together. Only years later, after their separation, did they became legally married at a point when André needed a visa to visit France.[27] The first in her family to attend college, Anita majored in French and earned a teacher's certificate at Hunter College. Having had Iris at age thirty, she was over forty by the time she was appointed to a school. Now considered too old to be a new classroom teacher, Anita was assigned to homeschool handicapped children.

By then Anita and André had gone to live with other anarchists in the Mohegan Colony, the oldest of several Westchester County leftist summer enclaves. Spending her childhood there, Iris attended the colony's progressive school until sixth grade. "We did plays and dances and things like that and caught frogs," recalled Iris, but she did learn to read. She immediately excelled. Despite being "awfully shy," she admitted to being "first in my class in everything." Her parents stressed culture at home, speaking French with each other and with Iris, who typically answered in English.

As a child, Iris witnessed the unraveling of Anita and André's relationship. André was "a good human being" with "very good values," Iris said, but he was "quite a stick-in-the-mud," extremely quiet and always reading. By the time Iris was five, Anita had separated from André and was living with the dashing Henri Dupré, also an anarchist, who liked to dance and was a protégé of the great French chef Auguste Escoffier.[28]

Figure 4.2
André Miroy and Anita Spiegel.

(Anita "would have liked to take me with her," Iris explained, but André said "No. You have Henri." So Iris continued to live with André in the Mohegan colony.) In their tiny Greenwich Village apartment at 5 Minetta Lane, Anita and Henri would dine on the fancy French food that Henri brought home from the Essex Club, the exclusive Newark men's club where he worked.

To spend more time with Iris, André worked at home as much as he could manage. He would leave at 7 a.m. to catch the train to lower Manhattan, where he worked at the Lawyers and Merchants Translation Bureau, returning in time to be with Iris when she

Figure 4.3
Henri Dupré, standing, with Auguste Escoffier.

came home from school. Iris's friends loved to visit with André, "the man who knew most about everything," but she herself felt the pressure of being his whole life.

It was comforting for Iris that Anita and André remained friends, considering themselves comrades. On Friday evenings, when Anita came by train from Manhattan to visit Iris, bearing "jelly donuts from the wonderful store next to Grand Central," the three would have dinner and spend the evening together doing "homework or whatever." Anita slept over before taking Iris with her the next day, keeping her through the weekend. Iris always felt a little depressed on Sunday nights, because "the excitement of being at my mother's was always more fun." All four would spend Thanksgiving, Christmas, "or any important thing," as a family. "My mother and Henri would have the dinner, probably, but they would invite my father." Summers were spent with Anita and Henri and "we'd go on picnics."

Academically, Iris was ready for high school by the age of eleven, but since she was so young she instead attended the progressive Putnam Valley Central School. When she entered Swarthmore at age sixteen, she "studied, studied, studied every minute." Before leaving Swarthmore at age nineteen, Iris married her childhood sweetheart, Andrew Dibner, "the boy that everybody thought was handsome and wonderful." The Dibner clan was also part of the Mohegan community.[29] Andy and Iris had decided when Iris was just thirteen that when they grew up they would marry, as they did shortly after Andy came out of the army.[30] "Everybody in those days got married pretty early," Iris said. It would be through Andy's three-years-older brother Dick Dibner, Stan's patent attorney at Hupp, that Iris later met Stan.

The newlyweds then moved to New York City, where Iris attended Brooklyn College and then Hunter College to make up the credits she needed to graduate from Swarthmore in the spring of 1948 with a degree in biology. They next moved to Michigan, settling in Wyandotte, 10 miles or so south of Detroit. Andy attended graduate school at the University of Michigan in clinical psychology and, supported by the GI Bill, did clinical work for his dissertation at the Veterans Administration hospital in Dearborn. When Iris wasn't accepted by the University of Michigan's medical school, she gave up the idea of becoming a physician and took eighteen hours of education courses to earn both a master's in biology and a teacher's certificate. She then taught biology for two years at Fordson High School in Dearborn.

After becoming pregnant, she stayed home and found "it was almost too quiet." Then in analysis three times a week as part of his professional training, Andy seemed distant.

Stan and Iris

Iris and Stan had met several times between 1952 and 1955 in the normal course of their lives. One of their first meetings was at the Workmen's Circle, where she was intrigued when he took out a notebook in which he had collected clippings about nerve physiology. A memorable exchange between them took place in the spring of 1952 at a party hosted by Dick Dibner and his fiancée Ursula. Introducing Stan to Iris, Dick said, "You have to meet this exciting man." For the rest of his life, Dick would regret making the introduction.

Andy and Stan got into a heated discussion at the party. As Stan recalled, Andy and his psychology colleagues were talking shop and voicing the conventional 1950s view that people have to "make a certain mental adjustment" to their society. Stan took issue, raising the example of Spartacus and the rebel slaves who were condemned to die. "Do you think that they should have adjusted to going quietly to their death?" he asked. As Andy and his colleagues had no ready response, Stan asserted, "I would have said, let's get together and fight these guys." Iris did not respond well to Stan's challenge, finding him "unpleasantly argumentative, arrogant, and generally rather conceited," certainly "not her type." But she also remembered being annoyed by his flirting with a "buxom blond dopey lady," and thinking that he was "extremely handsome."

Some months later, Stan had another argument with Andy when he and Iris visited Norma and Stan at their home and they all went out to eat at a delicatessen. They discussed the purges in Russia, which Andy interpreted as "social growing pains." Stan challenged this reduction of political conflict to psychological terms. He insisted that collective struggles shouldn't be seen as ordinary growing pains. This time Iris supported Stan's side of the argument, and Stan remembered feeling very pleased. Clearly, interest was developing between the two. On another occasion, in December 1954, Stan remembered sitting behind Iris at a Workmen's Circle event and patting her head. In Iris's memory, "I was sitting behind Stan," and "suddenly I was very interested. He seemed like a very super bright guy and seemed so handsome." It was not long after this that both realized they were in love.

Both remembered vividly when their worlds changed on January 1, 1955. They were at Dick Dibner's New Year's Eve party. Iris had had her long blonde braid cut off before the party, and her hairdresser had put stars in her stylish short hair. "When Stan walked in the door," Iris said, she and Andy were already there. "I felt sort of—wow, what happened? Boy did we hit it off. It was really madly falling in love." She remembered, "We talked and talked and talked. And I even said, 'Why don't we go into the bedroom?' It was crazy." Afterward, Iris "was thinking about Stan, Stan, Stan, Stan, Stan. And a few weeks later he called." They decided right then that "sometime, somewhere, some place, we'll get together," said Iris.

Stan and Iris each told their spouses about the sparks that had unexpectedly flown between them. Norma and Andy were both shocked. Iris became enraged when Andy remarked that he felt like "somebody took his watch. That's how important I was to him!" She decided that Andy "was shocked, but not, I don't think, that devastated." After that, Stan and Iris saw each other whenever possible. Both felt sorry to be causing pain to their spouses, but the feeling between them was stronger and deeper than any

Figure 4.4
Stan and Iris in upper Michigan, summer 1955.

other feelings. The Dibner family soon forgave Iris and blamed Stan for sweeping her off her feet. Dick Dibner "didn't talk to me for about ten years," Stan recalled.

Andy sent Iris to see a psychiatrist, and "she came back laughing," Stan recalled, because the therapist had suggested that to "get over this thing" she go away with Stan for a weekend. They went to the beach in upper Michigan and "had a great time," said Stan. "It wasn't anything we were going to get over." From then on, despite all the struggles and mishaps in his career, "I was a happy man," Stan said. "I felt that I had achieved the peak of what life was about."

Soon, Andy decided on his own to separate Iris and Stan. On completing his PhD he found a position in the psychology department at Clark University in Worcester, Massachusetts. Iris, Andy, and their two young children moved there late in the summer of 1955. At this point, Stan seriously considered divorcing Norma, but the speech psychologist working with Dale predicted that would have a "devastating effect" on the child. Stan decided against the divorce, but "from then on we were a couple," Iris said.

The next years were extremely difficult for all involved. Iris remembered, "We wrote letters every day and called every day and we each had separate lives." Stan managed to see Iris roughly half a dozen times a year, typically on business trips. "And we always knew we were going to get together," said Iris, "so it was pretty rough, but it was crazy. Every day I'd skip lunch and go talk on the phone. And I had my own mailbox. And he'd write ten-page letters every day. They were mostly 'I love you. I miss you.'" Stan remembered being "on the phone every day, several times a day, and writing every day, pages and pages, and she was doing the same." And they lived for the rare times when they could manage to get together.

Their letters and phone conversations were filled with all the things both cared about—work, family, love, science, politics, and how to make the world better for everyone. They shared a commitment to social justice, and both cared much more about fighting against exploitation than they did about money. "We thought we'd end up in a very small house with just bricks and wood for our book shelves," Stan recalled. "We were ready to live in utter poverty." Their many exchanges about their shared values over the years they were apart set the direction for their later work together. When Stan and Iris later started a new research company, they aimed to develop technologies that would help solve social problems (see chapter 5).

Stan recognized congenial social values when he met Anita and Henri. He and Iris would stay with them in their Greenwich Village apartment, where Stan felt at home, like he "was among old friends." He showed Iris a bowl filled with mail and said, "I get this same damn mail, Workers' Defense League." Stan appreciated that Anita was a "very bright woman" and liked the same Wobbly songs that he did.

Stan and Iris's romance put Herb Ovshinsky in an uncomfortable spot. He wanted to stay friends with Norma, as well as Andy and Dick Dibner, but that proved impossible. All broke off their relationships with him, and having been sworn to secrecy, he couldn't even tell Selma about Stan and Iris's romance. It was at the same time clear to Herb that Stan's and Iris's love "was stronger than all the strong family. And it was as strong the last day we were with her as it was in 1955." Herb saw how they complemented each other. Iris supported Stan's work in science and technology, while his brother's love and admiration caused the naturally shy Iris to blossom.

Schizophrenia and Epilepsy

Among the topics of Stan and Iris's daily communications during their years apart were the neurological questions that he was exploring as he continued to develop the cybernetic analogy of animal and machine control and communication (see chapter 3). Ovshinsky hoped to gain insight into the functioning of healthy nerve cells by considering cases where nerves fail to function properly, when signals were misdirected or distorted, or when people couldn't respond appropriately.[31] To do that, he began studying epilepsy and schizophrenia.[32]

Applying his reading of the medical literature on nervous disorders, Ovshinsky began to ponder whether both schizophrenia and epilepsy might have to do with changing electrical thresholds, building on the hypothesis that electrical waves from the brain control human actions. He suggested that, because of an injury or a genetic problem, "those waves are just the wrong ones" and result in neural disease. Driven by his interest in intelligence, Ovshinsky wanted to understand the difference between the wrong and right waves and contribute to medical science by learning how to control them.

Relating the behavior of nerve cells, intelligence, and electrical thresholds, he argued that epilepsy and schizophrenia are essentially "the same problem," both resulting from chemical, thermal, or physical interference with the ability to receive and act correctly upon messages from the brain. The resulting misdirection of messages in both diseases shows up, he argued, as lack of control. But while epilepsy arises from interference with the parts of the brain controlling motion, schizophrenia arises from the parts dealing with cognition. Schizophrenia, he maintained, was "epilepsy of the mind."[33]

As a machinist working on automation, it was natural for Ovshinsky to conceive of muscles as machines instructed by the brain and to try to understand the process of movement in terms of signals and feedback. He thought about the communications between the cortex and the cerebellum involved in any movement and how faulty feedback could produce irregular movements like tremors. In the early 1950s, he had begun to set down such ideas for himself in a number of papers, some of which he later published or delivered as talks between 1957 and 1961.[34] This line of thought, though it may seem to diverge from Ovshinsky's earlier work, not only emerged from his efforts to build smarter machines but would also lead to his most important inventions.

Iris, meanwhile, decided to further their ongoing discussions about epilepsy and schizophrenia by getting a PhD in biochemistry. She chose to study at the nearby Worcester Foundation for Experimental Biology, which had an advanced research program in the biochemistry of schizophrenia.[35] Her study proved more grueling than expected, however, because she was missing prerequisite undergraduate courses in quantitative analysis and physical chemistry. Worcester Polytechnic offered these, but she was denied admission because there were no facilities for women, so she had to commute to Boston and take the courses at Harvard. Stan "was so proud of her," Herb said, and "helped her every bit of the way." When the two were together, he would sleep on top of the desk while she worked.

Iris, for her part, urged Stan to publish his ideas on the synapse and nerve impulse in the scientific literature. "I wouldn't know really what to do about that," he replied. Iris decided to teach him how to write a scientific paper, drilling him on the technical jargon, typing and critiquing his writing, and teaching him about making indexes. "We used to have fun—what's an axon, what's a dendrite, etcetera," she recalled. From Stan's perspective, "She gave me my PhD, and we had a great time." On trips, the two often visited bookstores and libraries to learn about relevant research and find references.[36]

Neurophysiology at Wayne State

By the time Iris moved to Massachusetts, an important new opportunity had opened up for Ovshinsky. After helping him rewrite his paper on the nerve impulse, Iris helped him bring it to the attention of Ernest Gardner, a professor and the chair of anatomy at Wayne State University School of Medicine. Gardner found the article fascinating, as did his colleague Ferdinand A. Morin, who later succeeded Gardner as chair of anatomy when Gardner became associate dean and then dean of the medical school. Ovshinsky was surprised and delighted to receive Gardner's enthusiastic letter of June 17, 1955: "Dr. Morin and I read your paper with a great deal of interest. We both feel that it contains some extremely worthwhile and provocative ideas." The letter extended an invitation to work at Wayne Medical. Gardner asked, "Have you ever given any thought to the possibility of testing out some of these things experimentally? It seemed to us that experimental proof should certainly be sought for." He suggested that Ovshinsky and Morin get together for a discussion.[37] Overjoyed with the invitation, he soon joined them.

Gardner's letter profoundly changed the course of Ovshinsky's life by offering him the chance to add an identity as a scientist to his professional persona. Still working at Hupp, he now also conducted research at Wayne whenever he had free time. With Morin, he proceeded to explore the neurophysiology of the brain experimentally by implanting electrodes in the brains of cats and monkeys. Noting the twitching responses, he could tell which cells they were exciting or inhibiting. Herb helped by making micro-manipulator-controlled needles for implanting the electrodes accurately. "It was a little shocking," Herb recalled, but "I don't think it hurt the animals."

This first taste of membership in an academic community was "a wonderful experience" for Ovshinsky. "It sort of spoiled me. I was accepted immediately by these guys at Wayne, who were very good and did really valuable work. And what was so important to me was they accepted me. And so I thought that's the way science was, if you were making contributions."[38] He and Morin published on the relationship between the cerebral cortex and cerebellum, which Ovshinsky described as a servomechanism. He also lectured about his work, initially to the medical students. He encountered very little politics or backstabbing in this period, and he especially enjoyed interacting with the well-known Chicago-based neuroscientist Heinrich Kluver, who invited Ovshinsky to visit him in Chicago for scientific exchange. Unfortunately, he had no money for such travel. Morin and Ovshinsky did, however, regularly attend the Saturday scientific symposium on recent work in neurophysiology organized at the University of Michigan by Ralph W. Gerard, whose own interests included the biology of schizophrenia.[39]

The Ovitron: A Turning Point

Ovshinsky's work in neurophysiology may not have had a significant impact on that field, but it led to a pivotal development in his own inventive career. The paper on the nerve impulse that brought Ovshinsky to Gardner's attention proposed an analogy between the signals transmitted from one neuron to another and electrical circuits. Conceiving neural circuits cybernetically as organic equivalents of the sensing and feedback mechanisms of automation, he began to think of the nerve cell as a kind of switch that allows signals to pass once an electrical or chemical threshold is reached. He wanted to locate the threshold at which the cell fired and identify the electrochemical processes that allowed the impulse to pass between cells. To help him answer these questions he decided to build a physical model that worked the way nerve cells do, with a semipermeable membrane that allowed it to fire when the activating impulses reached a certain threshold.

Ovshinsky's model, which he referred to as his nerve cell analogy, was a new kind of switch. The Ovitron, as he called this invention, is arguably the pivotal achievement of his career because it motivated so much of his later work, as subsequent chapters will show. Working with Herb in the same storefront on McNichols Road that housed General Automation, he assembled a device that resembled an electrolytic battery, with tantalum electrodes immersed in a hydrochloric acid electrolyte containing a small amount of zinc. When an AC voltage was applied, no current flowed because of a thin insulating oxide film on the electrodes.[40] But when an additional low positive DC voltage was introduced by a third, nonreactive (palladium or platinum) electrode, the device switched on. Reversing the DC polarity switched it off again. Ovshinsky thought of the electrolyte as being like the fluids surrounding neurons, and, more importantly,

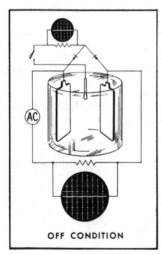 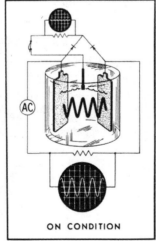

Figure 4.5
Operation of the Ovitron.

of the oxide coating on the tantalum electrodes as being like the cell membrane, a semipermeable barrier that allows transmission when stimulation reaches a certain threshold.

The Ovitron was an impressive switch. It was extremely fast, switching in 8.3 milliseconds, and it yielded an enormous power gain: the DC current of only a few milliamps at about 3 volts could control an AC current of as much as 20 amperes at about 100 volts. Because, unlike the transistor, it could handle such heavy AC loads, and because it worked differently from previous switches, it in time became the subject of several articles in electrical trade journals that repeated Ovshinsky's explanation of its principles.[41] It is now clear, however, that the analogy with the nerve cell is imperfect. Rather than acting like the neuron's semipermeable membrane, the insulating oxide layer on the tantalum electrodes simply becomes so thin that the current can pass between the electrodes; the electrochemical process of switching on and off is one of de-plating and re-plating rather than crossing a threshold.[42] Nevertheless, it was by pursuing the nerve cell analogy, boldly coupling biological and electrical processes, that Ovshinsky made this important discovery.[43] As with his later switching inventions, he was guided by his unique intuitions; scientific explanation came later, sometimes decades later.

Before continuing with the development of the Ovitron, we should pause to consider just what Ovshinsky was doing when he created it. Was he testing his theory of the nerve cell or trying to invent a new kind of switch? Was he doing science or technology? The answer seems to be "both," or maybe "something else," which may help us understand more about his creativity. He was certainly interested in pursuing scientific knowledge about neurophysiology for its own sake, but his interest first arose from thinking about automation, and it led to a series of technological innovations. Later in his career, when Ovshinsky began to reflect more on creativity and how he invented, a recurrent theme was his disregard for divisions between disciplines, including a refusal of the distinction between science and technology (see the interlude). Neither a biochemist nor an electrical engineer would have been likely to arrive at the Ovitron. Ovshinsky's omnivorous curiosity and willingness to go wherever his imagination led him were keys to his success.[44]

Support for the Ovitron

To develop the Ovitron, the Ovshinskys set up a new company, which they also called Ovitron, in 1958. They found support through Robert Allen, a schoolmate and friend of Ralph Geddes Jr., the son of Ovshinsky's former boss at Hupp.[45] Allen was very excited when he saw an early model of the Ovitron. Being the son of the legendary investment banker Charles Allen of the great Wall Street investment firm Allen and Company, he was also in a position to support its development.[46]

To look at the Ovitron, the Allen family sent their patent attorney, Leon Simon, from Washington, DC, to Detroit, where the Ovshinskys set up a demonstration on a Formica board, using a small transformer, a rectifier, and a light bulb. Simon was very impressed. He later told Herb "this was the only true invention he ever saw" and confessed that he had been "expecting to see this bunch of hustlers trying to take advantage of a young Bob Allen."

The Ovshinskys then traveled to New York to meet with Charles Allen and arrange financial support. Stan and Charles liked each other and began to talk on a personal level. Charles confided that he was very worried that his son Robert was headed for failure. He promised to support the development of the Ovitron if Robert could be a partner in the company. Acknowledging that his son could be difficult, Charles promised, "All you have to do is call me, and I'll take care of it." When Stan agreed, Charles loaned them $67,000 in cash and also endorsed bank loans for another $100,000. For developing the Ovitron, the Ovshinskys moved out of the storefront on McNichols into a different building several miles away at 14830 Schaefer Highway, rented from the same landlord. The work began officially in November 1958.[47]

At about this time, the Tanns, who in 1956 and 1957 had supported Ovshinsky's earlier work on switches, relays, and variable resistors, surmised that the Ovitron was an extension of the magnetic-ball switches they had supported, and sued the Ovshinsky brothers. As an electrochemical switch, the Ovitron was actually nothing like the electromechanical switch that Tann had supported, but the Tanns did not understand the distinction and proceeded with their suit.[48] Despite the pending suit, Ovshinsky worked tirelessly on the Ovitron. He would gradually learn how to make lawsuits work in his favor.

On July 8, 1959, after roughly eight months of development, the Ovshinskys publically disclosed the Ovitron. Organized by Ed Watkins, the disclosure took place at 10:30 a.m. at the Canadian Club on the eighteenth floor of New York's Waldorf-Astoria Hotel. Generously paid for by Charles Allen, the event included cocktails and a luncheon at noon. Ovshinsky expounded the theoretical basis of the Ovitron in the electrochemistry of the neuron, emphasizing its advantages over all existing control devices. Subsequent articles in trade publications, which repeated his explanations and claims, testify to the success of his presentation.[49]

At this point, the Ovitron was still a prototype, not yet a commercially viable device. Ovshinsky looked forward to using the concept to make electronic memories and to

Figure 4.6a
Research area in the Ovitron storefront.

Figure 4.6b
Ovitron shop area.

making it work with a better electrolyte. ("I don't like to have these electrolytes that can burn your fingers.") Indeed, he hoped in time to find a way to avoid the liquid electrolyte altogether and make the Ovitron into a solid-state device. Another disadvantage was that the device gave off small amounts of hydrogen gas, a potential hazard. In fact, after it had cycled countless times and had worked beautifully during the demonstration, Ovshinsky got a call during celebratory drinks at a country club: one of the screws of the hand-made device had come loose back in the lab and it had blown up. It was clear that there was more work to be done.

Nevertheless, some who were working on the forefront of electronics could recognize the importance of Ovshinsky's invention. Among those who read the articles on the Ovitron in the trade press was Willis Adcock, who had helped build the first silicon transistor at Texas Instruments. When Adcock visited Ovshinsky in 1959, he expressed his admiration for the Ovitron. "What you have here is remarkable, and you can do something the transistor can't do. You can handle tremendous currents, and you can handle AC." Adcock repeated the story years later when he spoke at the Institute for Amorphous Studies (see chapter 7).

Most academics ignored the Ovitron, however, perhaps because it had been presented in a public disclosure rather than an academic paper. At Wayne Medical the scientists did not see the relationship between the Ovitron and the nerve cell, and when Ovshinsky spoke about his nerve cell analogy at the Detroit Physiological Society in 1959, "there was absolutely no response." He soon found that he was no longer getting notices of the society's meetings. A year or so later, when he bumped into the president of the society, he learned that they had simply dropped him because he had no credentials.

Work on improving the Ovitron was going well, but working with Bob Allen was not. Ovshinsky had for some months been suffering in the relationship with Allen, who was driving him crazy with foolhardy ideas about science and invention, sometimes phoning in the middle of the night to propose them. He would insist on their validity even when Ovshinsky explained they violated the laws of thermodynamics. The situation became critical when Allen informed Ovshinsky he was going to buy Ovitron out and go public, hoping to make a killing.

Ovshinsky went back to Charles Allen to remind him of his earlier promise to intervene if his son caused problems. "I did promise," Allen replied. "But he's my son." The final straw came when Bob Allen decided to run Ovitron without the Ovshinskys and simply locked them out of the building. Ovshinsky knew that Allen could not succeed because he didn't understand the science behind the invention, but work on

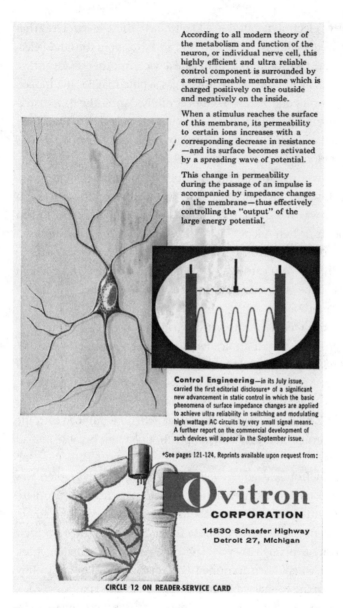

Figure 4.7
Ovitron ad (designed by Ed Watkins).

Figure 4.8
Ovitron promotional photo, with Herb, Bob Allen, and Stan.

developing the Ovitron had to stop. Another lawsuit ensued, and it took until March 1960 to reach a settlement (see chapter 5).

It was also at this point that the Ovshinsky brothers ended their collaboration. Herb would continue working on automation, but Stan's work on the Ovitron had made him redefine his goals. "From now on," he announced to Herb, "my future's in materials."

5 New Beginnings in the Storefront (1960–1964)

The early 1960s were a crucial period in both Stan's personal life and his inventive career, marking new beginnings in each that would set his direction for decades to come. These strands also now became more closely intertwined, so that in following them we need to keep shifting our focus back and forth between them.

A New Family

Iris had carefully timed her return to Detroit to occur on Stan's thirty-seventh birthday, November 24, 1959. She brought a cake, as well as both her children. When Stan met them at the train station, he and Iris were jubilant, but Robin, days away from her seventh birthday on December 5, and Steven, who had turned five on October 6, were both anxious. Robin remembered feeling as though she was falling into an "unknown and uncontrollable situation." Steven recalled being ill with flu-like symptoms during the long train ride from Boston. Decades later he remembered lying awake during the night "and just feeling terrible." They were too young to understand why they had left their father in Boston and why he had gradually moved out of their home in Worcester, by the end spending just weekends with them.

Both children had already met Stan when he visited their home in Worcester and at Anita and Henri's apartment in New York. They had been charmed by the smiling man who entertained them by drawing fanciful animals or people, building on numerals or letters they wrote. But they had no idea yet that he would become a second father to them, or that their parents were separating permanently. In the late 1950s, divorce was uncommon in middle-class American families. "All the way into high school," Robin recalled, "I was the only kid in my class who had divorced parents."

After the initial welcoming hugs and kisses, Stan drove his new family to the tiny rented house at 16818 Gilchrist Street, which they would call home for the next year and a half. In a lower middle-class neighborhood, where most of their neighbors

Figure 5.1
Robin and Steven Dibner, May 1960.

worked for the auto companies, their new home was hardly bigger than a garage, and a tree "blocked the whole damn house," Stan recalled. He had already begun to fix up the place, for it had been in very poor condition at the time he signed the lease. But since many Detroit landlords would not rent to Jews, there were few options. The location, however, was extremely convenient. The kids could walk to school, and the storefront was just five minutes away by car. And when Stan and Iris were done decorating, the house "was just beautiful," Stan said, "a small gem." Herb remembered it as "filled with love."

On arriving, Iris found a card that Stan had written on a shirt cardboard that said, "Welcome home, wife." (Iris and Stan did not marry legally until 1962, but their official wedding date was less significant to them than the day of Iris's move to Detroit.) Robin told how when Stan's shirts came back from the laundry he would save the cardboards and use them for writing such messages, especially on the anniversary of their getting together. "And every year after that on November 24, there would always be a big bunch of red roses, with one of these shirt cardboards propped up, and in some colored pen he usually would have written, 'Happy anniversary,' and some loving message, half in English and half in Yiddish."[1]

The new family arrangement was confusing for both sets of children. For Stan's boys, who continued to live with Norma, "suddenly, there were two Mrs. Ovshinskys," Harvey recalled. And because Iris and Norma both had Hudson's credit cards, there were frequent phone calls from the department store. "I remember my mother on the phone once, in tears, trying to explain who the real Mrs. Ovshinsky was." Norma did not hide her strong feeling of having been betrayed and made it clear she would need adequate compensation for letting Stan go. Inevitably, the boys absorbed some of her anger and resentment. Harvey recalled, "I remember screaming at Dad, and accusing Iris of being a prostitute. It was so ugly. But, to their credit, they both endured it. They knew it was the price they had to pay for falling in love."

Despite the initial obstacles, Iris was pretty sure the move would work out well for all involved. She had seen how Anita became noticeably happier after leaving André, and how they had remained friends. Things would indeed eventually work out for all five of Stan's and Iris's children, but their experience of having divorced parents would be much more complicated than Iris's had been. And despite her overall optimism, Iris also continued to fear she might be "ruining seven people's lives to make us happy." As for Stan, he experienced Iris's return to Detroit as "a turning point in my life, the start of real happiness." "We *really* fell in love," and he felt they were "falling in love every day."[2]

Nurturing Invention: The Storefront

In January 1960, Stan and Iris set up the new company they had often talked about over the years of their separation, a company guided by their progressive values. As Stan put it, their mission was "using science and technology to solve the world's societal problems." Iris described their goals in less exalted terms. "Stan and I never were that unrealistic to think that we could solve all the world's problems. We wanted to do what we could for society, and also we were very committed to making sure it was a successful company, not just a dreamer's company."[3]

For their work on the Ovitron, Stan and Herb had moved out of the modest storefront on West McNichols Road that had housed General Automation, but since the building was available after their break with Allen, they now moved back. Despite its drab surrounding neighborhood, which included a small drugstore next door, a flower shop, and a barbershop, the space became an oasis of creativity.[4] Family members contributed to launching the new company. Anita helped Stan and Iris move in, set up the files, and ordered stationery, while Mashie set up the bookkeeping. Stan assembled an oscilloscope from a Heathkit. Iris hung pictures, including a large chart of the periodic table, and she helped Stan furnish shelves upon shelves with books on all subjects.[5]

Those diverse books reflected Ovshinsky's insistence on the freedom to disregard conventional scientific disciplinary divisions. The pamphlet he and Iris produced describing their new laboratory expresses the conviction that "science is indivisible, that one must be able to utilize seemingly unrelated information from one discipline and apply it to another, to accomplish new and unique solutions to technological and scientific problems."[6] This declaration offers a notably early instance of a recurrent theme in his later reflections on creativity (see the interlude).

In defining their social mission, Stan and Iris had focused on the topic of energy, particularly the problems resulting from reliance on oil.[7] They began working on ways other resources could be converted into energy, and to signal this research direction they named their new company Energy Conversion Laboratories (ECL).[8] Later, as the focus expanded to include Ovshinsky's work on information, he would describe energy and information as "the twin pillars of our global economy," conceiving them as complementary, for to store or transfer energy requires information, and vice versa. Understanding information as both the means and the result of energy conversion, as "encoded energy," gave a larger meaning to the name he and Iris had chosen for their company in 1960.[9] Two decades later, they would celebrate its twentieth anniversary by issuing medallions whose two sides showed the brain and the sun.

Figure 5.2
ECD's two-sided medallion.

Ovshinsky had been thinking for many years about the geopolitics of energy. He remembered when the United States had cut off Japan's oil supply in 1941 and how that soon led to the attack on Pearl Harbor. "You can't cut off the energy supply to a country and not expect that there's going to be a war," he observed, adding that in ancient times there were wars over salt. That perspective, as well as his sensitivity to pollution from his personal history of asthma, contributed to his concerns, whether about the potential for future wars over oil or about increasing environmental damage, even before the growth of the environmental movement and the energy crisis caused by the oil embargoes of the 1970s.[10]

In addition to these concerns, Ovshinsky felt strongly that developing countries deserved affordable and up-to-date technologies. In talks he gave during the early 1960s, he pointed out that the out-of-date equipment being shipped to them by the industrialized countries was too expensive, inefficient, and polluting. He believed they should have "their own technology or their own branches of science." Here he was also ahead of his time, anticipating what would later be called appropriate technology.[11] In later years, when Ovshinsky and his company, ECD, developed thin-film solar panels, he would advocate their use not only in the developed countries, where the problem was dependence on fossil fuels, but also in the developing countries, where the problem was lack of infrastructure. He would often close his talks with an image of a young Mayan woman in the Chiapas rainforest carrying a box of thin-film solar panels on her back and a baby in front. "Look at this woman," he would say. "She is carrying the future on her back and in her arms."[12]

Figure 5.3
Mayan woman with baby and a box of solar panels.

For most of Ovshinsky's audiences in the 1960s, such ideas were too progressive (and perhaps too undeveloped), so he decided to bring them to the attention of the left-wing radical leader Fenner Brockway, an Indian-born member of the Independent Labour Party, a friend of Gandhi and Nehru, and at that time a member of the British House of Commons. In a letter to Brockway in May 1960, Ovshinsky expounded some of his ideas for using "solar energy, nuclear energy, ordinary forms of heat, or energy differentials of various types to create electricity directly," and explained that he and Iris wanted to offer them "to interested governments of underdeveloped areas," which could implement them using available natural materials such as wood or sand. Summarizing, he declared, "We must use our imagination to exploit technology rather than people, if the underdeveloped areas are to make great progress."[13]

When Brockway came to Detroit, Ovshinsky invited him to enjoy a home-cooked dinner on Gilchrist to discuss his energy ideas in a comfortable setting. Brockway congratulated Iris on her delicious vegetarian meal, saying that he had never had a better one. Suddenly an ashen-faced Iris called Stan over to speak with her in the kitchen. She told him she realized that she had used oxtail soup in preparing the dish and thought she had better let Brockway know. But when she asked him whether he was a vegetarian out of principle, he replied, "Oh no, I am long past that. Out of habit." A much-relieved Iris again called Stan into the kitchen to explain that she would not tell him after all. "If it was principle I would have to tell."

As for Ovshinsky's energy ideas, Brockway was an appreciative listener, and he arranged for meetings with a number of influential Indian politicians. But Ovshinsky could find no way to relate to their bureaucracies, and he was similarly frustrated when he tried to explain to a group of African ambassadors at the United Nations how to avoid exploitation by US industries by developing their own technologies from local materials.

Funding ECL's ambitious mission in the storefront was a continuing problem. Stan and Iris quickly spent the $10,000 that Iris brought from the sale of her father's house. They turned to writing proposals to government agencies, but ECL was a small and unknown company, and the subject of their work fell outside the existing categories, so their proposals were rejected. One preliminary proposal to develop a thin oxide-film computer, whose logic circuits would have extended the Ovitron nerve cell analogy, drew initial interest, but when they submitted a full proposal in late 1960 to build a prototype it was rejected in favor of one from Westinghouse.[14] Iris remembered being "so annoyed" when they later got "a call from Westinghouse saying that they got the contract and could we help them!"

The flair with which Ovshinsky pitched his mission brought in just enough funds to support ECL in this period, but the work of raising money was never easy. Nor would the company ever make much profit, for as one project succeeded, he would immediately follow up with a more ambitious and more costly initiative. In this period, some support came from Ovshinsky's connection with Alcoholics Anonymous, which his old friend Bill W. (William Griffith Wilson) and his partner Dr. Bob (Robert Holbrook Smith) had founded in Akron in 1935. Wilson told Ovshinsky, "All these drunks that I know, they run banks, they run big companies," and he convinced several friends who were recovering alcoholics to support ECL. Wilson, who would sleep in the attic when he visited Stan and Iris's home in Detroit, later served on the company's board of directors.[15]

A substantial addition to ECL's operating budget came in March 1960 with the settlement of the Ovshinsky brothers' dispute with Robert Allen about his takeover of the Ovitron Company. Since he had no case, his lawyers forced him to settle, and the Ovshinsky brothers won $96,000 to be divided equally between them. Stan divided his half with Norma as part of their divorce settlement, which was then being negotiated but not completed until two years later, on March 28, 1962. The $24,000 that he and Iris received covered ECL expenses (salaries, rent, equipment, and supplies) for the time being. When Stan asked Iris whether there was something she might like to buy with the money, she recalled saying "a set of matched towels would be my interest, not all these different old towels we have around." Iris insisted that financially "we were okay," if not flush. "We even went to Florida once for a vacation. And we'd go to New York occasionally."

Early Energy Conversion Work

In its early days, ECL seemed to lack a well-defined research program. With unlimited ambitions but limited experience and resources, Ovshinsky's initial experiments with energy conversion had inconclusive results, though some anticipated later successes. One yielded a high-temperature lithium battery, whose prototype was patented and gained support for a while from the Lithium Corporation of America.[16] But when that funding ended, so did Ovshinsky's battery work for nearly two decades. Another thermoelectric device attempted to get energy by splitting and recombining hydrogen molecules and at least succeeded in producing a mysterious spot of bright light without heat. Ovshinsky was unsure about the validity of his results, however, and didn't pursue the idea further.[17]

More generally, Ovshinsky thought about how to use hydrogen as a clean fuel, starting a quest he would pursue for half a century. He conceived of a complete energy system he called the "hydrogen loop," in which solar-generated electricity would electrolyze water and create the hydrogen to be used as fuel either in combustion, fuel cells, or other devices. The hydrogen would, of course, need to be stored safely; that problem would only be solved at ECD two decades later. It is notable, however, that at the outset of his energy work Ovshinsky was already thinking not just of individual inventions but of systems.[18] And eventually Ovshinsky's interest in hydrogen bore fruit when ECD's hydrogen storage program led to the nickel metal hydride (NiMH) battery, a hydrogen fuel cell, and a hydrogen-powered car (see chapter 9).

The Early ECL Family

ECL began as a family operation, and even when the company eventually grew to employ over a thousand staff members it kept some of that family feeling (see chapter 7). In its earliest days, Iris would join Stan at the storefront soon after getting Robin and Steven off to school, typically beginning her day's work with a visit to the library to find articles and books for Stan. Herb, now the sole head of General Automation, continued to work in the storefront too until December 1960, when he moved his company to Troy, but both before and after that he would help Stan and Iris any way he could.

The first paid employees were high school students who worked after school or on weekends and two young men from the nearby drug store who worked at ECL during their off hours and learned how to mix chemicals and run tests. Among the high school students was fifteen-year-old Harley Shaiken, whom Ovshinsky came to consider "like my son." Harley had met Stan and Iris through their early involvement in the civil rights movement, first on a picket line protesting Woolworth's segregated lunch counters in the South, then at a meeting of the Congress on Racial Equality (CORE), whose Detroit chapter Stan had helped organize. Harley could sense immediately how, as Stan talked about the larger context of a civil rights protest, his mind connected politics with everything else in life, with "a generosity and an openness and an engagement that I've never seen." The encounter had a lasting influence on the young man.

Max Powell, the first African American and one of the longest serving among ECL's and ECD's employees, also began working part-time in 1960, when he had a small

Figure 5.4
Explaining the hydrogen loop.

Figure 5.5
Experimenting with hydrogen conversion, 1960.

office-cleaning business. A few years later, Stan persuaded him to join ECL full time as the official driver, a role he performed for many years. Born in 1910 in Selma, Alabama, Max had moved north to get away from racial violence after seeing one of his friends tied up with fireworks that were then set off.[19] Max appreciated Stan's intelligence, but was just as impressed with Iris: "She was the glue that brought together whatever he had." He also noted that Stan and Iris were "never further apart than they could touch each other." All the children loved Max, who soon became a family member and friend who would take them out for candy and share his life experience, making them more aware of problems like racism.[20]

In July 1963, at about the time that Max began working full-time at ECL, another long-serving employee joined the laboratory. Lionel Robbins, an electrical engineer who had worked in sales at Perkin-Elmer, was able to understand Ovshinsky's technology and consider its commercial possibilities. As Robbins saw it, his job was to "to find people who could figure out a way to bring the technology into their product lines," and he had some success. With a conversational facility that Ovshinsky lacked, he also assumed the role of guide, leading visitors through the laboratory and explaining its work.

Family Life

Iris and Stan were committed to making their new family succeed, but there were predictable complications. Robin and Steven came to love their "Stan-Dad," who was happy to help raise them. At the same time, they continued to miss their father Andy, whom they would see some half dozen times a year and for a month in the summer; he would also call them once a week. In time, Andy remarried, but he never accepted Stan.

Also harboring resentment, Norma made it difficult for Stan's boys to see him and Iris in the years before they were legally married. Like Harvey, Ben admitted reflecting a lot of the "antagonism and some degree of hostility" that he absorbed from Norma back to Iris, and he later felt grateful that "Iris took it well. She was quite saintly." In turn, Robin and Steven heard Norma described in bitter terms at home, especially when she asked for more money or tried to curtail Stan's visits with the boys. Robin remembered being asked to ring the bell at Norma's house one day when they had come to pick up the boys for the weekend. She braced herself and was ready to run if a "witch-like person" appeared. Then "this very nice lady opened the door and said, 'Would you like to come in?'"[21]

On March 30, 1962, two days after Stan and Norma were divorced, Stan and Iris were married in Toledo by a justice of the peace. But in true anarchist fashion, they considered their formal marriage but a legal detail, "an unimportant event," Iris said. In later years they often forgot to celebrate their wedding anniversary, but neither ever forgot their two important dates: Stan's birthday, November 24, marking the day when Iris, Robin, and Steven arrived in Detroit, and January 1, marking the day in 1955 "when we fell in love." To have one less date to remember, they decided to celebrate their marriage each year on New Year's Eve.

After Stan and Iris's marriage, Ben, Harvey, and Dale were able to visit them on a regular schedule that included both a weekend and a weekday. Iris would cook "some wonderful meal that the kids would enjoy, like spaghetti." And, as she recalled, "We'd try to do things together as a whole family as often as possible. We'd go to an art museum, or to Greenfield Village." In summer, "we'd take the car for a picnic and go swimming." Whenever possible, Iris would separate the children, so that she or Stan could do something special with each of them. Harvey particularly enjoyed the times when Stan took him to buy magic tricks or monster masks, and he fondly remembers the time when Stan took him to see the monster film *Gorgo*. Stan was generous in supporting Harvey's interests, buying him a chemistry set and a telescope when he showed interest in science, and then "a great little printing press with rubber stamps and ink pads" when he began focusing on journalism, an aspiration Iris encouraged by typing what he wrote. One of Harvey's most prized childhood possessions is a letter he received from *Twilight Zone* creator Rod Serling, who wrote to him with advice on how to become a writer. "Unbeknownst to me, Dad and Iris framed the letter. They wanted to surprise me, and they did."

But the children's lives were also restricted by Stan and Iris's political beliefs. Stan prevented them from watching TV shows like *Zorro* because of its violence, or *Victory at Sea* because it glorified war. The boys were also not allowed to join the Cub or Boy Scouts because, as Ben recalled, "they were a 'para-military fascist youth organization.'"[22] The younger children were also kept from scouting; instead, Robin recalled, "Mom started an after-school science club for Steven, me, and friends."

As children, the boys did not share Stan and Iris's political passion and found standing on picket lines intimidating. "I probably knew segregation was wrong," Harvey said, "but I wanted to *swim* at the local swimming pool, not protest against it." Ben enjoyed the UAW Labor Day parades with their hotdogs at the end, but like Harvey and Steven, he found some of the CORE and SANE (Society for a Sane Nuclear Policy) protests "scary a little bit."[23] When younger, Ben would be disappointed when told on Saturday mornings that they would be going off to picket rather than test cars in Milford, but by the time he was eighteen, in 1964, he went south in an old VW Microbus and spent six months in Mississippi living in tar-paper shacks to help register voters for the Student Non-Violent Coordinating Committee. Being younger, Steven "enjoyed all of the songs and May Day celebrations without knowing completely what was going on." But despite any misgivings about joining their parents at demonstrations, all the children were proud that their family stood up against social evils.

Dale, who was ten at the time Iris joined Stan, continued to struggle developmentally. When he was fifteen, Norma and Stan sent him to a special school on an Arizona ranch, where he thrived. The school psychologist took him with her to the library at the local university, where she was studying for her PhD, and where Dale could read books that weren't available at the school. "I will always be grateful for that," he said, "and proud of the fact that in the Rorschach test she gave me I scored high on creativity." Stan and Iris were proud too, but not surprised. Still, during this time their relationship with Dale was uncertain and tentative. At one point, Iris recalled, he "didn't want to see us at all." But when they stopped off in Arizona to see him while traveling to California, "he was thrilled," especially when they took him on a trip in the desert. And then "he was very close again." After Norma and her husband moved to Los Angeles, Dale left the ranch school to live with them and attended Rexford Junior and Senior High School, a private academy in Beverly Hills.

An exceptionally sweet child, Dale eventually became articulate and well read about the things he found interesting. "Despite his struggles," Harvey said, "Dale always had the biggest and most generous heart of all the brothers, and in his own way, the most clever sense of humor." Much later, he moved to Florida where he became a born-again Christian, proud of his Jewish heritage but "a devoted believer in Jesus as my Messiah." Stan and Iris were happy that Dale now had a circle of friends, but they were not pleased when he passed out religious tracts at Robin's wedding. Every year Stan would bring Dale to Detroit for his birthday. On those visits, Max listened with interest to Dale's stories about his life in Florida. Marveling at the power of Dale's memory, Max suspected that Dale was "the *other* genius in the family."

All the children enjoyed visiting the storefront, typically on weekends or after school. Only later did Steven realize that Stan and Iris took them because "they didn't have enough money for baby-sitters." Harvey likened the storefront to Santa's workshop. "Drilling holes in shards of plexiglass was especially fun." Robin remembered being "given little jobs," like pressing the amorphous material for switches. She was pleased that her switch, dubbed the Robin Device, "worked the best." Steven, who often played his violin there, remembered the fun of going to the corner ice cream shop with Max and looking under the counter at "the incredible number of pieces of gum that people had stuck there." Explaining their father's livelihood was at times challenging for the children. Harvey at twelve told Stan he felt he couldn't tell his teacher that his father was an inventor, because from the teacher's viewpoint Stan was not a "real" inventor like Alexander Graham Bell or Thomas Edison. Stan replied, "Tell them I'm an engineer if it will make you feel better."

Stan and Iris tried to give the children some sense of their Jewish heritage, sending his boys to Shalom Aleichem, a secular Yiddish school, and all the children to a social democratic Workmen's Circle school. But unlike Stan, the children did not take to their Jewish cultural training. Ben recalled attending Yiddish school with Harley Shaiken at the Workmen's Circle. It was "too many afternoons a week, and I never learned a single word of Yiddish except 'schmuck.'" When Harvey at thirteen did not want a bar mitzvah, Stan and Iris gave him a used Nikon camera and thirteen volumes of an encyclopedia that he wanted. Norma was not impressed and complained that Stan and Iris were "buying his love."

In August 1962 the family was suddenly evicted from their house on Gilchrist after Stan and Iris invited several black CORE members to a meeting there. The landlord, who had been contacted by a neighbor, arrived the next day to say he would no longer be renting the house. Iris remembered him peering into their living room and muttering, "You read a lot, don't you?" She regretted having paid two months' rent in advance. Now the family had to find a new house immediately, and Iris needed to enroll the children in new schools so they could begin on time in the fall.

Moving was already on the agenda by then because ECL needed more space. They had been thinking about moving the company to Troy and so started looking at houses in Birmingham, a pleasant nearby town with good schools. After three or four landlords turned them down because they were Jewish, they found a lovely house to rent at 1692 Villa Road and again decorated it beautifully. For the next two years, until Stan finally did move the company to Troy, he and Herb, who lived in Detroit but worked in Troy, often passed each other as they drove to or from work.

As it happened, the family moved to Birmingham the week when Stan and Iris had organized a SANE meeting to discuss nuclear fallout, an issue that worried many Americans then. Fearing the radiation exposure from strontium 90, Iris would "give the kids powdered milk."[24] Stan and Iris advertised the meeting in the local Birmingham paper with a full-page ad sponsored by SANE and signed prominently by "Stan Ovshinsky, chairman," and "Iris Ovshinsky, vice-chairman." But instead of listing the Central Methodist Church as the address for the meeting, the ad erroneously gave the address of their new home. "We were so afraid that our landlady would say 'Go,'" said Iris, but she apparently never saw the ad.

Open housing was another issue Stan and Iris took up in that period. When one successful black man in the insurance business was denied a certain home, Stan bought it and sold it to him.²⁵ Max Powell remembered a huge gathering that Stan and Iris arranged at Birmingham's Society Hall on Woodlawn where Stan explained how the existing housing laws discriminated against blacks. At the same time he reassured the concerned whites, "If they can afford it, you don't need to worry." Even though many on the Birmingham city council opposed open housing, with Stan's advocacy there were enough votes to push an amendment through.

Robin and Steven both loved Birmingham, which Robin compared to "a little New England village, nicer than where we lived in Detroit." They attended the Adams Elementary School, and their neighborhood was filled with children. "We ran around with a gang of twenty-one kids," recalled Robin, who especially recalled playing on the railroad tracks, while Steven remembered meeting with kids in the street. Although still very shy, he recalled, "By the end of the first day, I was playing with everybody."

After the move, Stan's boys interacted with Robin and Steven more than they had earlier. Steven and Robin felt as though they were in a younger generation, but "we never had any problems of feeling that we didn't want them around," Steven reflected. "Harvey was the most involved with us. He would make up these wonderfully creative hide and seek kind of games, almost like treasure hunts." During one phase, sixteen-year-old Ben wasn't getting along with Norma and moved into the basement of the Villa house. It was about then that Ben won a scholarship to attend the summer program of the Choate school in Connecticut, and the school's Russian travel program took Ben to Europe in 1963.

In Birmingham, Stan and Iris entertained guests even more often than they had on Gilchrist. Iris always cooked, and the children found interacting with the guests educational and pleasurable. "One thing that was always very consistent," Steven recalled, was "this sense of the conversation around the table. Stan was always the dynamic leader, and the content always concerned important matters in the world." Their many dinner guests included the eminent biochemist Linus Pauling, the physicist Edward Teller, the Alcoholics Anonymous co-founder Bill Wilson, the pioneers in LSD research Abram Hoffer and Humphry Osmond, the great socialist and pacifist Norman Thomas, and Fenner (now Baron) Brockway, who in 1964 had been made a life peer and taken a seat in the House of Lords.²⁶ Whenever Thomas visited, he would sing the old labor songs with Stan and the children. The physicist Hellmut Fritzsche (whose relationship with Ovshinsky is introduced later in this chapter) would play violin with young Steven. A visitor from the Swedish company L. M. Ericsson, which became one of Stan's first licensees for his chalcogenide switches, arranged for Ben to work in one of their factories in Sweden during the summer of 1964.

It was only much later that any of the children realized how hard Iris was working. While Stan "was in charge of the company," said Harvey, she was "in charge of him and the family." Steven recalled, "My mom managed to cook these amazing meals" after working all day. She would still be at work when he and Robin called to let her know that they were home from school, and then "she so patiently heard every little detail of our day. It wasn't until decades later I thought about how generous that was and how wonderful," said Steven. He also recalled that when Stan and Iris occasionally went on a business trip without them, Iris sometimes took "one of her handkerchiefs, and put her kisses in it and tied it up, and gave it to us. What I remember is that sometimes those were soaked with our tears by the time they got home. So there were times when we definitely were missing them."[27] The family did not always eat at home. "Once in a blue moon they would order pizza," recalled Steven. But "best of all was going to this place called the New York Bagel Factory in the Jewish Detroit neighborhood known as Dexter-Davison. The four of us would be in the car, and we would buy a bag of hot bagels, a big package of Philadelphia cream cheese, an entire Hebrew National salami, and a half gallon of skim milk. And then like savages, we'd just eat the salami, and dip the bagels in the cream cheese, and drink the milk."

The Search for New Switching Materials

In 1960 Ovshinsky's development as an inventor reached a crucial juncture. His early work on energy conversion and his other ideas like the oxide film computer had not yielded any important results, and he was blocked by the settlement of the Allen suit from developing the Ovitron, his most important invention so far. The settlement provided funds to keep ECL going, but he could use neither the design nor the same materials. Yet as he sought better materials and a better design, Ovshinsky managed to turn this setback into an advance. His persistent search led to the breakthrough discovery of a new switching effect in the invention of his threshold and memory switches, the greatest achievement of his career as an independent inventor.

Tracing his path to these discoveries offers our best chance of understanding Ovshinsky's idiosyncratic, intuitive creative process. We have to depend mostly on his later recollections, in which some gaps necessarily remain, but we can still assemble an account showing how several strands in Ovshinsky's experience combined here. Ever since the time he built the Benjamin Lathe, he had been thinking about the limit switches used in lathes and milling machines that had the function of stopping or turning around the pieces being machined. These electromechanical switches often failed in an environment of metal chips, cutting oil, and constant vibration. Becoming dirty, they sometimes did not turn on when the contacts of the relay touched. Ovshinsky initially experienced this failure as an annoyance, but then he found that when he increased the voltage the contacts would conduct. He realized, as he later told Hellmut Fritzsche: "If that is so, I don't need the relay. If that part [the film of dirt] is nonconducting, but with the higher voltage it does conduct, it is by itself a switch."

Another strand came out of the nerve cell analogy that had resulted in the Ovitron. In his earlier thinking about nerve cells as switches, their semi-permeable membranes were the key to transmitting signals, just as the oxide layer on the tantalum electrodes was the key to the Ovitron's switching behavior (see chapter 4). He believed that both thin films changed from insulating to conducting when the nerve impulses or voltage passed a threshold, and at some point he realized that the oxide layer he needed to replace had an amorphous structure. Could he draw directly on such disorder in materials to build a better switch than the Ovitron? This was an important step. In thinking along such lines, Ovshinsky was boldly diverging from the work of nearly every other researcher in the area of solids, who considered amorphous and disordered structures as useless "dirt materials."[28]

Almost all solid-state physicists dismissed amorphous and disordered materials in favor of crystals, with their rigidly ordered lattice structure, and standard textbooks on solids like Frederick Seitz's classic typically began with a presentation of the different possible crystalline structures.[29] This bias was not an irrational prejudice. It followed from the history of the whole field, which began with crystallography and became systematic through the fundamental achievements of quantum mechanics, which enabled physicists to develop formal accounts of solids at the atomic level.[30] Furthermore, crystals had become the basis for the growing semiconductor industry that took off with the invention of the transistor in 1947, and it was supposed that their success depended on their regular periodic structure. There were thus both strong theoretical and practical reasons for preferring crystals, and we need to recall that consensus to appreciate just how independent and original Ovshinsky's intuitions were.

By turning to amorphous materials for switches, Ovshinsky opened up a new line of research that would be enormously productive. He made a major effort to learn what was known about amorphous and disordered materials. The only book with any information at all on the subject was Thomas James Gray's, *The Defect Solid State*, published in 1957.[31] But while the book covered the notion of materials having defects in their periodic structure, and spent many pages discussing glass, it did not deal with the electrochemical aspects that Ovshinsky was most concerned with. Indeed, the very designation of amorphous solids as "defect materials" presupposed the unquestioned

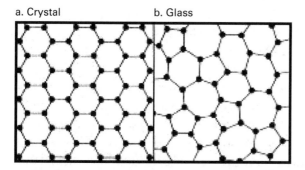

Figure 5.6
Regular periodic structure of a crystal (a) and irregular structure of an amorphous (glassy) material (b). Adapted from Zallen, *The Physics of Amorphous Solids*, 12.

crystalline norm that then prevailed, what he would later call "the tyranny of periodic constraints."

As Ovshinsky came to realize, even crystals are typically disordered at their surfaces, where their periodic structure is interrupted, and in fact the success of the field-effect transistor, the basis of the microchip, depended on its thin surface layer of amorphous silicon oxide.[32] Focusing on surfaces, as he had already done with the Ovitron, led Ovshinsky to work on creating thin amorphous films that, like the insulating film of dirt on the lathe relays, would act by themselves as switches. This was another adventurous step: making switches from thin films was radically original.[33]

Searching for materials to make thin films, Ovshinsky began working through the periodic table of elements, visualizing their electronic structures and looking for those that might offer the mechanisms he needed. Gray's book on defect solids briefly mentioned tellurium and selenium, two of the chalcogens, the elements grouped under oxygen in the sixth column of the periodic table (group 16 in the present numbering). He began to think that compounds of these elements, chalcogenides, might be promising replacements for the oxides he had used in the Ovitron, because like oxygen, the chalcogens all have a valence of two, with six electrons in their outer shells.[34]

Ovshinsky tried to learn what was known about chalcogenide glasses and other amorphous materials. He attended scientific meetings of those studying glass but found them uninformative, despite the fact that the technology of glass is thousands of years old. He already understood from his experience with metallurgy that when molten material is cooled rapidly onto a cold substrate it does not form crystals. Beyond that, the scientists studying glasses were thinking only about their optical properties, ignoring their electronic structure and surfaces. He learned little more in visiting the existing research labs. At Xerox in Rochester, New York, people were friendly, but when he asked to see their amorphous work, "they said that they had none." "But your drums are amorphous," he pointed out, a fact Xerox later acknowledged. (They were made of amorphous selenium.)

There is one other possible source of guidance for Ovshinsky's choice of the chalcogenides. In attending meetings and reading about amorphous materials around 1960, he may have come across the work of Boris Kolomiets, who with his collaborators at the Ioffe Institute in Leningrad had been studying amorphous chalcogenides and describing their properties in Russian journal papers since the mid-1950s. This, however, seems unlikely. Ovshinsky typically did not hesitate to get in touch with leading workers in areas he was interested in, but there is no record of any communication with Kolomiets or of any visit to Russia before 1967, and he never mentioned meeting Kolomiets before they met in 1967 (see chapter 6). In any case, although the Kolomiets group reported the semiconducting behavior of the chalcogenides in 1956, and extensively studied their electrical and optical properties, Kolomiets did not observe switching because his experiments used bulk materials instead of the thin films that Ovshinsky worked with.[35] Indeed, Ovshinsky's use of thin films distinguished his approach from those of nearly every other researcher in the field at that time and was crucial to his success.

Finding little help from others, Ovshinsky had to depend on his own intuitions as he considered the chalcogenides. He was especially intrigued by selenium and tellurium, which, unlike sulfur, have a molecular chain structure that reminded him of DNA, whose double helix chain structure was continuing to be widely discussed. (The discovery of this structure, which was published in *Nature* in 1953, would be recognized with the Nobel Prize in 1963.) To Ovshinsky the DNA discovery showed that a noncrystalline double chain structure with cross-links between the chains could carry information. As he later put it, "I chose tellurium because it was chained like that."[36] This was another fruitful analogy, which, like the nerve cell, bridged organic and inorganic structures.

Ovshinsky was already prepared to appreciate the significance of such chain structures, drawing on a strand of experience that started two decades earlier than his thinking about switches. While working at Goodrich in 1941 he had taken a seminar on the chemistry of rubber, which he learned was a polymer composed of long molecular chains connected by cross-links.[37] Reinforcing the DNA analogy, his recognition of the shared polymer structure engaged Ovshinsky's distinctive ability to visualize molecules, and helped to confirm his choice of the chalcogenides. He later described his search for new switching materials: "I wanted something that could have cross-links, and none of the elements except the group 6 really has that. I wanted something that has built-in plasticity and flexibility," presumably because he believed that would facilitate making thin films. He focused on tellurium and found that to get the switching effect he needed, "tellurium by itself doesn't work. It has to be able to have cross-links with other elements. And," he added, "because I was brought up in Akron, Ohio, I was familiar with polymers and cross-links. Because that's what rubber tires are made out of."

Ovshinsky would learn to create cross-links in tellurium by adding small amounts of other elements such as germanium or arsenic. The role of these cross-links was explained in microscopic terms only later, when he could collaborate with trained physicists to publish scientific accounts of his discoveries (see chapter 6).[38] But well in advance of theoretical explanations, Ovshinsky's intuitions led him to choose these elements and experiment with using them in different combinations. He could not afford expensive evaporation and deposition equipment for creating thin films, so he ground his materials into powders and experimented with combining them, like a modern alchemist seeking the magical formula. To help him test different ratios of elements, he built a small box he called his "universal tester," where he pressed together the various powders and probed them with the electrical leads of his homemade oscilloscope.

Trying to achieve greater stability and conductivity, he combined tellurium with elements like arsenic or antimony from neighboring groups. When he inserted his probes into these powders, he saw an erratic switching effect on his oscilloscope, but he did not yet have a reliable switch. Then a new possibility occurred to him. The powders in the stagnant air of the storefront had given Ovshinsky a mild case of arsenic and tellurium poisoning, but they also gave him an idea. Suspecting that the polluted air had deposited an invisibly thin film on the micrometer he would carry in his toolmaker's apron for measuring thicknesses, he attached a power source across the calipers and connected the leads to his homemade oscilloscope, which he had configured as a curve tracer, the X-axis showing voltage, the Y-axis showing current. He brought the calipers together and gradually increased the voltage. A dramatic "cross" pattern appeared on the screen. This could be called his eureka moment: Ovshinsky had in effect created his first amorphous threshold switch, for which the cross became the electronic signature.

Ovshinsky immediately sensed the importance of his discovery. The cross pattern indicated an extremely rapid, almost instantaneous switching (now known as the Ovshinsky effect) from insulating to conducting and back. Its symmetry showed that its behavior, unlike that of a diode or transistor, was reversible and bipolar.[39] No one else had seen this effect before. Ovshinsky's hunch about making semiconductor switches

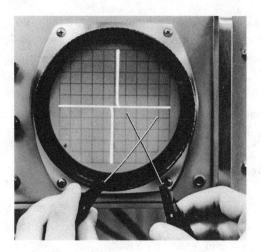

Figure 5.7
Oscilloscope trace of the "cross" pattern.

from thin films of amorphous materials had been confirmed. Now he needed to learn how to reproduce the effect reliably.

To repeat the experiment systematically, Ovshinsky had to control the composition of the film, both the elements in it and their proportions, which was a demanding process. Still working with finely ground powders, he mixed different elements, heating and pressing them down in thin layers on a substrate. "Then I'd put electrodes on it," he said, "and it worked." He had done related metallurgical work much earlier when he was still in high school, hot-pressing powders. "I was used to mixing different elements into powders to get different things, right out of trade school," he recalled. Even then, he would try to use as many different elements as possible that would hold together. Now again, he experimented with heating and compressing his combinations of disordered materials into thin layers. "I kept on making them thinner and thinner and they still worked," he said. "And these were new things."

Before following Ovshinsky's next steps, we should pause to consider the historical implications of his breakthrough discovery. The thin-film amorphous chalcogenide semiconductors that developed from his threshold switch have found their principal applications in advanced information technologies, in devices such as phase-change optical and electronic memories (see chapter 10). Yet the story of his discovery takes place in a world far removed from the sites where accounts of the birth and growth of the information age are usually set. From the advanced research facilities of Bell

Labs, where the transistor was created by highly trained physicists, to the rigorously controlled environments of the contemporary cleanrooms that are essential for microelectronic research and fabrication, semiconductor technology depends on resources and conditions that were notably lacking in the ECL storefront. And yet Ovshinsky's discovery was made not just in spite of those deficiencies but also in some respects because of them.

The key here is dirt. The insulating thin film of dirt on the limit switches that conducted when the voltage was turned up, the polluted air that left a thin film on the micrometer in his toolmaker's apron—all evoke the dirty environment of the shop floor and the tool room. Amorphous semiconductors have their material roots in the blue-collar working-class environment where their inventor had his social roots, just as his empirical research methods have their roots in the artisanal practices he began while in trade school.[40] And this impurity continued in the technology developed from these dirty materials: Ovshinsky's threshold switch did not require the crystalline purity and perfection of silicon semiconductors or the delicate precision required to make them, and as he loved to demonstrate, its functioning was unaffected by contamination or rough handling.

Early Promotional Efforts

Soon the Quantrol (Ovshinsky's original name for his threshold switch, because he believed that it worked by quantum control) began to receive attention and some support.[41] The British company Electronic Machine Control, Ltd. not only bought the first license (November 30, 1962) but also held a press conference in January 1963 at the Savoy Hotel in London to announce the new device.[42] One of the problems they encountered in the demonstration arose because the Quantrol was an AC device and the Savoy was one of the few DC hotels in London. Iris also remembered working in stadium boots "because it was so cold," and that "when Stan would come back to the Savoy Hotel carrying his oscilloscope, the doorman didn't want to let him in because he looked like a worker." A larger problem, shown by the transcript of the press conference, was the frequent frustration of the audience with Ovshinsky's explanations.[43]

By the summer of 1963, Ovshinsky had made some progress in promoting his new switch, but his efforts to attract major funding had failed. He realized that to succeed he would need an endorsement from a well-known scientist whose work was "beyond reproach or prejudice." He sought out John Bardeen, famous for the invention of the transistor and the discovery of the BCS (Bardeen, Cooper, and Schrieffer) theory of superconductivity. Bardeen had already won the first of his two Nobel Prizes, in 1956, for the invention of the transistor.[44]

Ovshinsky didn't know Bardeen, but he picked up the phone and dialed.[45] Bardeen's secretary came on the line and told him that Bardeen was in Pittsburgh. "Why don't I connect you?" Ovshinsky thought this odd, but soon a voice answered and said, "This is John Bardeen." Ovshinsky began, "You don't know me at all, but I know of course who you are." He explained that he had called to talk about "something that has to be seen to be really believed. It's new and there's nothing like it in semiconductors." Bardeen started to explain that he didn't do that kind of consulting, but when he heard that the device was non-rectifying he knew it was based on new physics. "Are you telling me," Ovshinsky recalled Bardeen asking, "that there is no PN junction, that it's not a rectifying junction?" "Yes, absolutely," he replied. "That's very interesting," Bardeen said. Ovshinsky could hear him flip through his date book. "I can make it in December," he said. At that, Ovshinsky blurted out, "Professor Bardeen, we'll be broke by then." Bardeen then suggested two physicists who might be able to come out sooner, either Hellmut Fritzsche, a young professor at the University of Chicago, or Nick Holonyak, who had been Bardeen's first graduate student at the University of Illinois. Ovshinsky chose Fritzsche because he had read a paper by him on tellurium in *Science* magazine.[46] He thought Fritzsche would be interested. As Ovshinsky recalled, Bardeen said, "Well, Hellmut is a fine physicist. You'll really enjoy him. And he's going to be tough."[47]

Fritzsche was indeed an excellent choice. He was already working with the disordered systems created by embedding impurities in crystals. Trained by the great Karl Lark-Horovitz, the head of the physics department at Purdue University, he had been studying the tunneling conduction observed at low temperatures in germanium with high concentrations of randomly distributed impurities, such as arsenic or antimony. Studying conduction in amorphous materials sounded to Fritzsche like a relatively small step, and he felt fully prepared. "All my instruments in Chicago were ready to be turned on to the electrical properties of Stan's materials."

Before arranging the visit, Fritzsche had to consult his wife Sybille, then expecting the third of their four children. Her due date was three weeks away, but she agreed to let him visit Detroit for one day. The visit, which probably occurred on Friday, August 30, 1963, "profoundly changed my life," Fritzsche recalled almost forty years later. It was "the beginning of a most fruitful and exciting collaboration and a deeply enriching friendship that includes all our family members."[48]

Fritzsche remembered Max Powell meeting him at the airport.[49] As they drove in, he eyed all the big buildings they passed and looked for a prominent sign that said Energy

Conversion Laboratory. He saw none. They eventually parked on a nondescript street in front of an unimpressive storefront without any sign. Ovshinsky greeted his visitor and led him into his office, where Fritzsche particularly recalled being impressed by Ovshinsky's many books.

Ovshinsky took Fritzsche back into the storefront's laboratory space and showed him the threshold switch with its crossed wires coated at their point of contact with a virtually invisible layer of amorphous and disordered material. When Fritzsche saw the symmetrical cross pattern on the oscilloscope, he realized he was dealing with something new. "I saw that the material was highly resistive, but at a certain voltage level it switched to a very, very conducting state. And then it switched back to the insulating state. That was completely miraculous."[50] He asked what was in the thin film coating the wires; germanium, tellurium, silicon, and arsenic, Ovshinsky told him. Fritzsche had worked extensively with germanium and tellurium and felt he knew their behaviors. "But a mixture of all of these would not form a crystal," he said. Ovshinsky explained that the material was not a crystal, and that, unlike a crystalline material, "the exact composition is not that important." This contrasted sharply with the extreme sensitivity of crystals to their exact composition, and Fritzsche was also amazed by how insensitive the amorphous film was to rough handling. The electrical characteristic did not change when Ovshinsky wiped the wire, or even when he stepped on it. It also made no difference whether or not the surface was dirty. "So after studying crystalline semiconductors where the surface is of extreme importance and the material composition is very delicately chosen," Fritzsche recalled, "I realized that this was a completely new phenomenon and a completely new material. And, of course, my interest was at a high point."

Fritzsche was also taken by the forty-year-old Ovshinsky's "immense intellect, exuberance, and personal warmth." As the two sat and spoke about the switch and amorphous materials, he realized that Ovshinsky had read extensively in the scientific literature on semiconductors and tellurium and had done a thorough literature search. Ovshinsky showed him his cabinets filled with scientific papers, and pointed to the files that covered all the surfaces.[51] Fritzsche noticed that he had even read "some of the Russian literature and some of the books on glasses. He knew much more than I knew at that time."

Ovshinsky meanwhile was forming his impressions of the thirty-six-year-old Fritzsche. The professor from Chicago seemed "very young" yet sure of himself as a scientist. It seemed to Ovshinsky that Fritzsche initially "thought that I knew nothing about the subject that I was working with." As he recalled, Fritzsche walked into his lab and said, "Mr. Ovshinsky, I am a detective. Just let me in the laboratory to see it, to work with it, and I will explain it to you at the end of the day. I will solve your problem." Ovshinsky was not so sure about that and decided just to leave him alone. "At the end of the day," he recalled, Fritzsche "was a very chastened human being," uttering expressions such as "I don't understand it. I can't explain it. All I know is it may be more important than the transistor." He finally asked, "What can I do to help?" Ovshinsky replied, "Just become our consultant." Fritzsche agreed.

As ECL's first physics consultant, Fritzsche visited often, helping on several fronts as friend, interpreter, collaborator, and when necessary a firewall between Ovshinsky and the scientific community. To begin, he devoted much time to translating Ovshinsky's accounts into language that scientists, engineers, and patent attorneys could understand. To help build a strong patent position Fritzsche worked with the Chicago-based patent attorney Charles (Chuck) Spangenberg, who began working with ECL at about the same time he did, and the two traveled widely in Europe to file applications.

Patents would become increasingly important as the company grew, and Ovshinsky focused increasing attention on his patent department. Much of his legacy would rest on his ability to protect his intellectual property through strong and broad controlling patents, as well as a policy of developing the next generation technology before others could. While in future years, many companies would attempt to use ECD's technologies without paying for them, Ovshinsky's patent coverage typically resulted either in their paying hefty fines or becoming an ECD partner or licensee—sometimes both.

Fritzsche also responded to commercial inquiries and met potential patrons. He helped draft nondisclosure agreements, which, at $10,000 each, kept ECL going for a while. By November 1963, he recalled, "I was already engaged in conferences and negotiations with West Bend, DuPont, Crystallite Division in Toledo, the National Cash Register Company in Dayton, Eriksson of Sweden, the North American Phillips Laboratories, and the North American Aviation Science Center."[52] Some companies had novel uses in mind for the switches. One Illinois manufacturer wanted to use them as sensors in coffee pots to turn off the power when the water boiled.

By October 1963, Ovshinsky had submitted his foundational patent application for the threshold switch, "Symmetrical Current Controlling Device," and was publishing data on his chalcogenide switches based on tellurium alloys (with arsenic, silicon, and germanium).[53] There were articles on the Quantrol in *Electronics* and *Control Engineering*. The one in the April 1964 issue of *Control Engineering* seems to be the first that received much attention. Ads for the new switch later featured a stunning photograph demonstrating the novelty and simplicity of the device: two crossed wires with the active film at their point of contact.[54]

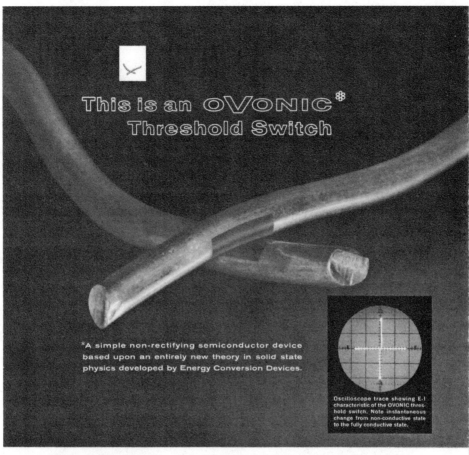

Figure 5.8
Ovonic threshold switch ad (designed by Ed Watkins).

Almost in passing, the last section of the article also mentions a second switch, a memory device that was a serendipitous offshoot of the threshold switch research. Its discovery came from an accident that Ovshinsky's prepared mind was able to interpret and seize upon. The assistant from the drug store, who was studying the characteristic cross pattern of the threshold switch's transition from high to low resistance and back, came to him very upset because he could not make the device he was testing switch anymore. Once it had switched, it remained in the low-resistance conducting state, so he threw it away. Ovshinsky retrieved the discarded device from the trash, gave it a strong pulse, and it again became a non-conductor. He immediately realized the importance of what he'd discovered: a bi-stable device based on a slightly different composition of the material used in the threshold switch, which could become the basis of a fast rewritable electrical memory.[55] Later known as the phase-change memory, this switch would eventually become recognized as one of Ovshinsky's most important discoveries.[56]

Leaving Birmingham, the Storefront, and Wayne State

One day in the fall of 1964 the doorbell rang at the Ovshinsky home on Villa. Robin came to the door. A deliveryman had "something for the people who are moving to Squirrel Road" then still a dirt road in Bloomfield Hills. "We're not moving," Robin said, but Steven, then in fifth grade, pushed past and said, "Yes, that's us." When Robin asked Steven about it, he replied, "I can't tell you." He had overheard Stan and Iris talking about the family's upcoming move to Bloomfield Hills, but Iris had told Steven not to tell anyone because they were trying to keep the news from Norma, in fear she might push for more money and reopen the divorce settlement. Stan and Iris were not actually sure they could afford the new house, but Richard Cummings, the senior vice president of Stan's bank (and later a director of ECD), had told Stan he was getting a very good price and urged him to buy. "This area will grow," he explained. "You're not taking any risks." When Iris worried that the location was isolated, the banker said, "Buy a dog." The kids loved Buffy, their huge white Great Pyrenees, but they couldn't keep the dog because it gave Stan serious asthma attacks.

Again, Iris and Stan furnished their new house to suit their life style, with books and study space everywhere. The ten-acre wooded property around the house, with its small lake where they could swim in the summer, would offer enjoyment to the family and to countless visitors. Later they added an indoor pool so that the swimming could continue year-round. Robin recalled how her friends loved to spend time at her house, with its many books and its Danish modern furniture, in contrast with the chintz and white fireplace décor then popular in the wealthy Detroit suburbs. Her friends enjoyed conversing with Stan and Iris, responding well to their liberal and international perspective, which was typically quite different from their parents'.

The company also moved some months after the family moved to Bloomfield Hills (see chapter 6). It had been clear for some time that the company needed more space, and the neighborhood surrounding the storefront "was getting much too developed," Iris commented, with "a lot of traffic." They kept delaying the move because Stan was reluctant to slow his work. He initially dealt with the space problem by adding an annex storefront next door on the same side of the street, plus an extension laboratory two blocks away across McNichols Road.

One summer day in 1964, however, he received a serious electrical shock in the extension laboratory.[57] Iris wasn't there and was simply told to come over, that Stan was ready for lunch. "They didn't want to tell me because 6 Mile is a very big street. They didn't want me to go running across." And so she took her time. "I combed my hair, put on my white gloves, walked across the street, and there's Stan lying on a stretcher." He was hospitalized for ten days, and there was talk about his dying, because the shock had passed through his heart.[58] Iris "kept saying, but you look so good, Stan." She was incensed when he replied, "I'll make a handsome corpse." Although he recovered, it was now completely clear that their working space was not only too small but also unsafe.

Lionel Robbins soon found an empty warehouse suitable for the company's work, roughly 10 miles north of the storefront at 1675 W. Maple Road in Troy. Energy Conversion Devices, renamed to reflect its increasing commercial orientation, would work in its new Troy space on Maple for four decades.[59] Close to the intersection with Crooks

Figure 5.9
The new facilities in Troy, Michigan.

Road, the warehouse was just one building away from Berz Airport, used for small propeller planes. A plane had once missed the field and crashed on the other side of the street. "Whenever I heard airplanes going overhead I always worried," Ovshinsky confessed.

At the time the lab moved to Troy in February 1965, Ovshinsky stopped his neurophysiological research in the wet lab that he had set up in the storefront. The work had been difficult to continue because Morin, now chairman of the Wayne State anatomy department, had been very ill for some time. Shortly after Morin died on November 26, 1964, Ovshinsky was suddenly dropped from the Detroit Physiological Society.[60] And when he tried to borrow a book from the library at the medical school, a librarian told him that this would now be violating university rules. It was a great disappointment to Ovshinsky, who had considered Wayne State his academic home. He would now create his own intellectual community.

II Inventing with Others

6   The Birth of ECD: An Invention Factory (1965–1979)

In February 1965, everyone working in the storefront packed up and moved into the new Troy space for the recently renamed Energy Conversion Devices (ECD).[1] Stan and Iris's offices were again side-by-side, as they had been in the storefront.

ECD's goals and culture remained much the same but Ovshinsky's role changed, especially after September 1967, when the laboratory became supported by publicly traded stock.[2] No longer the lonely independent inventor, he became the head of an ambitious research laboratory. His earlier inventions, from the Benjamin Lathe to the Ovitron and threshold switch, had been created directly by his own efforts with the help of only his brother Herb and a few employees, and from the barn on Chaffin Road to the storefront on McNichols Road, the places where they were produced had been necessarily modest. Now, with the move to Troy and ECD's subsequent growth, Ovshinsky had larger and better facilities plus a growing research staff. The many inventions that came from ECD over the following decades were still his creations, produced under his leadership, and directed by his vision, but others now did the hands-on work. Like Edison's lab, ECD became an invention factory with hundreds of scientists and engineers and supporting personnel.[3] And like Edison and many other inventors, as well as many academic scientists, Ovshinsky became an administrator—but one who was always intensely engaged with ECD's ongoing research programs and the science behind them.

Instead of working in the lab, Ovshinsky needed to devote more effort to representing ECD and raising funds. Meetings with licensees, accountants, and attorneys absorbed an increasing amount of his time. He now regularly reported on the company's progress to major investors and to ECD's new board of directors. For such activities, and eventually for all his work, Ovshinsky now dressed regularly in a handsomely tailored three-piece suit, a costume change that reflected this important transition in his life as an inventor.[4]

As Ovshinsky's role changed, so our narrative of his life also changes at this point. Here and in the following four chapters, ECD becomes a major character, sometimes diverting our attention from Ovshinsky himself. Its technological achievements all stemmed from his inventive genius, and its aims and character all arose from his and Iris's social vision. But tracing the history of the institution they created will occasionally require displacing them to make room for the many new figures who enter the story.

Physical Review Letters and the New York Times

On November 11, 1968, Ovshinsky's name suddenly became widely known to physicists working on semiconductors, when his paper describing his chalcogenide switches appeared in the prestigious journal *Physical Review Letters*.[5] Until then, Ovshinsky had described his switches only in trade magazines, for which the screening process of scientific peer review is unnecessary.[6] But the 1968 paper describing the experimental behavior of his threshold switch passed peer review, although with some difficulty.[7]

On the same day the *Physical Review Letters* article appeared, the *New York Times* published a front-page story about the switches written by the well-known *Times* science writer William Stevens. Ovshinsky was surprised and happy to see the article while the family was visiting the Interlochen Arts Academy in northwest Michigan, where as a treat for Steven they were attending a benefit concert on November 10 by the Soviet-born violinist and conductor Isaac Stern.[8] Iris recalled, "The next day we opened the *New York Times* and there was Stan's picture on the front page!"[9] They learned that an article on the switches had also appeared on the last page of the *Wall Street Journal*.

Iris recounted the steps resulting in the *Times* article. "Stan decided this was going to be quite earth-shaking" and called a press conference, to which he invited reporters from the *New York Times* and the *Wall Street Journal*. Unlike the Ovitron press conference (see chapter 4), this was a simple affair, held in ECD's new conference room. After being shown around the laboratory, the reporters asked Ovshinsky to comment about the future implications of his switches and memory devices. He predicted the development of thin-film computers and flat panel displays, while Hellmut Fritzsche and his fellow ECD physics consultant Morrel Cohen did their best to explain the science. When Stevens phoned the respected elder physics statesman Sir Nevill Mott for his comments on the switch, Mott said, "it is the newest, the biggest, the most exciting discovery in solid-state physics at the moment," adding that unlike the transistor, whose principles could have been worked out on the basis of existing knowledge, the

Figure 6.1
Article about Ovshinsky's invention in the *New York Times*, November 11, 1968.

discovery of the Ovshinsky effect was "quite unexpected" and represented "totally new knowledge."[10]

What appeared at first to be a triumph of recognition and publicity quickly backfired. Many academic and industrial scientists were outraged. The very publication of an important scientific claim in a newspaper seemed to taint Ovshinsky as a self-promoting charlatan. Fritzsche countered by pointing out that unlike Ovshinsky, "all the scientists who were criticizing Stan had fixed salaries from universities, from Bell Labs, from General Electric" and didn't need to advertise and seek support for their ideas. On the other hand, the contrary offense of a leading scientific journal accepting the work of someone who had no PhD was equally infuriating. Both the *Times* and *Physical Review Letters* soon came under strong pressure to withdraw their articles.

As the reaction unfolded, it became clear that besides Ovshinsky's use of publicity and lack of credentials, the very fact of his discovery incensed scientists at major research labs, where from their point of view such a discovery "should" more appropriately have occurred.[11] Some of the critics pressed Stevens to withdraw certain statements from his *Times* article, but Stevens showed them his notes and did not retract anything.[12] When they tried to get Mott to call the *Times* and retract his statements, Mott didn't either, but he felt extremely uncomfortable about the affair and expressed his concerns to Fritzsche in a private letter. As a young professor, Fritzsche also found the negative publicity embarrassing, but he promptly wrote back to Mott to explain the background of the press conference and to state his conviction that the criticisms of Ovshinsky were unjustified and rooted in jealousy.[13] Some physicists also urged Samuel Goudsmit, editor of the *Physical Review*, to withdraw the paper, but Goudsmit refused because it had passed peer review. He later told Ovshinsky that he had never experienced such craziness in his scientific editing career. Fritzsche reflected that the objectors must have recognized something important about the discovery. Otherwise, "it would have been dismissed by everyone."

Many critics seized on Ovshinsky's predictions in the press conference about the future implications of his invention, which they considered greatly exaggerated. It was to be a recurring theme of his career, for he could envision the successful consequences of his inventions without worrying about the practical problems that needed to be solved. In this case, the predictions would be borne out. Some also objected to the name Ovonics, which seemed self-aggrandizing and promotional.[14] But what most threatened researchers at Bell Labs or General Electric was the suggestion that technology based on amorphous and disordered materials would replace technologies based on crystalline semiconductors, then already the basis of a billion-dollar industry.

To make matters even worse, the notion that silicon was obsolete temporarily depressed the stocks of companies like Texas Instruments, Motorola, General Electric, and RCA, and caused ECD's stock to shoot up briefly to nearly triple its Friday price.[15] The incident, which looked like manipulation, "created very bad feelings on Wall Street that he was not to be trusted, despite that what he was saying was correct," noted the Stanford chemist John Ross. Of the companies that expressed criticism, the sharpest attacks came from Bell Labs, which claimed "nothing new has been disclosed" in the paper.[16] Bell Labs staff members were not permitted to visit ECD or invite Ovshinsky to visit for some years. That fact was driven home when not long after the appearance of the *Physical Review Letters* and *New York Times* articles he was invited and then promptly uninvited to give a talk at Bell Labs.[17] Having idealized scientists as among the purest and most civilized members of society, Ovshinsky could hardly believe the way they were behaving. "I was naïve," he reflected. "I wasn't expecting the rejection and hostility."[18]

Despite the uproar and enmity, publishing in a leading physics journal attracted many researchers to the area of amorphous and disordered solids, bringing the field "out of the backwater," as Morrel Cohen noted, and the article became widely cited. The previous scattered work, in Leningrad, Bucharest, and elsewhere, now coalesced into an exponentially growing, coherent international research area. Just for fun, the ECD physics consultant David Adler would often show in his talks a graph of the number of publications on amorphous and disordered materials in academic journals over time. The graph showed a steady rise after November 1968, when Ovshinsky's first *Physical Review Letters* paper appeared.[19] That increase had been anticipated earlier when Ovshinsky finally visited the Soviet Union in 1967 to speak in Leningrad about chalcogenide switching and memory effects at the fourth Symposium on Vitreous Chalcogenide Semiconductors. There he met Boris Kolomiets, the Russian researcher who had for over a decade been examining the optical and electrical properties of chalcogenides (see chapter 5).[20] After the talk, Kolomiets drew a slowly rising line on the chalkboard representing progress in the amorphous field until then and predicted that because of Ovshinsky's work, the line would now rise almost vertically. The prediction was fulfilled; the next international meetings saw between five and six hundred people attending, and the trend continued.[21]

The historical impact of Ovshinsky's 1968 paper in *Physical Review Letters* was amplified by a second paper in that journal in 1969, nicknamed CFO, for its co-authors Cohen, Fritzsche, and Ovshinsky.[22] This paper dealt with the electrochemical properties of amorphous semiconductors in relation to the switching effect. Eventually as much cited as the 1968 paper, CFO contained an account of the surprisingly sharp activation energy observed for the threshold switch, known as the "mobility edge," which, as Cohen put it, went "completely against everything that we were taught and that we were teaching at that point."[23]

Creating a Research Laboratory

As ECD gained resources from licenses and public stock offerings, Ovshinsky used them to turn the company into a laboratory that would not only develop new inventions but would also pursue fundamental materials science research. It thus entered the story of American industrial research laboratories, which begins in the later nineteenth century.[24]

In addition to the growing roster of scientific consultants (discussed in chapter 7), Ovshinsky hired several outstanding physicists as full-time staff members. Together with the consultants, they formed a group that came to be called the Physics Department, whose work aimed at developing a better understanding of Ovshinsky's discoveries and the nature of chalcogenide glasses, as well as developing applications of the Ovshinsky effect.

In many ways, the relationship between Ovshinsky and the physicists who worked for him mirrored his complicated relationship with the larger scientific community. Some of these physicists appreciated his unorthodox approach to research and valued him precisely because he worked outside the constraints of institutional science. Others, including many who had to take their research directions from him, were frustrated by his reliance on intuition and his difficulty with communicating ideas. The tension could also be felt within Ovshinsky himself. While he took pride in his independent, outsider position, he also craved institutional recognition for his scientific achievements.[25]

Optical Phase-Change Memory

Managed by Fritzsche, the Physics Department focused between 1968 and 1970 on understanding the science of the threshold and memory switches.[26] One of the early members was the colorful and brilliant Julius Feinleib, who had written his thesis on metal-insulator transitions and was then an assistant professor at MIT and a physicist at MIT's Lincoln Lab.[27] He contacted Ovshinsky after hearing about his work on amorphous materials, which appeared to relate to his own work. When Feinleib visited ECD, he found Ovshinsky rather impressive in his elegant three-piece suit and enjoyed the generous dinner and good wine he was offered. But from the first day, the two men were at odds about the physics of Ovshinsky's switch. Feinleib doubted Ovshinsky's claim that the mechanism was electronic, suspecting rather that it depended on heat. When Ovshinsky then said, "I want you to prove that it's electronic," Feinleib said, "Forget about it. I'm not interested. I have no feel for it."

Ovshinsky nevertheless made Feinleib a generous offer because of his expertise in the use of lasers. Ovshinsky had felt for some time that his phase-change switching could probably be induced by laser light and thus become the basis for an optical memory. Information could then be recorded and rewritten digitally using a laser to crystallize and amorphize precise locations on a diskette.[28] Agreeing that this might be both possible and interesting, Feinleib arranged a two-year leave of absence from Lincoln Labs, joining ECD in 1968. He convinced Ovshinsky to also hire the laser expert Sato Iwasa, then based at Honeywell, to help build the complex experimental apparatus. Having enjoyed Ovshinsky's 1968 *Physical Review Letters* paper, Iwasa was excited to come to ECD, arriving in 1970 with his bride Alice. He found ECD a lively place for research.

Ovshinsky's relations with Feinleib continued to be strained because of the two men's differences in both personal and scientific styles. Feinleib would annoy Ovshinsky by coming to work late and spending daytime hours taking flying lessons at nearby Berz airport or flying the old plane that he and Iwasa purchased together. Ovshinsky's official view was that "as long as they made contributions I really didn't care," but he still found it very irritating when he wanted to see Feinleib and learned he was out flying. The core problem, however, was their scientific mismatch. Feinleib did not believe that one could trust intuition in science and was irked when Ovshinsky lectured on results that matched his expectations before they were shown to be scientifically conclusive.

In spite of these differences, Feinleib and Iwasa soon succeeded in showing that a laser could crystallize a tiny spot on the amorphous material to indicate a one or a zero and, by applying another laser pulse to the same spot, they could cause it to change phase back to amorphous. This is the mechanism behind the rewritable optical memory, a technology that would later become widely used in CDs, DVDs, Blu-Ray, and high-definition disks.[29] Their resulting paper in *Applied Physics Letters* is often cited as the first optical memory paper.[30] Ovshinsky expressed his delight with the results characteristically. Alice Iwasa recalled, "One day, I was home from work, and all of a sudden someone knocks on the door with this big bouquet of flowers and a bottle of champagne."

Figure 6.2
Julius Feinleib and Sato Iwasa with the first rewritable CD.

Optical phase-change memory would become the first commercially significant application of the Ovshinsky effect. It was also the first instance of what became the typical pattern of Ovshinsky's later inventions, in which he came up with the initial idea and then got others to realize it. The change from working as an independent inventor to inventing with others made his creative efforts both more complex and more fruitful.

Growth of the Physics Department

The physics department was well supported in the early 1970s by a grant to survey the optical, electrical, and thermal properties of chalcogenide glasses. The support came from ARPA, the Advanced Research Projects Agency (now DARPA) responsible for developing technologies for the military. Bringing in roughly $300,000 a year, the grant allowed the group to flourish for some years. Ovshinsky soon hired other PhD physicists, including Ed Fagen, trained at the University of Pittsburgh, and Simon Moss, an x-ray radiologist from MIT who set up a structures lab, complete with an x-ray diffractometer as well as scanning transmission electron microscopes capable of seeing the phase changes from amorphous to crystalline and back.

John de Neufville had learned about Ovshinsky in 1968 while studying materials science under Harvard's David Turnbull. What most attracted him to ECD was Ovshinsky's question, "If you came here, what would you *like* to do?"[31] De Neufville outlined a systematic study of Ovshinsky's amorphous and disordered materials using the technique of sputtering to lay down combinations of elements. "If you came here, you could do that," Ovshinsky replied.[32] In addition to this De Neufville also contributed to the development of the optical memory. He remembered his five years at ECD as one of the happiest and most productive times in his career.[33]

Besides these PhD physicists, Ovshinsky also added a number of junior scientists Richard (Dick) Flasck, who joined the technical staff in 1970 with an undergraduate degree in physics from the University of Michigan, was attracted by ECD's recently added education program to support further training for staff members. By studying nights at Oakland University in Rochester, Flasck soon gained a master's degree in physics. He recalled this period at ECD as "an exciting time," but he also sensed the strain many scientists felt in dealing with Ovshinsky's claims. "They tried to be as flexible as possible without breaking their ethical backbone." Part of the problem, Flasck noted, was that "Stan did not think like a standard physicist or chemist—not in numbers and not in principles, but in pictures. And sometimes that gives insight that you can't get from standard mathematics."

At the same time that Feinleib and Iwasa were developing an optical phase-change memory, others were developing an electrical phase-change memory that would eventually prove more important. The work involved collaboration between ECD and the newly formed Intel, an outgrowth of the relationship that began when Intel's founders, Robert Noyce and Gordon Moore, who had been members of William Shockley's original team in his Shockley Semiconductor Laboratory, visited ECD in the summer of 1968.[34] They were attracted by news of Ovshinsky's switching and memory discoveries and were especially interested in his technique for etching "down to the atom." One of them said, "Stan, do you want to make a fortune? Use this for masks" (for etching computer chips). But at the time, Ovshinsky recalled, "I wasn't interested in masks, so I screwed that one up." Noyce and Moore visited ECD several more times, and in 1970, Moore became a co-author of one of ECD's early phase-change memory papers.[35]

By then, Ovshinsky had established a separate division run by Ron Neale to commercialize electrical phase-change memory. Neale's work with D. L. Nelson and Gordon Moore resulted in an important new kind of integrated circuit. Initially made in limited quantities by ECD, the 256-bit memory (the RM-256) was later manufactured by Intel and called the "read-mostly memory" (RMM). It avoided certain serious problems of the existing "read only" (ROM) and "random access" (RAM) kinds of memory. In particular, while the inflexibility of the ROM prevented its data from being changed and the volatility of RAM allowed data to disappear during power interruptions, the RMM could be programmed, read, and reprogrammed repeatedly, and it retained its data unless intentionally altered.

While the 256-bit RMM was not commercially successful, its development was an important moment in the history of nonvolatile memory, for it was the first time that a phase-change memory switch was integrated with a silicon chip. A long arc of development followed from it, leading through ECD's later improvements of its electronic phase-change memory (see chapter 10) and up to the present (2016), when a direct descendent of the RMM is entering production (see the epilogue).

As a growing research laboratory, ECD needed a machine shop to make experimental equipment and prototypes. In 1969, Harley Shaiken, a returning veteran from the storefront, set up the shop in a steel garage that Ovshinsky had erected in a corner of the parking lot to store the Bentley he had been given in England. Between working in the storefront for some months in 1960 and returning now to ECD for about five years, Shaiken had earned his journeyman's card as a machinist.[36] He recalled working with Herb to design and produce various machines as "the most satisfying work years I've ever had." He especially enjoyed the fact that everyone was treated equally at ECD, and he admired its "pliable" hierarchy, designed to draw out the best in people

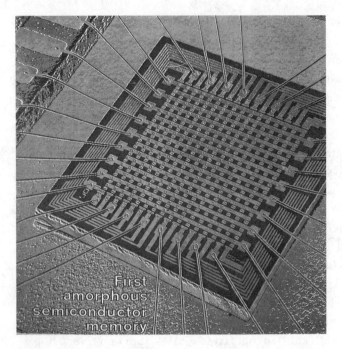

Figure 6.3
The 1970 256-bit RMM array.

by rewarding them "more on merit and engagement than any place that I've worked or studied."

As a fellow machinist Shaiken had a special rapport with Ovshinsky, who had not lost touch with the early experience that contributed to his discoveries. Shaiken recalled how Ovshinsky occasionally came by the shop, and they "would do things together." On one occasion when Shaiken was making a particularly complicated and time-consuming cut on the lathe, Ovshinsky suggested he use a different approach. Knowing that Ovshinsky hadn't worked on a lathe for many years, Shaiken said something like "that sounds good but I think maybe this would be a safer setup." Ignoring the suggestion, Ovshinsky proceeded to remove the part and put raw stock in the lathe. "Let me show you," he said. Turning up the speed much higher than Shaiken felt comfortable with, Ovshinsky cut the part in a way that the younger man considered "brilliant."

Another addition to ECD with a shared history was the anarchist-metallurgist couple from Connecticut, Laurence and René Pellier. Laurence was a metallurgical engineer, and her husband René was a machinist who prepared the specimens for her to study with their electron microscope. Longtime friends of Iris's family, the Pelliers worked for

ECD as consultants while living in Connecticut. The attractive glass models of atoms and molecules that René made were on display in Ovshinsky's downstairs office at home throughout his life, an embodiment of the way he himself visualized the structure of materials.

Theories and Models of Chalcogenide Semiconductors

In the efforts to explain the Ovshinsky switching in terms of fundamental scientific principles, one controversial issue was whether it was a thermal or electronic effect. The debates polarized the physics group. (According to Dick Flasck, "there were almost fist fights.") Ovshinsky felt that the mechanism behind the switching was electronic but was unable to explain why in scientific terms. Some believed that he preferred the electronic explanation simply because a heat-based phenomenon would be less reliable and so not suitable for commercialization, or because electronic switching would also be a more fundamental discovery. Feinleib, who on his first encounter with Ovshinsky had declined a challenge to prove the electronic explanation, would get annoyed when Ovshinsky argued by analogy for electronic switching, noting that the device's behavior seemed similar to that of electronic phenomena, and different from thermal

Figure 6.4
The Pelliers with Stan, Iris, and glass models.

Figure 6.5
Talking physics, 1976.

phenomena, where there is a time lag. Feinleib considered this way of arguing unscientific, but Fritzsche and Adler accepted it.

Not until the early 1980s was the controversy resolved. Melvin Shaw, a professor at Wayne State University, who joined ECD in 1970 as a consultant, showed with numerical simulation techniques that the Ovshinsky switching is first electronic but then also thermal. The initial event, Shaw explained, "the thing that breaks it down, is electronic. After it breaks down, it channels and gets hot."[37] Shaw recalled that when he reported his findings, "Stan thought I was going to show that it was thermal. When I told him it's electronic, he said 'I knew it all along,'" a characteristic response of Ovshinsky's that exasperated some of his colleagues.

Other ECD consultants worked to explain the Ovshinsky effect at the atomic level. An important contribution came from Marc Kastner, one of Fritzsche's graduate students. Drawing on his chemistry background, Kastner noted that of the four outer p-electrons of the chalcogen atoms in Ovshinsky's materials, only two are normally used in bonding. He suspected that the two remaining (normally non-bonding) electrons, which are called lone pairs, determine the special properties of Ovonic materials.[38] Ovshinsky was "enormously excited" when he heard about Kastner's idea. "Once

I knew it was a lone pair, it all came together," he explained. "In my mind, I saw exactly the whole form of it." Ovshinsky's enhanced power of visualization, seeing the positions of the electrons in the atomic structure of his materials, gave him an alternative to scientific calculation. "You figure out the bonds," he told Fritzsche. "In my mind I see them clearly, but I stutter when I try to describe them."[39] Introducing the concept of the lone pairs was a crucial moment in the process of science catching up with Ovshinsky's intuitions and then feeding back into their further development.

Fritzsche suggested that Kastner write a paper on his model under his own name, even though he was still a graduate student. That, Kastner said, "really helped my career get off to a start." Later, after Kastner found a position at MIT, Ovshinsky invited him to join Dave Adler on a visit to ECD for a workshop where Mott presented a model of the role of dangling bonds in amorphous semiconductors. Listening to Fritzsche point out problems in the model, Kastner "suddenly had a glimmer of an idea of how to make this work based on the idea of lone pair semiconductors." On the plane back to Boston, he explained his idea to Adler, "and by the time we landed, we had a draft of a paper." A week or so later, at a conference in Williamsburg, the two of them huddled with Fritzsche to work out the consequences, and their new model was soon published.[40] Adler and his colleagues developed this model further into a fuller account that became generally accepted.[41] The Adler model left some aspects of the Ovshinsky effect unexplained, however, and subsequent attempts have still not resolved all the issues. As the physicist Steve Hudgens recently observed, "The 'deceptively simple' two terminal devices that Stan described forty-four years ago still provide us with a fascinating mystery."[42]

Ovshinsky himself was struggling to convey his own conception of amorphous and disordered materials and decided that he needed to have actual physical models to show what he saw. The opportunity to create them arose in 1971 when a young Indian biophysicist, Krishna Sapru, moved to the area to be with her husband, who worked in Detroit. She had a research fellowship at Wayne State University and planned to work on DNA replication, but when her professor moved to California, she applied for a position at ECD. She felt an immediate rapport with Ovshinsky, who showed interest in her work on DNA, and she was especially impressed when he remarked, "There's no distinction between physics, chemistry, and biology." It was a view that few scientists would have expressed at that time but which reflected Ovshinsky's consistent disregard for disciplinary boundaries.

Sapru's first assignment was to create "some real models" of the amorphous and disordered materials. After studying the electronic structures, she and her young daughter sat on her patio and constructed the models using several hundred small Styrofoam balls, which they colored and connected with colored pipe cleaners. She told her daughter, "We will pretend that yellow is tellurium, brown is sulfur, green is germanium, and so on." Over a weekend, working with formulas that Ovshinsky supplied, they assembled thirty to fifty models of the switching and memory materials.

As Sapru worked on the models, she developed a feel for Ovshinsky's perspective on the "personality" of each atom and molecule based on their number of protons and neutrons, and particularly the distribution of the electrons in their quantum-mechanical orbitals.[43] She began to recognize that atoms always try to form the strongest bond. For example, when they combine to make lithium fluoride they are "two really happy atoms. And that's ionic bonding, when one electron goes mostly over to the other one, whereas in the case of covalent bonding, they share electrons." In chalcogenides, the lone pairs in the p orbital (represented by two pipe cleaners) "are not happy in the sense that they are dangling bonds, not paired up with anything," and so "anxious to make a connection." This anthropomorphic and visual kind of thinking, with which

Figure 6.6
Ovshinsky lecturing with Styrofoam models.

Sapru became adept in the course of her model-building work, was more like the way chemists, rather than physicists, typically think about atoms and molecules.

Ovshinsky was excited when he saw the models that Sapru had made; he told her, "We are writing a paper!"[44] He would keep her models in his office bookcase for the rest of his life, using them in presentations, and indeed whenever possible, in his efforts to explain his insights about the switching and memory materials to his befuddled listeners. In watching him talk about his work while manipulating his models, Joi Ito (whose parents Masat Izu and Momoko Ito were ECD employees) recalled, "Stan would talk about science in a sort of artistic way." He would be "holding these models of Styrofoam and pipe cleaners, shaking them and saying, 'See these dangling bonds. There's energy here.'" Ito added, "It took Nobel laureates to translate what Stan was feeling."

OIS and OMI

As ECD grew and developed new applications for amorphous and disordered materials, Ovshinsky found that the demands of managing its business made it hard for him to stay as closely involved as he wanted in the work of research and development. To get help, in November 1969 he asked the lawyer and accountant Keith Cunningham, then a senior executive at the accounting firm Touche, Ross, and Co. to become ECD's president and chief executive officer while Ovshinsky remained its chief operating officer.[45] As Chet Kamin, who would later become Ovshinsky's attorney and adviser, explained, "Stan brought Keith in because he knew he wasn't good at financial stuff. He wanted somebody that he could lean on to basically run the business part of the company so that he could spend more time and energy on research." By early 1971, Cunningham had raised enough money to set up two subsidiary companies aimed at commercializing ECD's technologies: Ovonic Memories, Inc. (OMI) in February 1971, and Ovonic Imaging Systems, Inc. (OIS) in April 1971, both located in Southern California.[46] Even with the funding Cunningham had arranged, the financial basis of the two companies seems to have been precarious.[47]

The aim of OIS was to commercialize instant imaging and non-silver films for microfiche records.[48] This technology seems to have been the first ECD application of amorphous materials beyond switching. The idea itself was not new: using photoconductive amorphous materials to copy documents was already the basis of xerography. In the Ovonic system, however, the image was reduced and transferred to a microfiche card, which could not only be read but also revised before being stored again. It was thus the analog equivalent of a digital rewritable memory, an ingenious technological advance.

The key to making the non-silver film work not only for copying images but also for writing and rewriting text was activating it with an electron beam, an idea proposed by the longtime ECD scientist Peter Klose.[49]

To help with manufacturing the film, Cunningham offered a position to Herb Ovshinsky, who was Cunningham's friend and neighbor. In Ovshinsky's earlier years in Akron, New Britain, and Detroit, he had often relied on his brother in designing and building machines; after joining ECD in 1971, Herb resumed this role. Working with his colleague, Al Adominis, Herb built a machine for roll-coating the new instant imaging film. It was ECD's first use of a continuous production method, the kind that would later be used for making thin-film solar cells (see chapter 8).

To work with Herb on the instant imaging technology, in 1972 Ovshinsky hired the young chemist Masatsugu (Masat) Izu, who had taken his doctorate at Kyoto University under the great theoretical chemist and later Nobel laureate Kenichi Fukui. Izu was then a postdoc at the University of Waterloo in Canada; to recruit him, Ovshinsky offered to double his postdoc salary. Izu decided to move to Michigan, along with his wife Momoko and their two small children, Joichi (Joi) and Mizuko (Mimi). Ovshinsky also hired the even younger chemist David Strand, who came to ECD at this time with a bachelor's degree in chemistry from Michigan State University. Strand, who over the following decades would become a mainstay of ECD's research programs, helped develop the coating of the film, a tellurium-based organo-metallic compound, which when exposed to light formed a latent image that became visible when heated and could be printed on paper or shown on a display.[50]

Unlike other microfilm systems, the Micro-Ovonic Fiche (MOF) allowed users to revise and save the stored information. It was therefore considered a revolutionary technology, and ECD, as the OIS parent company, entered an agreement with 3M to commercialize it. Despite this promising beginning, however, the timing was wrong. The venture with 3M never materialized, and OIS failed because a market for the microfiche retrieval technology could not be found.[51] In any case, the MOF analog imaging system, however sophisticated, would eventually prove unable to compete with the emerging electronic digital technologies.

The other company, OMI, also soon failed. It had been established to commercialize the optical memory developed by Feinleib and Iwasa. It aimed to manufacture and market a disc drive for IBM computers, promising a prototype with a capacity of 64 billion bits by spring 1972. Named the 4440, it would provide ten times the storage of IBM's popular 3330 with the same average access time of 30 milliseconds.[52] As with the microfiche, however, there was no market for the technology—in this case because computers did not yet require that much storage.[53]

Ovshinsky felt the failure of OIS and OMI also owed much to Cunningham's mismanagement and the inflexibility that made him a bad negotiator.[54] In any case, Cunningham's commitment of so much of ECD's resources to the two companies was clearly, as John de Neufville said, "a very high-risk approach," and their failure "almost broke the back of ECD." Over the next three decades Ovshinsky would continue trying to find people to relieve him of more routine management responsibilities, but he would never again delegate so much authority.[55]

After Cunningham left the company in late 1974, ECD's fortunes continued to decline, and it was necessary to reduce the staff to about twenty-five.[56] The few remaining employees were called on to juggle several tasks. Dick Flasck, one of those few, "more or less inherited the materials research lab/physics lab, the analytical lab, the bomb room and a number of other departments." In addition to his research, Flasck was sent on business trips so often and on such short notice that, he reported, "I kept a packed suitcase in the trunk of my car. He had me flying close to 200,000 miles a year." Flasck also recalled how Ovshinsky, refusing to be daunted by the downturn, continued to give one-hour talks every other day expounding his ideas. Sometimes, when there were no consultants visiting, Flasck alone made up the entire audience. Like many others, he was usually baffled by Ovshinsky's explanations.

In the midst of ECD's struggles to survive, Ovshinsky got some help for developing his electrical phase-change memory. Late in 1973, he and Iris came to see Robert Johnson, then senior vice president of engineering at Burroughs Corporation in Detroit, a business equipment company that had become an important computer manufacturer. They told Johnson they needed $100,000 quickly.[57] Johnson recalled having met Ovshinsky, who many considered "a wild-eyed inventor." Ovshinsky "didn't sound crazy" to him, although it was immediately clear that he "wasn't particularly good at explaining things." Johnson was able to arrange a $100,000 advance on a contract for licensing ECD's read-mostly memory (RMM) to Burroughs.[58] Collaborative work between ECD and Burroughs continued for several years, and in 1978 resulted in an RMM memory array of 1,024 bits, not only much bigger and better-performing than the 256-bit array but also faster and with a somewhat lower programming current.[59] A few years later, Johnson would join ECD's staff (see chapter 10).

The Beginning of ECD's Photovoltaic Program

Of all the programs ECD undertook in its early years, developing the means to produce cheap and efficient solar power was the most important for advancing Ovshinsky's goal to replace fossil fuels. The enormous amount of energy radiated by the sun made

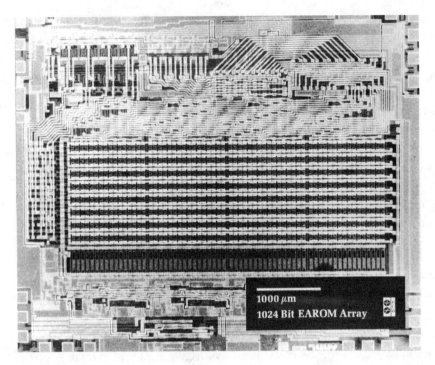

Figure 6.7
The ECD-Burroughs 1,024-bit integrated memory array.

it a promising source (in one hour the earth receives enough to meet all human needs for a year). But the photovoltaic program did not begin to make real progress until the later 1970s.

Understanding the problems ECD had to solve requires some understanding of the technology. A conventional solar cell is basically a diode, a one-way electron valve that converts light into electricity by means of the photoelectric effect, in which light dislodges electrons. Photovoltaic material is arranged in layers whose electronic structure has been altered (doped) by adding small amounts of other elements (boron and phosphorus) to create P- and N-type materials.[60] When these layers are brought in contact, an electric field forms at the P-N junction, causing dislodged electrons to move toward the N-layer, while the holes left behind behave like positive charges and move toward the P-layer, building up a voltage. Inserting the device in a circuit allows current to flow and do work.

Up to this time, solar cells had all been made of crystalline silicon because it was believed that only crystalline material could be used to make P- and N-type material.

Such cells, however, were costly, heavy, rigid, and fragile. Ovshinsky wanted to make amorphous thin-film cells that would be cheaper, light, flexible, and more robust. These had three layers, with P-type and N-type material separated by a thicker, undoped layer (an intrinsic, or I-layer) through which the electric field passes. Again, the dislodged electrons (now mostly from the I-layer) flow toward the N-layer and the holes again move to the P-layer. The difficult technical problem of making a thin-film solar cell was thus essentially that of using amorphous material to make an efficient PIN structure.

Ovshinsky knew that he would have a tremendous patent advantage in developing thin-film solar cells if ECD could use his amorphous chalcogenide alloys to create such PIN structures. That, however, depended on doping them to make P- and N-type material. But when his consultants analyzed the atomic structure of Ovshinsky's materials, they concluded that such doping would not work with them.

Further progress required switching to a different amorphous material, whose photovoltaic potential had already been demonstrated. In 1975, Walter Spear and his student Peter LeComber at the University of Dundee published a paper describing a process for making hydrogenated amorphous silicon with high photoconductivity. Using a new method (the glow-discharge plasma decomposition of silane gas pioneered by R. C. Chittick and colleagues), they found that amorphous silicon prepared with hydrogen could be doped like crystalline silicon. The addition of hydrogen furthermore neutralized the defects and dangling (unsatisfied) bonds that are numerous in amorphous silicon and would otherwise capture electrons and reduce the current produced in the deposited film.[61]

These results suggested that ECD should shift its photovoltaic research to amorphous silicon.[62] But Ovshinsky was not yet ready to give up on his chalcogenides. With new funding from United Nuclear Corporation between August 1976 and August 1977, he turned to what he would call "chemical modification," mixing larger quantities of elements from other groups with the chalcogenides to try to get the same effect as doping. Presenting this work at a meeting in Edinburgh in June 1977, Ovshinsky announced that it had increased conductivity by up to nine orders of magnitude.[63] He considered this feat of "atomic engineering" as "one of the most powerful things I've done," but it did not solve the problem of making chalcogenide solar cells.[64]

Ovshinsky was at last ready to turn to hydrogenated amorphous silicon. Fritzsche recalls persuading him to make the change at the March 1977 American Physical Society meeting in San Diego. (By then, David Carlson at RCA had announced making the first solar cells from amorphous silicon.)[65] Vincent Canella, a physics professor at Wayne State University who joined ECD in 1976, recalled, "Stan came back and said, 'We should do silicon.'" Ovshinsky then hired LeComber's student Arun Madan and directed the physics group to experiment with plasma-deposited amorphous silicon and germanium.

There was immediate tension however between Ovshinsky and the young Indian-born physicist. Madan wanted to do detailed, small-scale research aimed at fully understanding the physics, while Ovshinsky was eager to move forward and make complete cells that could be quickly brought into production. He therefore created a second photovoltaic group headed by Masat Izu to build and test cells. But the reorganization did not resolve the tensions, because Ovshinsky was still intent on developing a unique approach he could call his own. Having agreed to work with amorphous silicon, he wanted to substitute fluorine for hydrogen. In theory, this made a certain amount of sense, for like hydrogen, fluorine's outer shell lacks one electron, and Ovshinsky believed it would form a stronger bond with silicon. Fluorine would thus not only neutralize the dangling bonds but also give ECD the proprietary advantage he had failed to gain by using chalcogenides.[66]

But in practice, substituting fluorine was a failure. "It never really worked out, but we spent a lot of time finding out," Vin Canella said. The highly reactive fluorine would not only join with the dangling bonds but would also break desirable bonds, and as an etchant it would often break down the film as it was deposited. Stan's insistence on trying to use fluorine infuriated Madan. Dick Flasck recalled, "Arun was just frustrated as hell. His attitude was, 'I don't want to waste my time trying to find some way around the situation by not using hydrogen and using something else that's not quite as good, especially if, from the technical standpoint, there's no good reason to do that.'" Eventually, after these time-consuming efforts to develop an alternative, Ovshinsky went ahead with using hydrogenated amorphous silicon, and ECD's photovoltaic program became highly successful. But to the end of his efforts to make better and cheaper solar cells, he never completely gave up on using fluorine (see chapter 12).[67]

A Bold Concept and Major New Funding

Ovshinsky did not work directly on studying or designing the thin-film solar cells, but by the late 1970s he was very much engaged with producing them quickly and cheaply because he knew that was the key to replacing fossil fuels with solar power. Solar panels were then produced slowly and expensively, one at a time. Ovshinsky instead imagined a machine for manufacturing thin, flexible solar panels roll-to-roll, or as he liked to say, "by the mile," like film or newsprint. The basic concept for such production was not new (ECD had already made imaging film that way), but applying it to making solar cells was radically new. Ovshinsky envisioned a machine that would use plasma deposition to produce the thin layers of hydrogenated amorphous silicon on a moving stainless steel substrate, producing miles of solar panel. Based on only recently available concepts, the plan presented daunting technical problems.

Ovshinsky called a number of meetings with his scientists to explain his roll-to-roll concept. Fritzsche recalled those present shaking their heads with skepticism but also not offering much resistance to the concept, whose advantages were obvious: the thin and flexible panels could cover large areas, promising to bring down the price of solar energy. But the proposal was such a huge leap from the small experimental samples of less than a square centimeter the researchers had been working on that, as Fritzsche recalled, "it left us speechless." To the scientists, the problems involved in going from these small experimental cells to continuous rolls appeared overwhelming. Like his attempts to use chalcogenides and fluorine, it was yet another instance of Ovshinsky's asking for what seemed impossible. But in this case, his long-range vision would be vindicated, and the problems of building the roll-to-roll machine would be solved, as detailed in chapter 8.

Ovshinsky's search for support to fund this ambitious scheme succeeded, and succeeded in a way that dramatically increased the scope of ECD's operations. The support came from ARCO (Atlantic Richfield), then one of the seven or eight largest American oil and chemical companies. Such a partner might at first seem surprising, but in the later 1970s many major oil companies were developing alternative energy programs because the 1973 Arab oil embargo had made foreign supplies seem unreliable and because scientists had projected that the world's oil reserves would be depleted by the end of the century.[68] Indeed, this was to be the first of several such partnerships between ECD and an oil company. The energy crisis and economic slowdown of the 1970s encouraged an interest in new technologies that favored ECD's growth.

ARCO was already involved in solar energy, having in 1977 acquired Solar Technology International, which became ARCO Solar. It was (and, as its successor, SolarWorld, still is) primarily involved in making crystalline solar cells, but at the time it was also looking at other materials. The physicist Richard Blieden, who would later join ECD's staff, was working in ARCO's R&D program at the time he and Ovshinsky met in 1978 at the American Solar Energy Society conference in Boulder. Blieden had read Ovshinsky's 1968 *Physical Review Letters* article when he was teaching physics at Stony Brook and continued to hear about him and ECD later, when he directed solar programs at the National Science Foundation, the US Energy Research and Development Administration, and the Department of Energy. Ovshinsky's presentation at Boulder piqued Blieden's interest, and persuaded him to look more closely at what was happening at ECD. Blieden's visit after the Boulder meeting was followed by another from his boss,

Robert Chambers, the head of R&D at ARCO, and then by negotiations resulting in May 1979 in a $3.3 million grant for ECD's photovoltaic research.[69]

Less than a year after the initial ARCO grant, the opportunity arose for a much larger one. This time the connection came from Ovshinsky's old union colleague, Jack Conway, then on the board of ARCO Solar.[70] As an experienced activist and administrator, Conway appreciated the social implications of Ovshinsky's vision for solar energy, and he brought ARCO president Thornton Bradshaw to see him. A highly successful manager, Bradshaw believed in the social responsibility of corporations. Ovshinsky recalled him as "an American business executive who was interested in the great problems of the country and the world, a wonderful guy." Excited by Ovshinsky's concept for making solar panels by the mile, Bradshaw told him ARCO would fund ECD's energy research for three years. He could not guarantee support beyond that time because he was planning to leave ARCO then, but for three years ECD would have unrestricted scope for developing alternative energy systems.

Now Ovshinsky had to decide how much money to request from ARCO. He turned to Nancy Bacon, a successful accountant handling ECD's accounts at Deloitte Touche whom he had persuaded in 1976 to join ECD as chief financial officer. Bacon was to play a critical role in all of the company's financial transactions. She remembered Ovshinsky coming to her office after his meeting with Bradshaw and asking, "What do you think we should do?" She suggested asking for $15 million, which at first shocked him. But after much consideration of how they would structure the proposal, "Stan decided to ask them for $25 million," Bacon recalled, "and made the thing stick." The second agreement, in January 1980, provided not only an additional $6 million for the research and development of solar energy but also $19 million for other alternatives to fossil fuel, including thermoelectric and hydrogen.

The ARCO grants were a turning point in the story of ECD's ascendance as a major energy laboratory. Ovshinsky's boldness in envisioning roll-to-roll production and then in getting such a large increase in funding led to dramatic expansion. As Dick Blieden said, "For the first time at ECD they had enough money to build up the laboratory facilities they needed, to hire the patent attorneys to generate the IP, and basically to do all of the things Stan wanted to do to explore the opportunities in these new materials. ARCO really was a lynchpin."

The UNC Suit

The ARCO venture was, however, threatened even before it officially began. In May 1979, a New York lawyer contacted Chet Kamin, a Chicago attorney at the firm Jenner and Block, to ask whether he could help a Detroit client with an emergency in the Illinois state courts.[71] Kamin recalled that a day or so later the Ovshinskys showed up in his office "very agitated," because they had been served with an injunction preventing them from proceeding with their new ARCO contract. Challenged by "the interplay of science and technology, business, and law," Kamin took on the case.

The crisis had been incited by Keith Cunningham, who left ECD in 1974 to become CEO of the uranium mining and processing company United Nuclear. With plenty of money on hand from the uranium boom of the early 1970s, and still looking for a way to work with Ovshinsky, Cunningham gave ECD a one-year R&D contract for $.5 million starting in August 1976 "relating to the conversion of light, heat, or chemical energy into electricity."[72] Less than two years after the contract ended, in May 1979, UNC sued ECD on learning about its $3.3 million ARCO contract. It claimed that during ECD's one-year contract with UNC certain critical concepts relevant to the ARCO contract (in particular, "chemical modification") had been developed. In the trial, Kamin showed not only that the contract with UNC had expired before ECD's ARCO contract began, but also that the work under the UNC contract was not relevant to the ARCO grant.[73] ECD won and by June 1979 could begin working under the ARCO contract.[74] The UNC litigation was Kamin's introduction to ECD. Over the next twenty-five years, he represented ECD in many similar legal contests.

Figure 6.8
Chet Kamin with the Ovshinskys in the mid-1980s.

Expansion

With the UNC suit settled, ECD could move forward with several expanded research and development initiatives made possible by the ARCO agreement. Before considering the most important ones separately, we should pause briefly to note how their parallel and intersecting activities helped to make ECD a unique organization.

Just as Ovshinsky would say that a key to his inventive process was that "at any one time I have four or five deep things I'm thinking about simultaneously, and they feed upon each other," so a distinctive feature of ECD was that, once it had the resources, it maintained many simultaneous research and development activities. Conventional management wisdom urges a more focused strategy: identify the strongest, most promising or profitable activity, concentrate on it, and discard the rest.[75] But Ovshinsky refused to do that. His vision of the future was not confined to picking winners, because he believed that realizing the promise of his amorphous and disordered materials depended on pursuing multiple lines of investigation. And, like Ovshinsky's simultaneous thoughts, the different programs did repeatedly feed upon each other and yield new and unexpected discoveries, as we explain in the following chapters.

Despite complaints from some investors about what they saw as his "shotgun approach," Ovshinsky's unconventional strategy of maintaining many concurrent programs was highly productive, but it was also hard to manage and expensive. As Kamin observed, "What Stan was trying to do would have taxed the capacities of anybody you could think of. He was trying to start three or four industries at the same time." And yet, as we saw in the Cunningham episode, when he tried to rely on someone else to take over some of the responsibility it was a fiasco. As Kamin also noted, for a small company with no assured source of continuing revenue, "the financial requirements were enormous."[76] Ovshinsky's remarkable fund-raising abilities could usually meet those requirements, but there were times like the mid-1970s when many researchers were laid off and many programs cut.

It is clear, however, that the achievements of ECD depended on Ovshinsky's refusal to rank or separate its activities. "All through the history of the company," Kamin said, "he was taking something he learned or insight he got in one area and then applying it in a different area." And as we shall see, it was the ARCO contract that first made that possible on a large scale.

7 The ECD Community: A Social Invention (1965–2007)

Just as energy and information were two sides of the ECD coin, so were scientific innovation and social progress. As ECD gained resources and expanded its research programs, its growth allowed Stan and Iris to build a community based on their shared ideals, drawing on the examples of the Ferrer School, the Mohegan Colony, the Workmen's Circle, and political systems they admired. Building this community was itself an act of social invention. Ovshinsky had been working toward it since his early years in Akron and Arizona shops, where he had helped novice machinists and toolmakers develop their skills and had organized reading groups. ECD was designed to be egalitarian and supportive; people were rewarded for merit and given opportunities to develop their fullest potential. "This was how we believed society ought to be," he said. And just as he claimed "proof of principle" for pilot versions of his technological innovations, he would claim, "I proved a social thing," and would speak with pride about his success in changing and helping to mold the lives of ECD staff members.

In building the ECD staff, Ovshinsky wanted to "give jobs to people who had potential and didn't know it" and to create an environment where everyone could "live up to their potential, with ample opportunities, education, and culture." This aspect of the company initially surprised some on the staff, but it became widely appreciated, as in the course of their work at ECD members came to recognize talents and potentials they didn't realize they had. Mike Fetcenko, for example, joined the company as a vacuum technician, was able to attend college at night through ECD's tuition reimbursement program, and eventually rose to become senior vice president of the battery division (see chapter 9). He later observed, "Stan was more proud of the ECD culture than virtually anything else I can think of."[1]

Many were attracted by the shared goals and held by the close relationships. ECD felt like family, with Stan and Iris playing the roles of the benevolent father and loving mother. Fetcenko recalled, "I know of instances where an ECD colleague was going through some kind of personal hardship and Stan and Iris made a point of supporting them when they needed it most." "They would treat people very well," agreed Joe Doehler, a physicist who came to ECD from Bell Labs. In turn Ovshinsky demanded, and got, long hours of willingly committed work.[2] In bad financial times, many staff members worked even harder to help the company pull through. "It was," Ovshinsky said, "a participatory democracy based on merit and fairness and justice." Or, as the scientist Srini Venkatesan noted in less exalted terms, ECD was "just like a family. You know, if you make a mistake, they will take you to task."

Besides offering a supportive, collegial sense of community, ECD also offered some protection from the injustices of a world that didn't share its egalitarian values. For black staff members facing a segregated housing market, Ovshinsky would buy houses in his own name, as he had done earlier in Birmingham, a policy that sometimes resulted in death threats. (After an iron bar was thrown though a window of the Ovshinsky house, he slept for a time with a shotgun under the bed.) Growing up in the ECD culture, Joi Ito saw that "Stan was fighting for fairness, for equality. He wanted to save the world. And initially, when he was growing up, I think saving the world meant organizing unions and speaking up for the oppressed. On his wall it said, 'With the oppressed, against the oppressors.' That motto has influenced me deeply."

The progressive influence was pervasive. ECD avoided a hierarchical organization; conference meetings were held at a large round table that represented the equality of all participants, and Ovshinsky typically referred to staff members as "colleagues." His ideal (not always realized in practice) was to make their work fully and freely collaborative. "My kind of science is very much like jazz," he said. "You do your solos but you also interact with your fellow musicians. Everybody is a creator in a real jazz group."[3]

In addition to following such progressive principles within the community, Ovshinsky also encouraged ECD staff to support civil rights and become politically active. While he never insisted on their participation, most showed up for the first protest march against the Vietnam War in Birmingham, and ECD always celebrated May Day. In some cases, the influence made a clear difference. Charlie Sie, who worked on phase-change optical memory between 1969 and 1974, was born in China and as a boy fled with his family from the Communists in 1949. May Day parties were hardly one of his traditions, but he felt he got a valuable education in democratic socialism at ECD, particularly from Ovshinsky and Harley Shaiken, and as a result, he said, "I became a liberal."

Just as Ovshinsky believed people could develop new abilities if given the chance, he believed they could change their political outlook. He hired known anti-Semites and Ku Klux Klan members, expecting them to become less bigoted in a community where they were treated fairly and rewarded for their work. He was pleased at signs that being part of ECD could change attitudes, when "a Palestinian worked next to a Jew, a Kurd next to a Sunni; Chinese from the mainland worked next to the Taiwanese." He was also proud of hiring foreigners who learned English on the job. Rosa Young, for instance "had one team that was Russian, one team that was Chinese, and many of them didn't speak English." The international flavor appeared in other ways. For the ECD newsletter, *The Ovonic Link,* Iris contributed a regular feature called IRIS, or "International Recipe Ideas and Suggestions." In compiling these, she realized that there were thirty-five first-generation immigrants on the staff. "What was interesting was that we never tried for that. It just happened, which showed us that other people must discriminate against foreigners because why did we have so many?" "And the women outnumbered the men in vice presidencies and were equal on the directors," said Stan. "Then it was unusual," Iris added. "Stan said, 'Why waste 50% of the population?'"

ECD also provided generous benefits. Machinists got the same benefits as PhDs and were not necessarily paid less. What mattered was talent and motivation. "You can't overpay talent," Ovshinsky said, "or commitment." And, as Robin recalls him saying, "Nothing is too good for the proletariat." In addition to formal benefits, ECD staff would receive meals when working overtime, and sometimes gifts of flowers and champagne, or expensive dinners or theater tickets, as rewards for achievement or sometimes just in recognition of effort and attitude. And then there were the memorable Christmas parties, from the early days when the company was small, held in an Eastern Orthodox church hall "where there would be Middle Eastern food," Vin Canella recalled, "and Stan would walk around with a bottle of Metaxa pouring drinks," to later more lavish affairs, culminating in over four hundred people in the Detroit Institute of Arts, "where," he joked, "you could splash your drink on the Spanish armor."

Many staff members shared a sense of belonging to a closely connected community, and Stan and Iris also felt close to them. As he recalled, "We went to their births. We went to their marriages. We went to their graduations. We went to their funerals. And they knew that we really cared about them and would not take advantage of them." Both he and Iris were proud of how long they kept up and expanded their social experiment. "People would say you could build a utopia good for a couple people but once you get any larger, Stan, you'll find you can't keep it going. Well, I kept it going past a thousand people."[4]

Ovshinsky also kept this utopia going for decades without the company showing a profit. Many long-time staff members have memories of close calls with insolvency. As Canella said, "It was a strange company. You knew at certain times that there wasn't

Figure 7.1
Ovshinsky with his longtime colleague and friend Max Powell at an ECD party.

money for the next paycheck, and yet we never missed a paycheck. Stan would always go find another investor somewhere." As Ghazaleh Koefod, Stan's longtime assistant, recalled, "We used to kid around and say, 'Stan just pulls money out of a hat, like a magician.' Because we never made any money, but then there was always money." In lean times, there were layoffs and pay cuts, but the communal culture survived.[5] The considerable revenues from ECD's licenses, as well as the money from new investors, that went to fund research and development on the next, potentially world-changing technology also supported the social experiment. ECD was a utopian socialist enclave sustained by capitalism.

The Research Community

Stan and Iris's social experiment was not just an end in itself; it was also the means for pursuing the world-changing goals they had committed themselves to when they began. This intertwining of social and scientific aims is apparent in the story the physicist Jeff Yang tells of his brief job interview. Ovshinsky told him, "I have two conditions, and if you meet them I think I'm ready to make you an offer." The first was to agree to a general principle: "We have a social responsibility to keep people employed and to keep the planet clean." The second was to agree to try his ideas. It was easy to assent to both, though that could entail agreeing to seemingly impossible demands, as when Yang spent many months trying to find a way to use fluorine in solar cells.

The research that developed from Ovshinsky's goals and ideas covered a remarkably broad range. The main programs in energy and information will be treated in greater detail in the following three chapters; here the aim is to give a quick sense of the number and variety of areas ECD researchers explored. Their diversity and interconnections are an important part of what made ECD unique. At the end of chapter 6 we noted how this unconventional management strategy enabled cross-fertilization among programs, interactions that mirrored Ovshinsky's own creative process. Here we can get a sense of how it shaped the experience of ECD's research community.

The largest and longest program was in solar energy, beginning in the late 1970s and continuing until the year of Ovshinsky's death. He did not invent the material for amorphous silicon solar cells, but ECD researchers repeatedly discovered ways to increase their efficiency, and their success in realizing his idea for roll-to-roll production was revolutionary. Other ECD energy technologies included batteries, fuel cells, and thermoelectric devices; hydrogen generation, storage, and use; and superconductivity. In information technology, the capacity for making thin films developed for solar cells also yielded thin-film diodes and transistors that became the basis for flat panel displays. Other devices using Ovshinsky's materials included optical and electronic memories and the non-silver photosensitive films used in the rewritable microfiche and other photographic technologies. Finally, ECD used disordered materials for x-ray and neutron mirrors, for coatings on industrial products, and for high permanence magnets.

The diversity and scope of these efforts had a powerful impact on the physicist Rosa Young when she came to ECD in 1984. "I was so impressed. I had never seen so many projects under one roof, from photovoltaics and thermoelectrics to battery and hydrogen technologies, from flat panel displays to electrical and optical memories, from research to manufacturing. That's why I joined." What held all these together, she realized, was Ovshinsky's energy, passion, and extraordinary capacity for multitasking. "He would call people into one room, the optical memory team; in another room it's the photovoltaic team; in another, the battery team." Ovshinsky would walk from room to room and be involved in all the discussions at the same time. As she added, his frequent monitoring of all the various programs could even become annoying. "He will check with you twice a day: 'What's new?'" Sometimes she would object. "'Stan, how much progress can we make in four hours?' But that's Stan," she said. "He always wanted to discuss with you, and when I'd say 'OK, this is what we're doing,' he'd go pull some book from the shelf, open to a page, and say 'Read this. See whether this will help you.'"

Ovshinsky's pursuit of so many research programs at once was helped by ECD's democratic ethos. Instead of compartmentalizing his researchers, Ovshinsky kept the organization flexible. "Really, everyone was working on everything to a greater or lesser extent," Dave Strand said. Ovshinsky would transfer people from one project to another, wherever they could make the greatest contribution, and he encouraged them to be jacks-of-all-trades, to develop the capacity that he himself had for switching among research areas. The researchers worked long hours and often struggled with intractable problems, but they typically found their work both exciting and enjoyable. One of the most frequent words they use to describe their experience is "fun." For the physics consultant Mel Shaw, "ECD was the core around which you could do very interesting work in the amorphous field. I had the most fun ever, and when it all ended in 1985 because there was a financial crunch, I was just very sad that all that fun was ended." Shaw especially enjoyed consulting when David Adler and Marvin Silver were there. "We would laugh so much." Vin Canella recalled one time when there were about fifteen members in the conference room, and Adler was up at the board, "He got to a certain point and somebody said, 'The sign of your result is wrong.' And Dave looked up and said, 'Even better!'"

The Consultants

Shaw, Adler, and Silver were among the many outside consultants Ovshinsky added to a group that started with Hellmut Fritzsche in 1963 and continued to grow through the following decades. These distinguished scientists enhanced the research community in many ways, while their professional connections and public recognition helped to raise awareness of Ovshinsky's achievements.

Some of the consultants were or would soon become Nobel laureates. They visited because they recognized Ovshinsky as a peer and enjoyed their scientific exchanges with him, appreciating and profiting from his insights. The first of these to join ECD as an unpaid consultant was the celebrated physicist Isidor Isaac Rabi, who began in the mid-1960s.[6] Widely known as acerbic and unwilling to waste time with those he considered foolish, the respected Columbia University physicist connected with Ovshinsky on many levels. They had at times traveled parallel paths—both attending trade school while in high school and refusing to have a bar mitzvah. Rabi also shared Ovshinsky's intuitive approach to physics, for as he once said, "I am an intuitive person, and sometimes it is very clear to me that you must do this or that, but I can't explain it."[7] Rabi's support would be especially valuable to Ovshinsky when his work was attacked. After joining the ECD board in September 1966, Rabi warned him to protect his intellectual property, which, he predicted, "everybody is going to try to take away." Among much other sage business advice, Rabi suggested that Ovshinsky get a loaded vote for his ECD shares, a notion that he had not yet heard of at that point.[8]

Rabi would freely share his high opinion of Ovshinsky, even with relative strangers. B. J. Widick, Ovshinsky's Akron friend, who by the mid-1970s was an untenured faculty member at Columbia, recalled meeting Rabi on July 4, 1976, at a reception to celebrate Operation Sail, the bicentennial parade of tall ships in New York Harbor. "This short

Figure 7.2
Isidor Isaac Rabi.

guy comes up to me and he says, 'Are you B. J. Widick? I am Rabi.'" When they realized that both knew Ovshinsky, Rabi announced, "Stan is a genuine genius. His only problem is that they don't like him in the Ivy League because he doesn't have a degree, and they can't claim him." Years later, when Rabi was interviewed for a PBS NOVA documentary on Ovshinsky, *Japan's American Genius*, the interviewer tried to prompt Rabi to say that Stan was like Edison, but Rabi insisted that Stan was not an Edison. "He's an Ovshinsky, and he's brilliant," Rabi said. The producers edited out "he's brilliant," but the words are clearly audible in the outtake.[9] That Rabi made the comment meant a great deal to Ovshinsky.[10]

Interest in the science of Stan's work attracted the international elder statesman of solid-state physics Sir Nevill Mott, who like Rabi became a regular unpaid ECD consultant. Fritzsche had introduced Mott to Ovshinsky in the summer of 1967, when they were at a conference in San Francisco on metal-insulator transitions. Mott's subsequent visit to ECD was the start of both fruitful scientific exchanges and a close personal friendship.[11] Mott's interest in Ovshinsky's work helped make the study of amorphous solids "rather respectable," the physicist Richard Zallen observed, and it also gave Ovshinsky an immediate international audience. Ovshinsky, in turn, influenced Mott's work profoundly by introducing him to the field of amorphous materials, one of the areas for which Mott was later recognized when in 1977 he shared the Nobel Prize in physics with John Van Vleck and Philip Anderson for fundamental studies of magnetic and disordered systems. But while in private conversation, Mott would often credit Ovshinsky for his important role in building up the new research field of disordered and amorphous materials, the conservative scientist rarely credited Ovshinsky publicly.[12]

Another prominent physicist with whom Ovshinsky had a long-term scientific and personal friendship—surprisingly, considering their opposed political views—was Edward Teller, who became a prestigious advocate for ECD. On his many lecture tours, the famous physicist would describe Ovshinsky's work in glowing terms and deplore the lack of recognition he had received. And when Teller visited ECD, Fritzsche recalled, "all the scientists gathered around" to talk physics with the great man as he sipped his heavily sweetened strong coffee. Because Teller was the father of the hydrogen bomb, however, and the one most responsible for destroying Robert Oppenheimer's government career in nuclear physics, many of Stan's friends and colleagues considered Teller dishonorable, or worse. Rabi had described Teller, who was formerly a friend, as "the most dangerous man in America." Ovshinsky opposed Teller's conservative politics, but "when Teller showed honest integrity with me," he said, "I felt, well, this man you could work with." They "argued and argued" about physics, Fritzsche recalled, but when Teller spoke about politics, "Stan didn't give him an inch."

Among many other outstanding scientists who became ECD consultants, the Harvard chemist and Nobel laureate William Lipscomb visited frequently and consulted on Ovshinsky's work with fluorine and hydrogen. Linus Pauling, another Nobel Prize–winning chemist, offered insights into the structure of Ovonic materials. Like Ovshinsky, he often relied on intuition in his work, and the two enjoyed exploring their scientific and political affinities.[13] Robert R. Wilson, who had been a group leader at Los Alamos during the Manhattan Project and later became the founding director of Fermi National Accelerator Laboratory, was a later consultant and board member. As a physicist, machine builder, and artist, Wilson had much in common with Ovshinsky. One more consultant and board member was John Bardeen, who had earlier been instrumental in bringing Hellmut Fritzsche to ECD (see chapter 5). Iris remembered taking Bardeen to the London Chop House, where the hostess, a woman named Holly, with "two circles of rouge on her cheeks, like a clown," hugged Bardeen. The incredibly shy double Nobel laureate looked "like he was going to fall through the floor," Iris recalled.

There were also many gifted younger consultants who added to the culture of ECD and became more directly involved in its work. Two who began in the mid-1960s were Arthur Bienenstock and Morrel Cohen. Bienenstock, then an assistant professor at Harvard and later a professor at Stanford, developed an x-ray diffraction method for demonstrating that the Ovshinsky effect in the memory switch was a reversible transition between amorphous and crystalline states. Working with Fritzsche, he also explained why impurities, which have a large effect on the electrical conductivity of crystalline semiconductors, seldom affect amorphous semiconductors. Cohen, a solid-state theorist at the University of Chicago, played a leading role in the first major paper explaining Ovshinsky's threshold switching.[14] He continued for several decades to make theoretical contributions, serving, for example, as an adviser on Ovshinsky's cognitive computer project (see chapter 10).

One of the most important consultants was the MIT electrical engineering professor David Adler. Known for his brilliance and his ability to present extremely complex physics issues very clearly, Adler was especially useful to Ovshinsky, who lacked the ability to explain his physics insights. Adler's colleague Brian Schwartz (introduced later in this section) noted that if Adler did not fully understand one of Ovshinsky's ideas, he would press him to explain it again and again until he could explain it back. Julius Feinleib recalled Adler's "encyclopedic" knowledge of physics, his "great memory for every fact in everybody's experiments." That knowledge and his patiently acquired understanding of Ovshinsky's ideas enabled him to translate them into publishable form; the papers he redrafted would pass the referees.[15] Despite their fifteen-year age

Figure 7.3
David Adler and Ovshinsky.

difference, the two also became personally close. Adler came to view Ovshinsky as a loving father figure he wanted to visit two or three times a month. The two would talk about everything, and Adler's interaction with Ovshinsky led him to focus his own research at MIT on amorphous semiconductors.

Tragically, because of a family history of premature coronary artery disease, Adler did not expect to live much beyond forty, although by running and losing weight he managed to extend his life by about a decade. He decided to make the life he had as rich as possible, often taking his family to France and eating there only in three-star restaurants. When he died in March 1987, most of those who spoke about him at his memorial did so with humor and wit. Mel Shaw showed a picture of Adler, who didn't seem to react to cold, on a snowy mountaintop wearing just a shirt, Doris Zallen recalled. But when Ovshinsky rose to speak, words failed him and he broke into tears.[16]

John Ross, a physical chemist, was an MIT department chair at the time he joined ECD as a consultant in 1976. Initially skeptical about Ovshinsky when sent to check out ECD for John Deutsch, director of research at the Department of Energy, Ross "came

Figure 7.4
Physics staff and consultants, 1977. Rear left to right: J. T. Chen, Marc Kastner, Arun Madan (with sunglasses), Krishna Sapru, Larry Christian, Dick Flasck, Iris, unidentified, Ed Benn, Wally Czubatyj, Arthur Bienenstock, Hellmut Fritzsche. Seated left to right: David Adler, Bill Paul, Mel Shaw, unidentified, Ted Davis, Ovshinsky, Nevill Mott, Michael Shur.

away converted." He realized that one reason Ovshinsky appeared to be "exaggerating so much," was that "he was living twenty years ahead of himself." Ross decided to help him communicate more effectively.[17] And Brian Schwartz, a condensed matter physicist and colleague of David Adler at MIT, became a consultant in the early 1980s and soon played a major role in running ECD's Institute for Amorphous Studies.

The Institute for Amorphous Studies

To enrich ECD's energetic intellectual culture and share it with a larger community, Ovshinsky created an organization designed to attract unusually talented and creative scientists who would gather for discussions about the new science of amorphous and disordered materials. "His paradigm would be something like the Institute for Advanced Studies," said Iris. The plan to build the Institute for Amorphous Studies suddenly

Figure 7.5
Institute Board of Directors. Rear left to right: Heinz Hennig, Hellmut Fritzsche, John Ross. Middle left to right: David Adler, Brian Schwartz, Stanley Stynes, Robert Johnson. Front left to right: Mary Beth Stearns, Ovshinsky, Iris, Marc Kastner.

became feasible in 1982 when an attractive school building set on a ten-acre campus across the small lake behind the Ovshinskys' house became available. It had previously housed a private girls school, Iris explained. "Stan used to think, 'Wouldn't it be nice if I could have that place?' And then all of a sudden it went up for sale." Ovshinsky bought the building to use for seminars and colloquia and turned part of the space into a library. He asked Brian Schwartz, who was known to be good at managing events, to help create and run the new institute.

Besides his teaching and research at MIT, Schwartz was at the time working with the American Physical Society and had many professional connections: "It was very easy for me to pick up the phone and say, 'Can you come and give a talk?'" Ghazaleh Koefod, serving as the coordinator of the institute, helped Schwartz invite speakers and guests. They designed an attractive brochure listing the many distinguished members of its board of directors and advisory committee as well as the lectures in the initial 1982–83 series on Fundamentals of Amorphous Materials and Devices. The quality of the colloquia was exceptional, and the talks continued for about four years, allowing Schwartz to produce, together with Adler, eight edited books published by Plenum Press.[18]

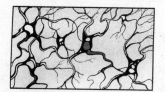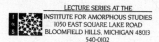

Figure 7.6a
Institute lecture poster, Leon N. Cooper.
Figure 7.6b
Institute lecture poster, Linus Pauling.
Figure 7.6c
Institute lecture poster, Robert R. Wilson.

The Institute for Amorphous Studies became the site of numerous scientific conferences as well as a series of popular lectures offered every three weeks or so aimed at educating the community on a range of topics, such as cosmology, superconductivity, and even political or artistic issues. "There were always 120, 150 who came," Hellmut Fritzsche recalled. Among the famous speakers in this series were Edward Teller, Linus Pauling, Maurice Goldhaber, John Deutsch, Leon Cooper, Nevill Mott, Robert R. Wilson, George Porter, and the ambassador to Japan, Edwin Reischauer. Many who came to hear the talks were academics and business people, but others, Koefod said, were housewives or bright high school students. "It was free. The institute got some money from the sale of the books, and the people who gave the lectures got an honorarium, expenses, and Stan's generous dinners. It was part of wanting to educate the community, to get them more interested in science." Overall, as Schwartz observed, the institute and its activities "added credibility to Stan's operation at the time."

Changing Lives

Vin Canella paid tribute to the qualities in Ovshinsky that enabled others to flourish. "What was so great about him was his ability to recognize talent in people and to bring it out and foster it." Canella noted how that often resulted from Ovshinsky's heavy demands. "He demanded stuff that was, in most cases, impossible. But through that very act you achieved way more than you could have. You kind of ran to your limit."

There are many instances of such personal development, but perhaps the most remarkable is the story of Momoko Ito, who entered the ECD community as a housewife and quickly rose to the level of vice president. Married to chemist Masat Izu, she had never worked before Ovshinsky hired her as a secretary. It was when he asked her to help as a translator in dealings with Japanese firms that he recognized her potential. "Everybody has within them something that is genius," Ovshinsky used to say. In Momoko, as in many others, he found and nurtured it.

He initially brought Izu along to translate when he was negotiating agreements with large Japanese corporations like Sony. But Izu "didn't feel that was his job. So Iris and I decided we'd ask Momoko. Well, it was obvious when we got there. I would speak for three minutes, representing what we were doing, and then she would speak for ten minutes, and everything would be very smooth." The patent attorney Marvin Siskind recalled, "She told me thank God he had her, because she would never translate what he told her to say. First, probably because of some coherency problems, but second of all, because you just can't say a lot of the stuff he'd say there. He'd act like he's talking to another American where you can get away with being insulting, and she would never translate that. She'd translate it into the most genteel Japanese, instead. Because she was respected in Japan, and because she knew how to do things correctly, a lot of these relationships were made." John de Neufville explained that Momoko would initially merely translate, but it was the kind of translating in which "she'd say, 'This is what he said, and this is what he meant.' And the next trip she said, 'This is what he said. This is what he meant. This is what you should say.'"

De Neufville also told a story about Momoko's skill as a negotiator during one of ECD's periodic financial crises. Seeking funding from the Asahi chemical company, which had a license for some of ECD's technology in Japan, they proposed to extend the agreement to worldwide rights, and Momoko was sent to Japan to negotiate. Soon after she left for the airport, a telex from Asahi arrived saying she shouldn't come because they'd decided that since they weren't using the technology in Japan there was no point in licensing it worldwide. Either she didn't get the message or just ignored it,

Figure 7.7
Momoko Ito.

but she flew to Tokyo. At the airport she was met by a senior Asahi executive who was supposed to tell her about the decision. "Well, after one evening talking to Momoko he had not only failed to inform her, he had actually changed his mind and decided that they should extend the license." The papers were quickly drawn up and signed the next morning. "And when the ink was barely dry, Momoko said, 'Could you please draw me a check?' They said, 'This will take a couple of weeks.' She looked at them and said, 'I need the check today.' And she left with a check," said de Neufville, who added, "That kind of nerves of steel combined with infinite charm was characteristic of Momoko."[19] As they continued working together, she and Ovshinsky developed deep mutual respect and affection.

In time, Momoko became a vice president of ECD. When the Japanese learned that Momoko was an executive officer, they put up initial resistance. "It was very unusual to have a Japanese woman executive in that generation," her daughter Mimi explained. As Ovshinsky recalled, "A roar went up from all these companies: 'You can't put a woman in there. You'll lose everything that you've gained here in Japan.' And I said, 'I'm sorry, it's against my principles to discriminate against anybody.' So she went and she won them all over." Ovshinsky stressed how important the connections Momoko

Figure 7.8
Edwin Reischauer.

made were to him and to ECD. "The Japanese saved me. In the 1970s when nobody wanted to pay any attention but to attack me and my work, the Japanese gave me open arms. And Momoko helped with doing that."[20]

Momoko had other resources besides her interpersonal and intercultural skills. As Mimi explained, "When she first went to Japan, my mom had one great physical asset going for her, which was that she had this amazing liver. She could drink a whole bottle of gin without batting an eyelash. And so she would go out drinking with these Japanese businessmen, and she would literally drink them under the table. And that was how she kind of earned her chops in the Japanese business world because it was such a big part of the culture."

In 1981 ECD formed a subsidiary company in Tokyo, Japan-ECD.[21] Momoko was president; she also recruited Edwin Reischauer, the former US ambassador and a leading scholar of East Asian studies, as chairman. At Stan's suggestion, Momoko had sought him out. "She just walked up to him at a cocktail party and introduced herself," Chet Kamin recalled. She soon became close friends with Reischauer and his wife Haru, and persuaded him to join ECD.[22] Reischauer had extremely high prestige in Japan, and recruiting him was a coup.

Now divorced from Masat and resuming her maiden name of Momoko Ito, she moved to Tokyo with Joi and Mimi.[23] Starting with a one-room office in her little apartment, she eventually ran a large office and had a large house where, as Mimi recalled, "She loved having parties every time Stan and Iris came. She was a real networker; she knew how to bring people together." Not only did she establish herself as an executive in Japan's masculine business culture, she also hired several other "really bright, underappreciated women. She knew that women in Japan were an untapped labor pool."

Momoko accomplished all this while being seriously ill with breast cancer that in time metastasized to her bones and liver. Ovshinsky considered her one of the bravest people he had ever met. Joi feels it was "partially because she had sort of faced death and conquered it. She had tuberculosis when she was five and was in a wheelchair for maybe half or a third of her childhood, and then she got cancer. And even as children we'd been told two or three times, your mother has only got a week, or a month. But then she would live, and so she really had no fear, and I think that built her character quite a bit." In Japanese, "momoko" means "peach." When Momoko died, the ECD staff planted a peach tree in her memory outside the institute. As Dave Strand recalled, "We all took turns using the shovel. It was a touching ceremony. Momoko was good to everybody; everybody liked her."

Shadows

As a utopian social experiment and as a diverse, productive research and development laboratory that both generated innovative technologies and fostered individual growth, ECD was a remarkable place, but it was hardly perfect. Its shining achievements were darkened by aspects that conflicted with its ideals, shadows cast by traits in Ovshinsky's character. For Vin Canella, "Stan was a person that I loved very dearly, but he was not always easy to love. He was a person with greatness and skills and talents, and he was a human being. He had faults in many ways equal to his greatness."

Perhaps the crucial fault was Ovshinsky's need for control. ECD and its achievements would not have been possible without his strong will and determination, but the ideal of ECD as an egalitarian participatory democracy whose members made voluntary sacrifices was sometimes belied by his demands for subservience and loyalty. A telling example is the recurrent "Christmas crisis," recalled by several researchers. Year after year, as Christmas approached, Ovshinsky would tell the leader of one group or another that some important work had to be done over the holiday. As Dick Flasck recalled, that had already happened to him once, when he spent Christmas Eve in his lab doing superconductivity measurements. "I'd been told, 'Dick, the future of the company depends on this. I hate to ask you, but you've got to work over Christmas.'" Flasck resented the imposition, which he believed was unnecessary, and was determined not to repeat the experience. A few years later, he "felt in his bones" that it was his turn again. Anticipating the project he'd be asked to work on, a version of the microfiche with a keyboard for entering text, he got his group to work on it secretly during the fall. By the time of the Christmas party, when Ovshinsky indeed told him he'd need to spend the break working on the device, it was already done. Flasck enjoyed the holiday at home, while Ovshinsky kept calling to ask impatiently how the work was going. Since no one had been in the lab, Ovshinsky was sure his orders had been disobeyed. He called a meeting right after New Year's and demanded to see the device, which Flasck calmly produced and demonstrated. Ovshinsky was furious. "But," Flasck said, "I didn't get the Christmas crisis again."

Ovshinsky probably did believe that each Christmas crisis really was urgent. He operated at a high level of intensity and frequently displayed a crisis mentality when problems arose. (His family remembers the repeated mantra heard around the dinner table, "This couldn't have come at a worse time," whenever there were economic difficulties or family issues that demanded his attention.) But whatever he believed, the Christmas crises were clearly instances of his exerting control and testing his subordinates' loyalty, and his anger at being outmaneuvered shows how important that control was to him.

Ovshinsky frequently indulged in anger. He could go into a rage when projects weren't moving along fast enough, or when he felt he wasn't being understood. "You don't know what you're talking about," he would shout, pounding his desk in frustration. Sometimes his anger seems to have been not so much a loss of control as a way of achieving it, something he could turn on and off. Steven recalled how he "could be screaming at someone on the phone, and he'd put the phone down and he's ready to go swim or have lunch."[24] Ovshinsky may have used anger to assert control, but he also showed the ability to let go of it. He might get enraged with those who opposed or failed him, but he did not nurse resentments. As Subhash Dhar recalled, "A number of very senior people threatened Stan and quit; almost all of them were welcomed with open arms when they decided to return."

Family

Iris's influence moderated Ovshinsky's intensity. Mike Fetcenko remembered "being in meetings where Stan needed to be restrained, and only Iris could do that. She would be his sort of calming influence, or his rock." She was also the seemingly inexhaustible source of nurturing and bonding energy for the whole ECD community. "Everybody fell in love with Iris," Ovshinsky said, using an image that appears in many accounts of her: "She was the glue that held everything together. They came to her with every kind of problem." Iris kept track of all the staff members' birthdays and the names and ages of their children. "She would carry around these note cards," Mimi Ito recalled, "and a big Mont Blanc fountain pen. We'd be having dinner and something would come up, and she would whip out her pen and a card and jot it down and tuck it into her purse. She was just the most together person, and she made it seem so effortless." Mimi marveled at all the tasks Iris managed, like the holiday cards and the Rosh Hashanah honey loaves she baked and sent each year to a huge mailing list. "She had this perfect kitchen and this perfect house, and she was managing this guy who was probably incredibly difficult to manage. She just managed everything down to the last exquisite little detail, and everything was done tastefully and on time and with style."

Just as Iris's energy and attention to detail reached out into the ECD community and helped make it feel like family, so the wider life of the community entered into

Figure 7.9
Books in Ovshinsky's library showing pages marked with pipe cleaners or sticky notes.

the family life that she and Stan nurtured. Growing up in the house on Squirrel Road, Robin and Steven were always included in the dinners for the many visitors, both at home and in the London Chop House. Robin recalled, "When I was applying for college, one of the questions was 'What was your most extraordinary educational experience?' And I said it was having dinner at home, because of the people that were invited night after night—really, really interesting people from all over. We were included in that a lot." Even after a full day at the office, Iris would produce one of her delicious meals, and the conversation would range over history, politics, music, literature, and art. Stan held his own comfortably, often consulting passages marked with pipe cleaners (later sticky notes) from the books in his large and remarkably diverse library.

The family circle expanded to include others, like Joi and Mimi Ito, for whom Stan and Iris were like parents. The ECD community "was really our life," Joi recalled, "and we learned just about everything from the Ovshinskys." Like Robin, he recalled the intellectual energy of dinnertime conversations and only later realized how extraordinary that experience was. Mimi explained that while their parents had social and cultural status in Japan, it didn't help in America, so becoming part of the extended Ovshinsky family "was really our socialization into a certain sort of sophisticated elite intellectual American culture that we wouldn't have had otherwise."

Stan and Iris were even more concerned with Robin and Steven, but their parental concern combined generous support with the pressure of high expectations. When Steven had shown early signs of being musically gifted, they immediately arranged lessons, first piano and then violin.[25] Despite her busy work schedule, Iris always drove him to his lessons and sat through them.[26] Steven was offered scholarships to music conservatories, but Iris and Stan insisted he have a liberal arts education and sent him to Indiana University, where he majored in music and languages before going on to earn a master's at Julliard. Music formed a lasting bond between Steven and his parents. He recalled that some of the best times they shared were when they would visit him in college and he expressed his feelings about different pieces; he was also pleased when his music could help them to relax.[27]

But there were also times when Stan and Iris's appreciation and support became a source of uncomfortable pressure. Even though Steven had chosen not to have a bar mitzvah, they celebrated by giving him a facsimile of Stravinsky's notebooks from the time when he was composing "The Rite of Spring." It was inscribed, "To the man, our son, who will honor Stravinsky by exceeding him." To the sensitive thirteen-year-old this loving gesture carried the weight of excessive expectations. Later, he felt pushed again when Stan suggested he "go on to be a conductor and composer." Steven did study conducting and proved good at it. After joining the San Francisco Symphony as a bassoonist in 1983, he and some colleagues organized a chamber orchestra, which had a highly successful first season, but he grew anxious when asked about plans for the next season. He realized what mattered most to him was connecting emotionally with his audience through performing—not organizing and raising money, which Stan of course shone at.

Robin also experienced a blend of parental support and pressure in pursuing her career. When she expressed an interest in medicine at Careers Day, her sixth grade teacher gave her brochures about becoming a medical technologist. This was not well received at home. "You're not going to be a medical technologist," Iris declared. "You're going to be a doctor and tell the medical technologist what to do."[28] Later, as an undergraduate at the University of Michigan, Robin became quite interested in philosophy and politics. "Stan thought it was great." But when Robin told Iris she had stayed up all night talking about Nietzsche with her friends, "Mom said, 'You think those conversations are important? They're not important. You should be studying. You're getting terrible grades. You're never going to be able to do what you want.' And she was almost right."

Years later, after Robin had graduated from Wayne State University School of Medicine, her relationship with Stan and Iris changed. Now the pressure came from their turning to her continually for medical advice. "For years I would say, 'You've got lots of doctors. I'm your only daughter. I would like to keep it that way.'" But "when he got older it was like I couldn't hold out any more, and I just tried to do my best with it. There were times when I actually cried and thought, 'If I wasn't a physician, I would have a totally different relationship with them.'" Still, she accepted her role. "If he was ever in the hospital [which was often], I came and stayed with him overnight. I would stay for days, and it was always a good thing I did because there were so many disasters and near disasters" (see chapter 13).

The relationships between Stan and his sons Ben, Harvey, and Dale, were considerably more complex, partly because they rarely lived with him and Iris. For years, Ben was away in Europe, but his experience there eventually reconnected him with ECD. He worked as a management consultant for several years, and in 1981 became ECD's European representative. But when Norma became fatally ill, Ben moved to Florida to nurse her. After her death in 1985 he moved to Berkeley, California, where he taught college and university courses in management, developed a management consultancy, and worked as ECD's West Coast representative.

Dale moved into his own apartment after living with Stan and Iris between 1969 and 1971. He enrolled in a community college for those two years, studying psychology and sociology and earning an associate liberal arts degree. He recalled the help and encouragement he got from Stan and Iris: "I will always be grateful for their support. It was Iris who encouraged me to take typing in school, which helped me express my thoughts." Dale also moved to Florida to be near his mother. There, at her prompting, he found employment for the next twenty-eight years in what Stan called "the food industry." "I bagged groceries at Publix," Dale said, "but for me the best part of the job was talking to the customers."

Harvey, however, remained nearby. In 1969 he met and married Cathie Kurek, then a graduate student at Lafayette Clinic, part of Wayne State University. She became a psychiatric nurse and they stayed in the Detroit area, where Harvey pursued a long

Figure 7.10
The Ovshinsky family, gathered in 1999 to celebrate Stan's sister Mashie's 80th birthday. Left to right: Steven, Ben, Sylvie Polsky (Robin's daughter), Robin, Iris, Stan, Dale, and Harvey.

and successful career in journalism and broadcasting. His work ranged from the underground newspaper the *Fifth Estate*, which he founded at age seventeen, to becoming director of production at Detroit Public Television as well as a storyteller and producer of many national and regional award-winning documentary films. Like Ben, he also used his professional skills on behalf of ECD, recording and publicizing its work. His journalism and storytelling reflected the progressive political values that motivated Iris and Stan and formed another link with the ECD community.

8   Solar Energy: Working at the Edge of Feasibility (1979–2007)

Ovshinsky's vision of solar energy focused on making efficient and affordable solar panels "by the mile." While he could clearly imagine this goal, ECD scientists were faced with the enormous practical problems of continuously passing a flexible stainless steel substrate, the web, on which thin layers of amorphous silicon would be deposited, through a series of gas-filled chambers without any cross-contamination. As the physicist Joe Doehler observed, the challenge forced them to work "at the edge of feasibility."

Over a hundred million dollars was eventually required to prove the concept, but Ovshinsky was equal to the task. His passionate belief in his vision enabled him to persuade the leaders of large corporations to invest the substantial sums needed for ECD's research, displaying his gift for what Mike Fetcenko called "reeling in whales." ARCO Solar was the first, investing tens of millions of dollars in Ovshinsky's solar vision; others included Standard Oil of Ohio, the Japanese electronics giant Canon, and the big Belgian-based wire company Bekaert. All the whales swam away after a few years, usually because changes in management led to their withdrawing support.[1] Yet the end of each funding agreement left ECD with greater resources for research and development, and over the next two decades Ovshinsky's bold plan for mass-producing the solar panels was indeed realized.

ARCO (1979–1983)

Solving the problems of mass-producing amorphous silicon solar panels began soon after Ovshinsky had negotiated ECD's second, much larger contract with ARCO Solar (see chapter 6). Early in January 1980, he called a series of staff meetings to announce the new effort and to share the ideas he had been developing. Jeff Yang remembered one such meeting on the first day after the holiday break, to which Ovshinsky had brought a pile of books with relevant pages marked. Gesturing with his hands, he began by reminding the group of how a solar cell works and then described his plans to make thin-film cells in large quantities. Yang was impressed when Ovshinsky proclaimed that ECD's goal would not be to make the best solar cells but rather to change the world by making them cheaply. Only by achieving economies of scale, he argued, could they reach the goal of making fossil fuel obsolete.

But when Ovshinsky went on to present his concept for mass-producing the solar cells by roll-coating them in large volume, those hearing it for the first time were highly skeptical because he had extrapolated from measurements made on tiny fingernail-sized experimental devices to miles of material.[2] At another meeting held at Ovshinsky's home on a Sunday night, Dick Blieden recalled that "virtually everyone there" had serious doubts, and that Steve Hudgens, who had earlier been one of Hellmut Fritzsche's PhD students and had just joined ECD, muttered something like "this guy is crazy. Maybe I'd better find something else to do." At the same time, even to the skeptics Ovshinsky's concept was powerfully attractive. Hudgens remembered appreciating the boldness of "thinking about building a 1-megawatt production plant for thin-film solar when the world market for solar cells was three megawatts." Ovshinsky's argument was characteristic of the way he thought about technology: by the time the proposed 1-megawatt machine was built, the market would be 20 megawatts, and when that was achieved the market would be 100 megawatts.[3]

Figure 8.1
Ovshinsky and an Ovonic solar panel, late 1980s.

To appreciate the challenge of building the roll-to-roll machine, we need to review the design of the thin-film solar cells, as discussed in chapter 6, and then consider the manufacturing process. Each cell consisted of three layers of hydrogenated amorphous silicon produced from silane gas (SiH_4) by plasma deposition. The P-type layer (doped with a small amount of boron) and the N-type (doped with phosphorus), sandwiched a thicker undoped, or intrinsic (I) layer; these were deposited in sequence on a stainless steel substrate. (Later, two and then three of these PIN cells would be stacked to increase efficiency.) Light shining onto the cell is absorbed mainly in the intrinsic layer and is then reflected back by the substrate. In the electric field created inside the cell by the PIN structure, the electrons freed from the I-layer by the light move to the N-layer, while the holes they leave behind move to the P-layer. When the cell is inserted in a circuit, current flows.

Previously, the layers of N-, I-, and P-type silicon were each deposited separately in a slow and expensive batch process. To deposit them continuously on a moving substrate with plasma-enhanced chemical vapor deposition (PVD) was another story.[4] Each layer required a separate gaseous environment, and while the stainless steel web passed through the successive vacuum chambers, the silane gas in each chamber had to be isolated from its neighbors, so that the P and N dopants didn't get into the I-layer. Preventing such contamination was extremely difficult. Joe Doehler, like Steve Hudgens, first thought, "This is impossible," and he became even more pessimistic when he went back to his office and did some quick calculations.[5] Considering the possibility of using a "gas gate" of differential gas pressure between the chambers to prevent diffusion of impurities from one to the next, he realized that "the degree to which you needed isolation was incredible." But when he shared his results with Ovshinsky, he recalled, "Stan looked at me and said, 'Joe, you're going to prove to me that it doesn't work or you're going to make it work.'"[6]

Doehler then worked on the gas gates with his collaborators in Masat Izu's group for about a year, making elaborate calculations to understand the fluid dynamics of the gases and adjusting the design of the chambers.[7] When they finally succeeded, "it was an intellectual thrill," he recalled. Keeping the pressure in the intrinsic chamber higher than in its neighbors created a "silane wind" flowing in both directions through the slots that kept out boron or phosphorus atoms. The gas gates were crucial in making the roll-to-roll machine work, though there would be many other problems to solve as ECD built successively larger machines over the years, from the first 1.5-megawatt machine built for Sharp in the early 1980s to a huge 30-megawatt machine by 2005.

Even before the first roll-to-roll machine was completed, Ovshinsky moved quickly to patent it. The job fell largely to the attorney Larry Norris, who joined ECD in the spring of 1980 and worked on patent applications as well as business and license agreements.[8] Within a few years more than a dozen basic patents were issued, most with the help of a second patent attorney, Marvin Siskind, who joined Norris in March 1982. Together they built up a patent department at ECD. In the early days, "Larry typically worked on the big picture, and I did much more of the nitty-gritty," Siskind said. An important objective was to obtain broad enough coverage to control secondary patents dealing with improvements.[9]

Over the next three decades, Siskind took on the main responsibility for converting Ovshinsky's confusing explanations into clear patent descriptions. As Mike Fetcenko recalled, "While others would pull their hair out after listening to Stan describing his inventions, Marv was the guy who could turn it into strength." Siskind realized that Ovshinsky skipped thoughts along the way, often jumbling them, so that what he said was not necessarily what he meant. When Siskind took notes during their meetings he would write what he thought Ovshinsky meant, rather than what he said. "If you couldn't do that, you would never be able to deal with Stan," he said.[10]

In April 1981, when the first of the two ARCO grants covering solar work ended, Ovshinsky attempted to extend and increase the R&D and cross-licensing agreement to a full-fledged joint venture.[11] But, unlike Bradshaw, the new executive vice president of ARCO Solar didn't share Ovshinsky's vision, and ARCO "really didn't think big company–little company joint ventures work," Nancy Bacon explained. Soon Ovshinsky was talking with both Sohio and Sharp in his efforts to raise further support. Making the new agreements was aided considerably by the fact that when the ARCO agreement ended, ECD was able to keep the technologies that had been developed, a coup that John de Neufville attributed to "the fine negotiating skills of Nancy Bacon."[12] These favorable terms set a precedent for ECD's negotiations with later patrons. With each successive partner in the solar program, as well as with most of those in the other energy and information programs, ECD was able to keep both the intellectual and physical property that had been created under the collaboration. As we shall see, these cumulative gains eventually enabled the company to grow from a contract R&D lab into an independent manufacturer.

Sohio and Sharp (Early to Mid-1980s)

Ovshinsky negotiated agreements in 1981 with the Cleveland-based Standard Oil of Ohio (Sohio) and at the same time with Sharp of Osaka; the result was overlapping

Plate 1
Stan Ovshinsky, with the same mischievous grin he had as a young boy. (Figure 01.b)

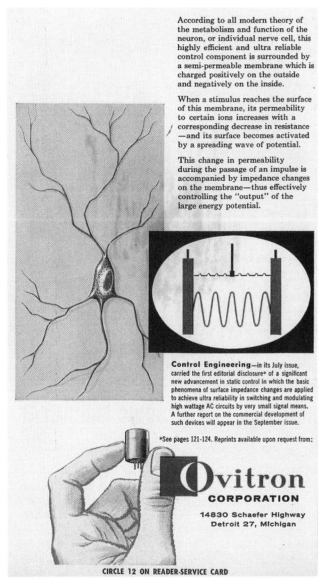

Plate 2
Ovitron ad. (Figure 4.7)

Plate 3
ECD's two-sided medallion. (Figure 5.2)

Plate 4
Mayan woman with baby and a box of solar panels. (Figure 5.3)

Plate 5
Ovshinsky's double portrait of Sacco and Vanzetti. (Figure 10a.10)

Happiness can be everywhere
Where two people share love

Life is so much easier on a porch
On a beautiful summer day

It is a pity that a beautiful summer day
Is not reflected in a peaceful beautiful world

Maybe an answer for our earth and its strife
Is a global porch and people sharing their love

With warmest wishes,
Stan and Iris

Plate 6
Image and text from holiday card, "A summer evening on our porch / Photo by our good friend, Dr. Takeo Ohta / Poem by Stan." (Figure 10a.11)

Plate 7
Ovshinsky speaking in Chile, October 2009. (Figure 12.3)

Plate 8
National Inventors Hall of Fame banner. (Figure 15.1)

joint ventures.[13] ECD's initial contact with Sohio concerning photovoltaics occurred late in 1980, when Ovshinsky's old Akron friend Harvey Leff put Bacon in touch with Sohio's Dick Smith, who in turn negotiated with Ovshinsky. Sohio was at this point rich from its oil fields on the north slope of Alaska, and like other oil companies was hedging its bets by building a broad portfolio that included a number of alternative energy approaches. To help him with the negotiations, Ovshinsky brought in Dick Blieden, who had been instrumental earlier in arranging the ARCO Solar contract (see chapter 6). Blieden worked initially on a consulting basis and eventually joined ECD's staff.

In negotiating with Sohio, Ovshinsky said he wanted to build a 4-megawatt machine, but Sohio negotiated it down to a demonstration 2-megawatt pilot production machine, the TA2 (Tandem Two) machine, designed to create double (i.e., six-layer) cells, for reasons of efficiency discussed later in this chapter. ECD's contract with Sohio ultimately contributed some $80 million of funding for solar, as Bacon recalled, supporting considerable research aimed at improving the efficiency of the solar cells.

Before the negotiation with Sohio was concluded, ECD also agreed to build a 1.5-megawatt machine for Sharp, the TA1, which created single three-layer cells.[14] At the time Sharp joined with ECD, it had one of the biggest solar cell manufacturing operations in Japan, but their cells were made of crystalline silicon. When Tadashi Sasaki of Sharp learned about the pilot production machine that ECD was building, he wondered whether using a roll-to-roll machine to produce amorphous solar cells could also fit Sharp's needs. He negotiated the deal that resulted in the Sharp machine, built between the summers of 1982 and 1983 during the early years of ECD's joint venture with Sohio. ECD then disassembled and delivered the machine to Japan in August 1983, where it was used by Sharp for some years.

The TA1, ECD's first commercial roll-to-roll machine, produced small solar cells for use in hand-held calculators, the first products that used amorphous silicon. Ovshinsky would proudly hand out the $10 solar-powered calculators to employees and visitors. To commemorate the historic achievement, he had a poster made that juxtaposed a picture of the Sharp machine with one of Ford's Model T, the first affordable, mass-produced automobile. Just as Ford's assembly line replaced the slower, more expensive process of making cars one by one, ECD's roll-to-roll machine replaced the slower, more expensive batch production of solar cells. As we discuss in chapter 11, this experience with amorphous silicon solar panels later enabled Sharp to make thin-film transistors (TFTs), eventually used in their work on flat panel displays.

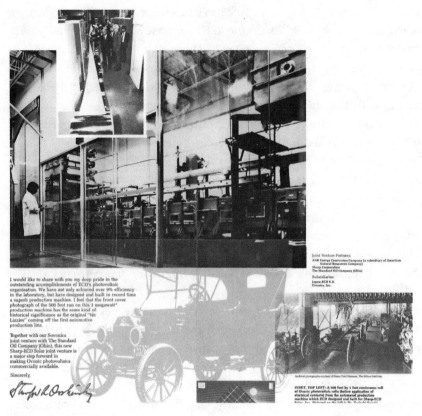

Figure 8.2
Poster comparing Ovshinsky's roll-to-roll machine to Ford's Model T assembly line.

The efforts to increase the efficiency of the amorphous silicon solar cells proceeded simultaneously along multiple tracks. While contamination had been the largest problem in designing the roll-to-roll machine, now the largest problem was the Staebler-Wronski effect. This had been discovered at RCA in 1977 by David Staebler and Christopher Wronski, who found that exposing amorphous silicon solar cells to sunlight over a six-month period causes a 10–30% decrease in efficiency.[15] While this degradation can never be completely eliminated, ECD developed various strategies to minimize it. The most important was making the intrinsic layer so thin that nearly all the electron-hole pairs created when light impinges on the cell can be collected by the electrodes before they recombine and create defects. Thinner solar cells, however, absorb proportionately less light than thicker ones, so the team stacked first two and

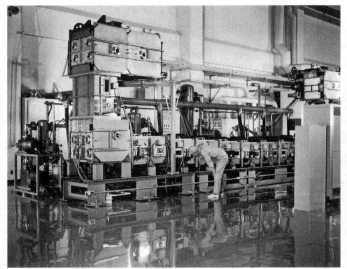

Figure 8.3a
The Sharp machine in Japan.

Figure 8.3b
Sharp solar-powered calculator.

later three thin cells on top of each other. A reflective layer at the back sent the remaining light up through the cell to generate more current.

Stacking two or three thin solar cells (i.e., increasing from three to six or nine layers) also opened the way to "spectrum splitting," developed in the early 1980s by Steve Hudgens, in which the stacked cells are not compositionally or optically identical. Incorporating different proportions of germanium in the silicon allowed the layers to absorb different light wavelengths.[16] The concept did not originate with ECD, but the material produced elsewhere was poor. ECD perfected the process. "Nobody else knew how to make silicon germanium alloy better than ECD," Jeff Yang said. In the triple cell, the top cell captured the blue end of the spectrum; the bottom cell captured the red, and the middle cell all the frequencies in the range in-between. The result was that over the course of the day, the newly designed solar panel used more of the solar spectrum than do ordinary crystalline cells; it "wakes up earlier, goes to bed later, and performs better in high heat and diffused light," explained Steve Heckeroth, an architect and builder who installed the solar panels in several of his projects.

Other important contributions to improving efficiency came from Subhendu Guha, who arrived at ECD in February 1982. The respected physicist had been working on amorphous hydrogenated silicon for about five years.[17] Finding a paper by Ovshinsky on using fluorinated amorphous silicon "of interest," Guha wrote to him asking whether he might come to ECD for his upcoming academic sabbatical leave from the Tata Institute of Fundamental Research in Bombay (now Mumbai).[18] Ovshinsky replied that he had heard about Guha's work and suggested he come to ECD for two years. Guha joined the group then headed by Hudgens.[19] After some months, Guha accepted Ovshinsky's invitation to stay on indefinitely, and within two years he had risen to be manager of the Advanced Research Group.

Soon Guha and Yang were collaborating on further increasing the efficiency of the solar cells, in which they made two major advances. The first was finding a way to invert the cell structure, from PIN, where P is on the bottom, to NIP, where P is on top. Basic physics principles indicated that it would be better to have the light enter through the P-layer, but the amorphous P-type material is very absorbent and so lowers the efficiency of the whole device. Guha suggested making the P-layer of microcrystalline rather than amorphous silicon, which made the NIP structure possible because microcrystalline silicon has higher conductivity and is less absorbent.[20] The second advance, known as "band gap profiling," increased the efficiency by varying the germanium/silicon ratio continuously over the thickness of each intrinsic layer.[21] Both advances substantially increased the efficiency, raising it to a record-setting 13%.

When Yang announced ECD's new efficiency record at a conference, however, he found himself in a bind because of Ovshinsky's commandment, "Thou shall not disclose the device structure." So when, in response to Yang's talk, someone asked him, "What is your device structure? Is it NIP or PIN?" he replied, "It depends on which way you look at it."

Sohio's generous support had led to remarkable progress in the design of the cells, but after five years British Petroleum, which had taken over Sohio, ended the agreement. "BP just wanted to focus on oil," Blieden said. But once again, with the help of Bacon and the attorneys, ECD benefitted from the dissolution. When Sohio-BP withdrew, they "left ECD with more technology, more understanding and in complete ownership, which allowed us to go forward and make a new deal," said Herb Ovshinsky, who added that "every back-out turned out to be for the benefit of ECD. Every single one." "It was sort of like swallowing an elephant," Blieden added, "because we had this small company."

But after ECD had swallowed the elephant there was little more to eat for some years. Ben Chao remembered "big layoffs in 1986 and 1987. We lost at least 40% of our people; maybe three hundred people left."[22] Those who remained at ECD during those dark days were sustained by their belief in Ovshinsky's goal "to make the planet a better place to live," Chao said. And once ECD's financial situation improved, "a lot of people came back."

In the midst of the lean years, a new funding opportunity arose from one of ECD's many patent disputes. Canon, a major manufacturer of copiers, was trying to make its copier drums with amorphous silicon instead of such toxic materials as selenium and arsenic. Because they were using the technique of plasma-enhanced chemical vapor deposition developed by ECD, the patent attorneys Marv Siskind and Larry Norris went to Japan, as Siskind said, "basically to charge Canon with patent infringement. They were doing it with radio frequency, but nonetheless they were doing it." The discussion eventually took a more positive turn. "After a while," Siskind recalled, "we convinced them we're not just here to say hey, you have a patent problem, but we think we can help you out with your commercialization. And so things changed quite a bit." As a result, Canon agreed in 1985 to pay ECD $10 million to develop the drums. Steve Hudgens solved the problem of depositing the required 30 microns of amorphous silicon at a high rate by changing to microwave frequency. Developing the license agreements for Canon's copier technology began what would become a close business relationship, leading eventually to a 1990 joint venture in solar energy. Meanwhile, as Siskind said, ECD "lived off of Canon in the last part of the 1980s."

Canon: Creating United Solar Systems (1990–2000)

After ECD's work on the copier drums, Canon expressed interest in buying a research machine for making solar cells.[23] This initiative became the occasion of another financial drama, which began on a Monday morning when Ovshinsky told Jeff Yang that ECD might not be able to meet the payroll at the end of the week unless they could demonstrate the machine's capacity to make a 12% silicon cell while a visiting Canon scientist watched. "It was kind of like having a gun at your head," Yang said. At this point the band-gap-profiling concept was not yet fully developed, so making a triple junction cell with the required efficiency in a few days was "like running before you were walking." But on Thursday they crossed the 12% threshold, and Canon agreed to pay a million dollars to have the machine dismantled and rebuilt in Japan. Then another plot twist came when it turned out that the money couldn't be moved from Japan by the next day. Nancy Bacon suggested selling the machine to Canon USA instead; the lawyers and accountants worked through the night, and by Friday the money for the payroll was in the bank.

The pressure continued even after ECD convinced Canon to begin supporting solar work while they were still arranging the joint venture. Yang recalled the grueling negotiating sessions: arriving in Tokyo in the evening, starting early the next morning and continuing for twelve hours, then dinner followed by an invitation to karaoke. "Subhendu was polite enough to say yes, but I said I really can't. Then at 7 o'clock the next morning, it starts all over, with the minutes of yesterday's whole discussion on a white board, day after day after day!"

The work with Canon in the pre-joint venture period was uncomfortable for other reasons, too. Guha recalled that he and Yang were sent to Japan to meet Hiroshi Tanaka, Canon's aggressive and imaginative senior managing director, known as "Tiger Tanaka." Before the meeting Yang and Guha met Momoko Ito for breakfast. She greeted them with "Congratulations on getting 15% efficiency." "I beg your pardon?" Guha responded. "You didn't get 15%?" she asked. He explained that they were then at roughly 13%. "Oh my god," she replied and promptly went to work to straighten matters out. Besides the embarrassments caused by Ovshinsky's tendency to exaggerate, he was also sometimes challenging for Canon's representatives to negotiate with. When Tanaka complained that Ovshinsky had misled him, Ovshinsky fired back with "What do you know about science?" Ovshinsky's tough negotiating style and characteristic demands for "more" could create tensions with his partners and sometimes contributed to the breakdown of collaborations. That did not happen in this case, but it does seem to have contributed to a shift in his role in ECD's solar program.

When the agreement with Canon was finally concluded in 1990, it was for a 50/50 joint manufacturing venture separate from ECD, to be called United Solar Systems (USSC).[24] Since the funding was coming entirely from Canon, they controlled the new company, including, crucially, its operational management. Ovshinsky had thought Guha would continue doing research in ECD while United Solar handled production, but Canon preferred to bring Guha into the joint venture. He thus became responsible for the operations of the new organization, becoming vice president for research and technology and later executive vice president in charge of both research and production. Canon was represented by a series of executives who typically came for a week each month.

The birth of United Solar was an important turning point in ECD's history. On the one hand, it greatly advanced Ovshinsky's vision of mass-producing thin-film solar panels, of which the new company eventually became the largest manufacturer. But in the process, he gradually lost control of the solar effort.[25] Under Canon, ECD's solar program was transformed from an R&D into a manufacturing operation. "We learned a lot from Canon," Guha remarked. "We learned manufacturing. They taught us you run it as a real company, not as an R&D company anymore. To be truthful, at the time ECD did not have a clue as to how to do manufacturing. Manufacturing is a different discipline. You don't make big changes. Canon's philosophy is what is known in Japan as *kaizen*. Make small changes—don't make big changes." That disciplined conservative approach was the antithesis of Ovshinsky's bold decision to go directly from solar cells measured in centimeters to making continuous panels by the mile. He could imagine and inspire that dramatic effort, but he was not so well suited to directing its day-to-day operations.[26]

By 1997, United Solar had a 5-megawatt plant in Troy producing triple junction panels in 12-inch- and later 14-inch-wide rolls. The coated webs were cut into panels in Mexico in a maquiladora operation just south of San Diego, where the labor-intensive finishing processes of laminating and adding connections could be done at lower cost. But outsourcing the last stage of production caused problems when it came to installing the panels. Some got bent during shipping from Mexico and had to be expensively rewired by electricians after installation, and the adhesive initially used for attaching the panels melted at 180 degrees and even, as the talented Indian-born engineer and technician Arun Kumar recalled, "caught fire big time." Steve Heckeroth encountered many of those problems in using United Solar roofing on projects in northern California. He tried to raise these issues after Ovshinsky appointed him ECD vice president of building-integrated photovoltaics, but that just created conflict with United Solar management, who resisted Ovshinsky's interference. Heckeroth found this all "hard to watch" because he felt ECD's solar product was outstanding and "they could have been so far ahead of everybody."

Such problems were in time corrected, and United Solar became very successful, even winning an endorsement from the White House. "You remember those solar panels, how big they used to be?" President Bill Clinton said, holding up a United Solar panel. "Look at this. I want everybody to look at this and consider this for your home. Look how thin they are. It's really an amazing thing, and most Americans have not yet seen these, but they can make a huge difference in what we have to do."[27]

Ovshinsky's role in all this, however, was reduced, and his attention shifted to other research areas such as hydrogen and fuel cells (see chapter 9). Eventually, as with the ARCO and Sohio joint ventures, the partnership with Canon ended with a change in its management. Tanaka, the greatest proponent of photovoltaics, lost his battle for power in the company, and the new president, as Robert (Bob) Stempel later put it, "took the view that nothing should be Canon's business unless it had a lens."[28]

Sovlux (1990–1996)

In 1990, while the Canon joint venture was underway, ECD also formed Sovlux, a joint venture with Kvant, which was a part of the Soviet nuclear industry. ECD built a huge

Figure 8.4
President Clinton in the Oval Office, holding an Ovonic solar panel.

2-megawatt photovoltaic machine in Michigan, then disassembled and shipped it to Moscow to be reassembled and installed. By the time it arrived, however, the Soviet Union had dissolved, and both installing it and getting payment from the Russians took longer than expected. A large ECD team, headed by Roger Woz, spent a long time working in Moscow to get the Sovlux machine running. Arun Kumar was the main troubleshooter, the one who "made sure the machines ran properly and the electric work was done well," Ovshinsky said. "If there was a noise, he'd stay up all night to find out where the noise came from." Kumar spent so much time there that he learned to speak Russian. He found that the machine had sat on the docks for months, and most of the electronics had become corroded by the salt air. "So the moment we got there, we took a lot of time cleaning that stuff out and getting it to work." By 1996, the ECD team had the machine running and turned it over to the Russians.

Bekaert (2000–2005)

Around the time when Canon was preparing to pull out of the United Solar joint venture, Guha received an inquiry from Bekaert, a large Belgian-based manufacturer.[29] He first thought they wanted to sell something to ECD, but when their representative arrived and expressed interest in investing, "that certainly got my attention," he said. Bekaert wanted to get into alternative energy, and after looking at all the current solar programs had chosen United Solar as the most promising. They initially proposed a tripartite venture, but Canon preferred to leave, so Bekaert bought out Canon's share, and in June 2000 ECD and Bekaert formed a 50/50 joint venture. As with the Canon venture, all the money—over $80 million—came from Bekaert, so they were in control, but they left the R&D and manufacturing with United Solar while putting in their own sales and marketing staff. Also like Canon, Bekaert stipulated that Guha remain in charge, and he now became president of United Solar.

The new investment led to another giant step in production capacity. From the 5-megawatt capacity of the Canon machine, Bekaert agreed to the ambitious goal of building a 25-megawatt machine.[30] The key to this advance was Herb Ovshinsky's suggestion to turn the stainless steel web on its edge, running six rolls vertically at once.[31] Like Ovshinsky's original roll-to-roll concept, it was an extremely complex engineering task, requiring a high degree of teamwork, but again the ECD staff succeeded. The ribbon-cutting ceremony in July 2005 for the plant in Auburn Hills, Michigan, was attended by US secretary of energy Samuel Bodman, who gave the keynote speech. Over the next few years, the first plant was joined by a second, and then two more were built in Greenville, each with four production lines, which gave ECD's solar program an enormously increased capacity.

Partnering with Bekaert also helped ECD find new markets for solar panels, especially in Europe, where there was increased interest in solar power after the Chernobyl nuclear accident and generous subsidies in Germany. United Solar's thin-film laminated roofing panels gained increasing acceptance, and by 2009, 80 percent of their greatly enlarged production was being exported.[32] By then, however, the partnership with Bekaert had ended. Once again, a change in leadership led to the breakup. With the declining European economy, Bekaert's new management became increasingly worried about the high cost of the new machine and whether it would really work. Eventually Bekaert decided not to continue, and in 2005 ECD bought them out for a fraction of what they had invested. "In retrospect," Bob Stempel noted, "that's one of the great things that happened for United Solar, because now the machine was ours." Several months later the machine began running, producing good material. "The sad part," Stempel added, "was the people from Belgium who helped us bring the machine up to speed. When they found out their company was going to leave, those guys were actually in tears some of them, because they knew the machine was ready to run and produce."

Figure 8.5
The ribbon-cutting ceremony for the 25-megawatt plant, July 2005.

Figure 8.6
Herb and Stan Ovshinsky at the ribbon-cutting ceremony.

Once again, ECD had benefited from the ending of a major joint venture. United Solar became a fully owned subsidiary and the largest US manufacturer of thin-film solar panels. It was now primarily oriented toward increasing production, though it also continued research on improving efficiency.[33] Guha stressed how much the company had gained from its large corporate partners. "We learned a lot from Canon in terms of manufacturing. We learned a lot from Bekaert in terms of financial discipline." But, he also emphasized, "the innovation came from ECD."

It appeared by 2005 that a tipping point had been reached in the world's appreciation of the need for alternative energy. United Solar's large area solar panels were selling so well that they had a backlog of orders. Even President George W. Bush, who as Ovshinsky said had been "brought up as an oil person," took notice of ECD and its impressive solar energy efforts when he visited the United Solar plant in Auburn Hills on February 20, 2006.[34] Despite having very different politics from Bush, Ovshinsky welcomed the visit: "First of all, he's President of the United States, and he came to see something that he normally would not."

For several years after ECD bought out Bekaert, United Solar dominated the amorphous silicon solar panel market. Over the course of twenty-five years, ECD had developed an outstanding product, cheaper than crystalline solar cells, adaptable to more kinds of installation, and much less fragile.[35] Ovshinsky, however, wanted more. Still intent on making solar energy cheaper than fossil fuel, he proposed another huge jump in production capacity, from the 25- to a 1,000-megawatt (1-gigawatt) machine. The concept depended on depositing the amorphous silicon at a much higher speed and would require an enormous investment of new funds. Guha and others at United Solar didn't think this was a feasible or affordable goal, or that it was needed. In time, Ovshinsky would proceed with the gigawatt concept on his own (see chapter 12).

Figure 8.7
Subhendu Guha showing President Bush the new solar plant, February 2006.

9 Hydrogen and Batteries: The Genie and the Bottle (1980–2007)

As early as 1960, Ovshinsky believed that hydrogen could be a key to solving the world's energy problems. His concept of the hydrogen loop, using solar-generated electricity to obtain hydrogen from water, offered a general scheme, but at that time he had no way to store the notoriously combustible gas safely (see chapter 5). "Everybody talked about how hydrogen was the volatile genie in the bottle," his son Harvey recalled, "but that's not how Dad saw it. From his perspective, the problem was never the genie. It was the bottle." By 1980, solving that problem had become possible. Ovshinsky had included the concept of a hydrogen-based energy economy in the second proposal he presented to ARCO. In response, ARCO allocated roughly $10 million for hydrogen research over the next three years as part of its sizeable energy grant to ECD (see chapter 6). These funds allowed Ovshinsky to create what was then the largest hydrogen research program in the United States.[1]

A Discovery in the Hydrogen Research Group

To direct the new hydrogen program Ovshinsky appointed Krishna Sapru, who divided the group into three sections: (1) hydrogen generation, which focused on electrolysis; (2) hydrogen storage, which focused primarily on hydride materials; and (3) hydrogen utilization, which focused on fuel cells.[2] These teams worked in a new hydrogen research building on Barrett Street off Maple, within walking distance of ECD's main lab at 1675 W. Maple Road. Although the group's work was ultimately aimed at producing practical energy devices, its culture was entirely that of a research lab, where the aim was understanding the nature and behavior of the materials they were studying.[3]

Sapru proceeded to hire and blend into a "close-knit family" some twenty-five to thirty well-qualified researchers. Among them were the Indian electrochemist Srinivasan Venkatesan, the Israeli electrochemists Benjamin (Benny) Reichman and Arie

Reger, Kuochih Hong, a materials scientist from Taiwan who had been educated in the United States, and Mike Fetcenko, "a local boy" who in his sophomore year of college was attracted to ECD because it offered a tuition reimbursement program.[4] The hydrogen program helped all its members educate themselves to a level where "chemists could understand materials science and materials scientists like me could understand chemistry and electrochemistry," Sapru recalled. Outside electrochemists came to lecture, and "everyone was teaching everyone else."

In an early instance of such mutual education, Hong's storage group initially used the slow traditional high-pressure and vacuum method for hydriding and dehydriding alloys. Venkatesan, in the hydrogen utilization group, suggested that electrochemical methods would be much quicker. Experimenting with one of Hong's alloys, he found that it could indeed absorb hydrogen during electrolysis, and also that its absorption capacity could be precisely measured electrochemically.

Meanwhile, working on nickel titanium alloys for fuel cells in the utilization group, Reger also found that hydrides were formed during electrolysis and that the amount of hydrogen absorbed was greater than had previously been observed. When he reported this discovery to Sapru, she called all three sections together to discuss the result. At this meeting Hong added that his research showed that certain of Ovshinsky's disordered materials could store up to 10% hydrogen. Hearing that, Reichman, in the hydrogen generation section, immediately thought this material would be a good candidate for a battery electrode and asked Hong for some of his material to set up an experiment. As Reichman recalled, he "didn't have to do much" to make a battery. Simply placing the material in a beaker of potassium hydroxide solution with a piece of nickel hydroxide to serve as the positive electrode, he formed an electrochemical cell.

To understand this experiment, it may help to note how any battery works. Like a solar cell, a battery is an energy conversion device, but instead of converting light into electricity it converts chemical energy to electrical energy. Chemical reactions at the electrodes free ions and electrons; the ions flow through an electrolyte between the positive and negative electrodes, while the electrons flow out of the battery in an external circuit and can perform work. Reichman recognized that, with its high capacity for storing hydrogen, Ovshinsky's disordered material was a promising source of hydrogen ions.[5]

The next day, Reger saw Reichman working on his experiment and asked about it. When Reichman told him about his battery idea, Reger said that he had thought of exactly the same thing at the meeting. The two friends decided to work together developing the rudimentary battery. Reichman recalled sensing at the time that this work might be important, but he didn't invest too much hope in it. "It was just something in a beaker, just a curiosity." When they brought it to Sapru, she was excited about the discovery but also felt awkward because their ARCO support had been allocated for work on hydrogen, not for making a battery. When she spoke with Ovshinsky about the experiments, however, he instantly recognized their potential and strongly encouraged the research. "Stan can see things," she said.

Ovshinsky called Reichman and Reger into his office for more explanation and, as Reichman recalled, asked with "a very straight and serious face" whether they could make a device. "I want to see if it can do something," he said. The team then worked feverishly over the next several weeks, coming in on the weekends. By the end of 1981, they had scaled up their initial experiment by sputtering Hong's material onto a larger area and constructing encased cells that looked more like ordinary batteries. Testing and comparing them with the predominant nickel-cadmium batteries showed they had a much higher capacity.[6] Creating a new battery had not been the aim of the hydrogen group, but sharing information and collaboration among all three sections had resulted in an important discovery.

It did not take long for Ovshinsky to announce the discovery, and in a way that was typical of his forward vision and sense of drama. On a Saturday morning early in February 1982, he convened one of his regular meetings of the heads of ECD research groups and other leading staff members. In the center of the large round conference table they found a black cardboard box, from which two wires were connected through a switch to a toy electric fan. Nothing was said about this mysterious centerpiece while each of the staff made their reports and silently wondered about it. After the discussion had ended, Ovshinsky asked Krishna Sapru to throw the switch, and the little fan began to turn.

Ovshinsky then lifted the box, revealing a small laboratory beaker holding some liquid and two metal plates from which the wires ran. This, he announced, was the prototype for a new kind of battery. ECD would create a subsidiary to manufacture and market batteries based on this discovery, batteries that would outperform all current types for a wide range of devices and that some day, he confidently predicted, would power an electric car. It was a bold declaration, the epitome of his claims to see tomorrow, but as we shall see, his predictions were fulfilled.

Ovshinsky now moved quickly toward developing the battery, asking Venkatesan to make a prototype that could work like a conventional battery in everyday devices. Venkatesan replaced the negative cadmium electrode of a tiny commercial nickel-cadmium battery with one of the new metal hydride electrodes. To demonstrate that this battery could run something, he took his toddler's toy train, which ran on 1.5 volts. "I stole his toy, put it in there, and it worked," he said, and "they liked it so much that they kept the train. Fortunately my son didn't realize that he lost it, so I bought him another one, and I quietly replaced it."

As was soon discovered when Ovshinsky pushed to patent ECD's new battery, the basic ideas behind it were not new. There had been earlier versions of the battery—and earlier patents, too.[7] Indeed many working in the battery field did not initially consider ECD's battery a true innovation, and some were appalled that the patent was granted. The battery expert Dennis Corrigan became convinced that ECD's patent was valid only later, in the mid-1990s, when scientists at Philips, who had previously dominated this field, wrote a paper describing the ECD battery as new. "Well, if it's good enough for them, it's good enough for me," Corrigan said.[8]

The patent recognized that ECD's was the first nickel metal hydride (NiMH) battery that could be commercialized. The earlier batteries had too little power and too short a cycle life, soon becoming sluggish. They would stop working because the oxides that formed in the harsh alkaline environment prevented them from recharging. As Mike Fetcenko explained, "Every hydrogen storage material is more stable as an oxide than as a hydride." The key to making a commercial NiMH battery was to use Ovshinsky's disordered materials in its negative electrode. Seven to eleven different metals

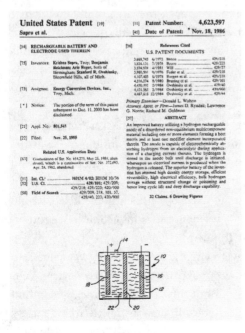

Figure 9.1
ECD's original nickel metal hydride battery patent.

were melted together to create an alloy that could succeed where others had failed.[9] "It all comes down to materials," Ovshinsky insisted. Fetcenko added, "Stan's multi-component, multi-element, multi-phase materials provided a resistance to oxidation that had never been seen before."[10] That was precisely why ECD's battery was a major invention. Corrigan compared Ovshinsky's invention of the NiMH battery to Edison's invention of the light bulb: Edison did not come up with the original concept but deserves the credit for making it work.

Not until 1986 was ECD granted its patent for the NiMH battery. As the inventors, it listed Krishna Sapru, Benny Reichman, Arie Reger, and Ovshinsky.[11] Reichman considered it entirely appropriate for Ovshinsky to be listed on this patent, for he and Reger were simply doing interesting experiments to satisfy their scientific curiosity, while Ovshinsky created the hydrogen program and recognized that they had made an important invention. "Nothing would have happened without him," Reichman said, adding that Ovshinsky "pushed, he pushed, he pushed" to turn the experiment into a battery with significant commercial potential. He was often so demanding "that it was sometimes annoying. After a meeting with Stan, you'd come back and you'd start working, and you'd get a telephone call. He wanted us back there, in one minute." And when Ovshinsky asked when something could be done and was told in two weeks, he'd typically say, "Can't you do that in three days?" Reichman also noted that when Ovshinsky spoke to others about the invention he would often say, "Here is Benny Reichman, who should get the credit for this." Reichman in turn emphasized that other members of the team like Reger, Venkatesan, Fetcenko, and Hong also deserved much credit for the battery.[12] The NiMH battery, which was probably the most commercially successful of Ovshinsky's inventions, is also the outstanding example of collaborative invention that came from ECD. As with phase-change optical memory or the roll-to-roll production of solar panels, others did the work of research and development, but it was Ovshinsky whose vision guided their efforts and whose energy drove them.

The Ovonic Battery Company

Within weeks of the invention of the new battery, Ovshinsky was arranging its commercialization. A leader in this effort was Krishna Sapru's brother Subhash Dhar, who joined ECD in October 1981. Dhar had encountered the battery for the first time at the February 1982 staff meeting when the invention was demonstrated. He remembered feeling skeptical about Ovshinsky's prediction that the little beaker battery would someday power an electric car and asked who would lead the development.[13] Ovshinsky looked straight at him and said, "You!" Dhar was taken aback. "What did I know about batteries? Nothing." Giving important assignments to people who weren't trained for them might seem risky, but Ovshinsky trusted his intuitions and they usually proved right. As with Momoko Ito and so many others who rose through the ranks, he could see potential in people who weren't aware of it themselves. After spending four days studying batteries, Dhar accepted the assignment.

Ovshinsky now moved most of the scientists in Sapru's hydrogen group (including Venkatesan, Reichman, Reger, and Fetcenko) into a new division, the Ovonic Battery Company (OBC), headed by Dhar.[14] Largely for marketing purposes, they opted to call their invention a "nickel-metal hydride battery," a term already known in the industry.[15] Dhar and his staff then turned a small Troy warehouse at 1826 Northwood Drive into a laboratory for developing production methods, which was up and running by the second week of January 1983. By then, ARCO was no longer funding ECD, and the OBC entered into a joint venture with the local natural gas distribution company, American Natural Resources Company (ANR).[16] The R&D efforts first focused on developing prototype batteries under the direction of Venkatesan. Ovshinsky was eagerly involved and was soon pushing Dhar and the OBC team to go beyond lab prototypes toward practical production processes, for which Fetcenko's chemical engineering background was very helpful. By 1987, they began production on a small scale and in September 1988 entered a development and marketing agreement with Hitachi Maxell.

Meanwhile, in 1985, ANR was bought by Coastal Corporation, a Texas-based natural gas supplier that soon withdrew funding from OBC. Ovshinsky tried but failed to find a new funding partner. In 1987, a time when other areas of ECD (such as solar and displays) were also struggling financially (see chapters 8 and 10), OBC turned to a licensing business model. This made sense because by now OBC knew how to manufacture the batteries but could not afford to build a large factory. While continuing its small-scale production, OBC sold licenses to the German battery company VARTA, Hitachi Maxell in Japan, and Gold Peak in Hong Kong.[17] In this way, ECD's NiMH batteries found their way into many commercial products, including digital cameras, laptops, and calculators, and eventually, with new manufacturing partners, into electric and hybrid cars.

GM-Ovonic and the EV1

The story of the General Motors innovative electric car, the EV1, has been told before and need not be repeated here in much detail. In *The Car That Could*, Michael

Shnayerson gives a full account of its successful development from the perspective of an embedded observer at GM, while the California drama of its eventual destruction is the focus of the film *Who Killed the Electric Car?*[18] Ovshinsky appears in both, but not as the central figure. From his perspective, the story began much earlier, with the decision he and Iris made at the start of ECL to seek alternatives to oil.

The possibility of building electric cars, however, depended on the fluctuating political support for alternative energy sources. The energy crisis caused by the 1973–1974 oil embargo led to government-funded initiatives in renewable energy like photovoltaics, but by the time ECD had begun developing the NiMH battery in the 1980s the Reagan administration was dismissing the very idea of an energy crisis. As Shnayerson observes, "There seemed to be about as much need for EVs [electric vehicles] in 1982 as for cars made of Swiss cheese."[19] That changed in 1990, when the California Air Resource Board, hoping to reduce smog caused by automobile emissions, adopted a Zero Emission Vehicle mandate (applied to 2% of new cars sold in California in 1998, 5% in 2001, and 10% in 2003). Now there was a strong incentive for auto manufacturers to develop electric cars, and there was also funding for developing the batteries to power them. In May 1992 ECD received a grant of $18.5 million from the United States Advanced Battery Consortium (USABC) to develop and scale up their NiMH batteries for electric vehicles on the basis of tests in which their batteries outperformed other types, such as lead acid batteries, in terms of power, durability, and recyclability.[20]

ECD was already making substantial improvements in its batteries even before this new interest and support emerged. The batteries based on Ovshinsky's original patent were excellent for slow-rate applications such as radios, cameras, and computers, but they couldn't charge and discharge quickly enough to be used in cars, where the battery needs both to absorb the car's braking energy quickly and to release it quickly. A later advance, the catalytic oxide, made that possible.[21] In this version, nanoparticles of nickel alloy suspended in the porous oxide layer that forms on the hydrogen storage materials catalyze the reaction and greatly speed the charging and discharging process. This breakthrough made ECD's NiMH battery feasible for electric vehicles.[22]

In the year after receiving the USABC grant, ECD used this improved technology to build prototype batteries, which were first successfully tested (and publicized) after being installed in a Chrysler van. "That first test drive was a great moment for him," Harvey Ovshinsky recalled. "Dad called me up and asked me to capture it on video. I said, 'Sure, who's going to drive the car? I'll need a release.' He laughed. 'Where should I sign?'" The test was a triumph from Ovshinsky's point of view but an outrage from the battery consortium's, which insisted on controlling all publicity and enjoined

Figure 9.2
Ovshinsky in the electric Chrysler van.

him from announcing any further results until the USABC approved them.[23] To move ahead, Ovshinsky clearly needed help.

Help came with the entrance into the ECD battery story of Robert Stempel, the General Motors engineer and former CEO who is often considered the father of the EV1. Stempel had long been engaged with reducing automobile pollution. In the 1960s his early work on exhaust emission control had made him realize "how much bad stuff was coming out of a gasoline engine." Adding catalytic converters to change hydrocarbons and carbon monoxide to water and carbon dioxide started him on his long quest to "take the automobile out of the air pollution problem."[24] One milestone occurred in 1988 when GM ran its solar powered car, the Sunraycer, across Australia at 45 mph.[25] "That really opened my eyes to the fact that with relatively little energy you could do an awful lot of work," he said. Before he was ousted as CEO in 1992 for failing to restore the troubled company's profitability, Stempel had authorized the program to build the electric car, first called the Impact, later the EV1.

At the time he was pushed out, Stempel felt "really too young to walk away" from the effort to build electric cars. Serving at GM as a consultant, he maintained his contacts and used them to pursue the project, whose current design depended on lead acid batteries that limited the car's range to between 75 and 100 miles. Seeking possible alternatives, he spoke with Walter McCarthy, the chairman of Detroit Edison, who suggested seeing Ovshinsky about his battery. Stempel sent some of his former colleagues to investigate. He recalled, "A report came back: well, he's a hard guy to understand, but he does have this nickel metal hydride battery and he's talking about seven, eight, eleven elements all put together. We think this is alchemy." But when Stempel asked how well the battery worked, they told him, "It's got power and it's right now about twice as good as the lead acid." That was enough to bring him to talk with Ovshinsky himself. At that pivotal meeting in 1993, Ovshinsky convinced him to try ECD's batteries free of charge. "Our very first test," Stempel recalled, "the GM engineers took the car, and Stan in his usual fashion said to them, it will go 200 miles with that battery. A couple people looked at me and said, 'Bob, do you really believe that guy?'" Ovshinsky proudly recalled, "The first test was 201 miles on a single charge."[26]

That memorable result suddenly made the car look feasible for driving around town, and it also made Stempel and Ovshinsky firm allies. Stempel remarked that with Ovshinsky's background as a machinist and toolmaker, "he immediately understood what we were talking about." Stempel became, as Chet Kamin said, "the partner that Stan had been looking for." He was "someone who really understood Stan, who was sophisticated technologically, and could trust Stan, and talk to Stan. They really worked quite well as a team."[27] Ever since the Cunningham fiasco (see chapter 6), no one had succeeded as Ovshinsky's second in command until Stempel joined ECD.

Before severing his ties with GM, Stempel brokered a joint venture in February 1994 between GM and ECD called GM-Ovonic, owned 60% by GM and 40% by OBC, whose

Figure 9.3
Robert Stempel and Ovshinsky.

aim was to make batteries for GM's electric car. GM, which had operational control of manufacturing, provided the funding; ECD/OBC provided the materials, the battery electrodes, and other components, such as the separators, as well as its know-how and the use of its machine shop.[28] GM engineers then began producing the Ovonic batteries, initially working in Troy and later in a production plant in Ohio.

The GM-Ovonic partnership was moving forward, but while GM was working to bring the EV1 to market, it was also spending much more money fighting the California Air Resource Board mandate. And the USABC also continued trying to restrain Ovshinsky from promoting his batteries. As in other instances like solar power, the technology was advancing but the political and economic support for it was unsure.

Despite this uncertainty, by the late 1990s GM did bring out the EV1, which it promoted as the car of the future. "No car company will be able to thrive in the twenty-first century if it relies solely on internal combustion engines," GM CEO Jack Smith told the press. "Issues such as global climate warming, clean air, and energy conservation demand fundamental change from all industries and all nations." The first generation version (1997) was powered by lead acid batteries (GM considered the Ovonic batteries to be not yet ready; Ovshinsky, of course, disagreed), but the second, two years later, did have Ovonic batteries, which more than doubled its range. "It's a nickel metal hydride battery," boasted Vice President for R&D Ken Baker, when GM rolled out the new version at the Detroit International Auto Show. "That makes EV1 the undisputed leader in range and performance of any electric vehicle in the world. And Bob Stempel, Stan Ovshinsky, and Iris Ovshinsky certainly have been close partners in this process."[29]

But this triumph was short-lived. For ECD as well as for many enthusiastic drivers, the EV1 was a great success, but when the lobbying efforts and legal challenges mounted by the auto manufacturers succeeded in overturning the California Air

Figure 9.4
The General Motors EV1.

Resource Board mandate, GM recalled all the EV1s, which had been leased rather than sold, and crushed them. Ovshinsky and others at ECD considered this a disastrous mistake—a judgment that recent history seems to confirm.[30] The outstanding engineering of the EV1 could have put GM many years ahead of both its American and Japanese competitors. Instead, only in recent years have they reintroduced electric and hybrid vehicles.[31]

Battery Litigation

To build a battery business, ECD had to defend itself when unlicensed companies tried to appropriate its technology. But with ECD's strong patents on both electrodes, litigation with major companies that infringed them also proved an important source of income, especially when the litigation was with large companies such as Matsushita, or Toyota, which had big plans for its Prius. As Marvin Siskind recalled, from roughly the mid-1980s to the mid-1990s litigation over batteries brought in about $10 million, plus a number of licenses with other battery companies. "The 1990s were the big era of the battery supporting the company, instead of the solar." In 1994, ECD brought an action at the International Trade Commission against Sanyo, Yuasa, and Toshiba, who quickly settled. ECD again prevailed in its important case against Matsushita, the parent company of Panasonic. All the companies infringing Ovshinsky's patents had to pay substantial settlements and take out licenses.[32]

A later case turned on the scope of those licenses. ECD held that the license agreements with Matsushita from the previous cases applied only to small batteries, not the large Panasonic batteries in the Toyota Prius and other hybrids, which also used ECD's technology. "We had them dead to rights on patent infringement," Chet Kamin recalled. "Ultimately the case was settled. The terms are confidential, but they were very favorable to ECD." As a result, from 1997 to the present, the Prius has continued to use ECD's NiMH battery technology. In such ways, despite the debacle of the EV1, Ovshinsky's NiMH batteries both helped sustain ECD and advanced his goal of replacing fossil fuels.

Baotou

With ECD's control of the battery patents came other opportunities. One notable example began in the spring of 1999, when Subhash Dhar received inquiries from Chinese representatives about a joint manufacturing venture. Baotou Rare Earth Manufacturing was interested in acquiring OBC's NiMH technology and building a manufacturing plant in Inner Mongolia.[33] After a series of tough negotiating sessions in Beijing, in which Dhar resisted the Chinese efforts to beat down the price, an agreement for $123.5 million was successfully concluded, which actually put ECD in the black for a while. Like Sovlux, the earlier agreement to set up a solar panel plant in Russia (see chapter 8), this joint venture depended on importing machines built in the United States, and a large new plant was constructed in the open Mongolian grassland. A team of about twenty OBC technical and engineering staff installed and debugged the production machinery and then turned it over to the Chinese. The plant had a production line for making metal hydride powder, a key battery component, and another for manufacturing electrodes and assembling batteries, mainly for electric bicycles.

The production methods OBC established in Baotou had been developed during the 1990s. The heart of the process was the metallurgy for making the special alloys for both the electrodes by melting the components in a furnace, casting huge ingots, and breaking them into powder to be compressed into sheets and cut to size. Ovshinsky came up with an ingenious method for breaking the ingots by using their high capacity for absorbing hydrogen. Flooding the massive 3,000-pound ingots with hydrogen gas caused the metal to expand and crumble, after which it could be mechanically ground into a fine powder and then made into sheets in another roll-to-roll process. In charge of the foundry where this took place was Meera Vijan.[34] (As she recalled, "Stan always says, 'I put a woman in charge of a foundry.'") She took on the same role in Baotou, working in the harsh conditions of the Mongolian winter without heating or electricity while the plant was being built. In four years, Vijan and the other OBC staff who went back and forth between Troy and Baotou managed to complete the plant and get it running.

Hydrogen Cars

In the late 1990s, when the prospects for the EV1 were still bright, Ovshinsky also began working to develop a car with an internal combustion engine fueled by hydrogen, another clean, non-polluting energy source. He had conceived this possibility in the early 1960s as part of his "hydrogen loop," but by the time he was ready to pursue it there had been several other attempts.[35] These converted cars all used either gaseous or liquid hydrogen, neither efficient nor safe for ordinary use. Ovshinsky, however, aimed to store the hydrogen in the form of a solid hydride, as ECD had learned to do much earlier. Placed in the tank of the car, this material would release only as much hydrogen as was needed, and it would be as safe as the dangerous elements sodium and chlorine become when combined in table salt.[36] Taking this ambitious step toward a hydrogen economy was the next episode in Ovshinsky's endless effort to achieve "more."

To lead this new project, Ovshinsky turned to the physicist Rosa Young, who had played an important role at ECD since 1984 and who would later become an even more important part of Ovshinsky's life (see chapters 12 and 13).[37] He had learned about her from Edward Teller and invited her to visit ECD in 1982. He remembered thinking when they met, "She has these very intelligent eyes." When he called Iris in, she agreed: "We really ought to hire her." That was not so easily done, however. Young already had two good jobs, working as a staff physicist at Oak Ridge and consulting at Helionomics on the commercial use of lasers in photovoltaics. Besides, she pointed out, her experience was with crystalline, not amorphous materials. Ovshinsky asked, "Can you learn?" "Of course," she snapped. "That's why I need you," he said. After two years of steady pressure, she accepted his offer, and in later years Ovshinsky would boast about his wisdom in hiring her. "I put her anywhere in the battlefield of advanced physics, and we always got results."

The first stage in creating a hydrogen car was a three-year cooperative research project on solid hydrogen storage funded from 1997 to 2000 with $18.9 million from the Advanced Technology Program (ATP) under the National Institute of Standards and Technology (NIST) in the Department of Commerce.[38] Ovshinsky asked Young to head the project.[39] ECD was the lead institution in a collaboration that included universities (such as the Colorado School of Mines), national laboratories (including Oak Ridge and Ames), and the industrial laboratory Crucible Research. As program manager for the collaboration, Young would lead brainstorming sessions every week. Her small group of six or seven people developed an experimental foundation for understanding hydrogen storage in a large variety of materials.

This work soon attracted oil companies, who in the later 1990s were worried about losing their market to electric vehicles. Supporting hydrogen was one way of protecting themselves.[40] Shell was the first of these to approach ECD, the only company working on solid hydrogen storage. In the middle of 1999 Shell Hydrogen funded ECD's hydrogen effort at about $700,000 a year in a joint venture. But disagreements about which company would play the leading role (e.g., whether Shell or ECD would get its name on refueling stations) led Ovshinsky to end the joint venture after only six months and seek a more congenial partner.

Texaco

In the late 1990s Texaco, rumored to have a billion dollars to invest in clean energy technology, also approached ECD. Impressed with the technology ECD had developed under the NIST program, Texaco was eager to form a joint venture. To the Ovshinskys, Texaco appeared to be the ideal partner, declaring, in effect, "We don't want to change you. We want to become an energy company, not just an oil company." Iris recalled, "We were so happy. We thought we were finally getting a really wonderful deal." And for two years, it was.

In October 2000, the two companies formed joint ventures for developing fuel cells and hydrogen systems, and ECD leased a 75,000-square-foot building in Rochester Hills to house the work. A third joint venture was formed in 2002, when GM-Ovonic became Texaco Ovonic Battery Systems. Texaco also made a substantial equity investment and bought 20% of ECD's stock. With a value of over half a billion dollars, the alliance with Texaco, Subhash Dhar said, "was the largest single transaction that ECD had ever made."

The hydrogen storage and battery agreements with Texaco were relatively easy to develop on the basis of ECD's previous active research programs, but the fuel cell agreement was tricky because ECD had little to show in this area. Texaco wanted to fund a fuel cell program because at that time all the automotive companies were touting the hydrogen fuel cell as superior to the battery for zero emission vehicles. As a way to counter the pressure to develop electric cars, both the auto makers and the oil companies were claiming fuel cell cars were "just around the corner" and would be available in four or five years. This misleading publicity helped kill the electric car, but it also helped generate support for fuel cell and other hydrogen research.[41]

In response to Ovshinsky's request for a fuel cell proposal, Rosa Young suggested building on ECD's work on the nickel metal hydride battery to make an alkaline fuel cell.[42] She chose not to work on developing it herself but rather to remain in charge of the hydrogen storage program, while Subhash Dhar took responsibility for the fuel cell program, transferring some of the electrochemists from OBC. Among them was Venkatesan, who confirmed that the metal hydride electrode of the NiMH battery could also work in the new device, since hydrogen ions are crucial in both systems. Using the Ovonic materials had several advantages. It was now unnecessary to use expensive catalysts to split hydrogen molecules into atomic hydrogen, because metal hydrides store and release atomic hydrogen. Furthermore, using the stored hydrogen removed the starting delay with conventional fuels cells. (Ovshinsky therefore dubbed the Ovonic version the "instant start" fuel cell.) Dennis Corrigan explained another advantage to ECD's design: "The first fuel cell cars didn't have batteries, and they were not as efficient as everybody expected because they couldn't capture the regenerative braking energy. But the Ovonic fuel cell had a built-in battery function, and I just thought that was a really neat concept."

While ECD's hydrogen fuel cell does not seem to have progressed beyond the early stages of development, the hydrogen storage program, on the other hand, reached its goal of successfully powering a car. Its first task was to develop a vessel to hold the hydrides, as well as an efficient heat exchanger.[43] This was a considerable engineering task, and it was hard getting help with it. The US job market was so tight that Young could find only a few US-educated scientists from Germany and China but no suitable engineers. Earlier, through Ovshinsky's connections, she had recruited an outstanding Russian metallurgist named Vitaliy Myasnikov, through whom she then hired a number of talented engineers from Russia, Ukraine, and Belarus. Young also enjoyed a cordial working relationship with Gene Nemanich, the manager Texaco assigned to the project, who helped her set and meet milestones.

Ovshinsky was characteristically impatient for the team to get its new hydrogen fuel tank installed in an actual car, but there were still several problems to be solved. The team of scientists had to experiment with several different metal hydride alloys before settling on one that would release hydrogen using only waste heat from the radiator. Once that was decided, the engineers had to design a lightweight container that would hold both the 3.5 kilograms necessary to achieve a 200-mile range and the heat exchanger. Myasnikov's team produced an ingenious design, a carbon fiber-wrapped vessel with all the components inside. It was, Young observed, like building a ship in a bottle.

Before being installed in a car, the tank had to be rigorously tested for charging and discharging. That required a facility that could handle large quantities of hydrogen, but the local fire department was reluctant to provide a permit. Chevron, which was then taking over Texaco, allowed ECD to set up the testing lab in Richmond, California, where they had a refinery.

The next steps were converting a gasoline engine to burn hydrogen, installing the tank, and testing it in a car, but ECD had no capability for doing all that. Through the retired Chrysler engineer Dick Geiss, Ovshinsky got in touch with Quantum Technologies, a company in Lake Forest, California that had the necessary equipment and expertise. Bruce Falls, the director of the alternative fuel program there and a longtime hydrogen proponent, was excited to help ECD achieve its hydrogen car. "That was exactly what I'd wanted to do all the way through high school and college!" he said. The deal was made in early 2002, and a few months later, Young and her team of engineers brought Falls their fuel tank.

Ovshinsky and Bob Stempel had chosen to convert a Toyota Prius, which, Falls explained, was the most reverse-engineered car ever made and so very well understood. Replacing the gas tank of a blue 2002 Prius with the ECD hydrogen storage vessel proved to be relatively simple; the main problem was the sophistication of the Prius's controls, which were highly computerized. But Falls and his team were able to make the modifications and adjustments needed for the Prius to burn hydrogen. The demonstration model had no safety problems, and, at a time when other experimental hydrogen cars could go no more than 80 miles without refueling, it could go 200 miles.

When Falls reported his achievement to ECD in August 2002, Ovshinsky immediately wanted to visit Lake Forest and drive the car. He brought along Stempel, Young, and Iris, and after he and the other visitors drove the car, he announced, "I'm so appreciative of what your team did, I want to take everyone to dinner." "Nobody will ever forget that dinner, that night," Falls recalled. He remembered thinking, "This is what leadership is about."

Chevron

But the headiness of achieving the hydrogen internal combustion car was not sustained for long after Chevron took over Texaco at the beginning of 2002. Dhar recalled that "one of the high-ranking Chevron people personally called Stan and Bob to assure

Figure 9.5
Rosa Young, Stan, Iris, and the mobile hydrogen refueling station funded by the Army NAC. Used at ECD for publicity, it was also driven to other sites for refueling demonstrations.

them that this acquisition of Texaco in no way, shape, or form was going to interfere and change the commitments and agreements or the direction that had been agreed upon by ECD and Texaco. That promise, I think, lasted only a few weeks." Chevron did help with ECD's major battery litigation against Matsushita and Toyota in 2003, but it soon became clear that Chevron had no interest in developing alternatives to oil, and the partnership became increasingly troubled. "We went from having the best partner in the world," Mike Fetcenko observed, "to very soon thereafter one of the worst partners."

Signs of trouble came first in the battery division, renamed once more and now called Cobasys (Chevron Ovonic Battery Systems). Marvin Siskind said, "Chevron put a quarter of a billion dollars into that venture. But Chevron, I believe, put in management with the wrong background. The guy who ran it was a marketing guy, a zero technology guy, a guy who hated Stan and did his best to drive the biggest wedge in the world between Stan and Chevron." Dhar, who had been running the division, left in November 2003 in the heat of "a lot of disagreements" with Chevron.[44]

Another sacrifice of the Chevron takeover was Gene Nemanich, who, Young said, "understood and believed in the technology and wanted to help us move forward." He was persuaded to retire at the end of 2003 with a verbal promise from Chevron that he could become president of the joint hydrogen program. He waited for the position and was ultimately disappointed when Chevron replaced him with someone who knew and cared nothing about the hydrogen technology.

Meanwhile, the hydrogen car was benefiting from Ovshinsky's gift for promotion. TV audiences learned about the car after it was entered in the Michelin Challenge Bibendum, a large demonstration of clean energy vehicles held in Napa Valley in September 2003. The car was a huge success. There Ovshinsky met the actor, producer, and science enthusiast Alan Alda, who was so taken with him, Iris, and ECD's hydrogen car that he featured them all in two episodes of his PBS show *Scientific American Frontiers*, "Future Car" (2004) and "Hydrogen Hopes" (2005).

Despite this enthusiastic publicity, Chevron continued to oppose the program, for it did not believe that Ovshinsky's hydrogen economy could ever be commercially feasible. In 2004 Chevron notified ECD that they would not put any more money into the hydrogen work and intended to shut down the testing lab in Richmond. When ECD proposed renting the facility from Chevron, they refused. The point, said Ben Chao, was that Chevron "didn't want to leave any fingerprints or have any liability issue." By November 2004, both ECD and Chevron were eager to end their joint hydrogen venture.

Before formalizing the conclusion of the joint venture, Chevron told Ovshinsky, "We're not going to sign unless you give us the car." Immediately planning to build another car, he attempted to bargain: "'Give us back at least our hydrogen tank, because you have no use for it.' And they said, 'No.' And I said, 'Well then, give us our turbocharger.' And they said, 'Hydrogen went through it.' And I said, 'For Christ's sake people, you're in the fuel business. Don't you know that gasoline is polluting the world and causing wars?' But they wouldn't give me back the tank and they wouldn't give me this damn little turbocharger." Then Chevron crushed the successful blue prototype built in 2002 during the Texaco and Chevron joint ventures, the one that said "Hydrogen" on its side.[45]

Ovshinsky experienced Chevron's crushing of the hydrogen car as a very aggressive act. Falls and Chao agree, but they also understood it as a defensive move on Chevron's part, since the car had not gone through all the required federal safety tests and so posed a potential legal liability. Deferring such tests, however, is standard auto industry practice for prototypes, and Chao felt "a meanness behind" Chevron's legalism, reflecting personal animosity toward Ovshinsky.

From a research point of view, the hydrogen internal combustion car was extremely successful. With funds included in Chevron's $4.6 million severance payment, Bruce

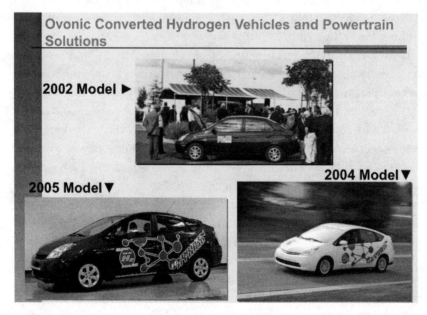

Figure 9.6
ECD's three hydrogen cars.

Falls and his team went on to build two more hydrogen cars, a white model in 2004 and a red in 2005. These were not sponsored by Chevron and were shipped back to Michigan when the Richmond testing facility closed. With small design changes, each model performed better than the previous one. Falls was proud that they achieved a 200-mile range with very little pollution, "orders of magnitude less than the gasoline one and no CO_2 at all—no climate change gas," thus comparing well with the electric car. They hoped to do even better. Ovshinsky hired Falls and his team to continue the program, but unfortunately by this time (2005) he and Stempel no longer had control of ECD (as explained later, in chapter 11) and could not get the board to approve funding for further development.

In the end, however, it was not only Chevron's or the ECD board's lack of support that killed the hydrogen internal combustion car but a growing sense, at least among those who had created it, if not Ovshinsky, that it was impractical. In spite of their success with the demonstration models, both Young and Falls concluded that the technology was not ready.

Ovshinsky's original concept of the hydrogen loop, proposed back in 1960, had evolved into a three-part scheme for powering vehicles: (1) use solar-generated

Figure 9.7
Promotion for Ovshinsky's hydrogen loop concept.

electricity to produce hydrogen through electrolysis of water; (2) store the hydrogen in the form of hydrides; (3) either burn the hydrogen in an internal combustion engine or use it in a fuel cell. Young pointed out that, despite the theoretical attractions of this completely clean energy system, it would be inefficient in practice because energy is lost at each stage. Using the solar-generated electricity in a battery-powered electric car would be much more efficient and would not require large new investments in hydrogen refueling stations. She tried to convince Ovshinsky that, given the current state of scientific progress, even with more research funding it would be hard to compete with battery-powered vehicles. Ovshinsky disagreed, and he found new hope of continuing to pursue his hydrogen dream in the place where his career as an inventor began.

The Akron Hydrogen Dream

This new venture began when Ovshinsky accepted an invitation from Frances Seiberling Buchholzer, the great-granddaughter of F. A. Seiberling, the founder of the Goodyear Tire and Rubber Company, to speak to the Akron Roundtable. His talk, "Can the Hydrogen Economy Solve Akron's Problems by Building New Industries," on February 17, 2005, at the Tangier restaurant attracted more than 540 people.[46] He was so moved by the turnout that he "spoke from the heart and just mesmerized everybody," recalled Jeff Wilhite, then working for Mayor Don Plusquellic in Akron's Office of Economic Development.[47]

About a week later, the phone rang in Wilhite's car while he was driving through a blinding snowstorm. It was Fran Buchholzer, who said, "Jeff, do you have a minute, Stan would like to talk to you." When Ovshinsky came on the line he said, "I'm still on such a natural high from the reception I had at Akron, I told our folks we're going to move the hydrogen testing facility from Richmond to Akron." For Wilhite it was "as if the heavens had parted," since the mayor had asked him several years earlier to head a new effort to attract small business interests to the city.[48] Wilhite was exuberant when the mayor's office quickly approved moving ECD's hydrogen testing facility to Akron. "It is without question one of the strongest opportunities the city has had, probably in forty years, because the rubber industries have left," Wilhite remarked.[49] "Lightning can strike three times," he exulted. Akron had "moved the nation's economy in the 1800s with our canals. We moved the world in the 1900s with our rubber industry, and we'll be moving the world with our hydrogen economy, and it's happening because of Stan's love for the city." Love certainly played a role, but it was need on both sides that fueled the scheme. Ovshinsky was desperate for some way to advance his stymied hydrogen program, and Akron was desperate for some way to replace its abandoned industries.

Wilhite proceeded to contact important people in town in an effort to raise support for the project and turn Akron into the world's hydrogen city. He made the most of the parallel that he saw between Ovshinsky bringing his hydrogen project to Akron after his positive reception there and the legendary story about Goodrich deciding to move his rubber business to Akron after his reception there (see chapter 1). One of Akron's attractions for a hydrogen industry was its fine municipal water system, and the proposal included converting all the Akron buses to run on hydrogen.

Ovshinsky was also ecstatic about moving the hydrogen program to Akron. He enjoyed talking with Wilhite about Akron's heritage: for example, about the Alcoholics Anonymous movement co-founded by Dr. Bob (Robert Holbrook Smith) and Ovshinsky's old friend Bill W. (William Griffith Wilson). Wilhite meanwhile enticed Ovshinsky with his plan to turn Bill Wilson's Akron home into a hydrogen institute ("H_2 and You") where visitors could learn about Ovshinsky's hydrogen economy, how to store hydrogen safely, and so forth. The enthusiasm also infected Iris, who, after going with Stan to visit his parents' graves in the Workmen's Circle cemetery, decided that Akron was the place where they should also be buried. It was a peaceful and beautiful spot. As Ovshinsky recalled, Iris had said, "You always loved Akron, but Akron paid no attention to you. Now that Akron loves you it's all right for us to be buried there." He agreed.

Combining a nostalgic return to his beginnings with a way forward toward the hydrogen economy, the Akron plan was an idyllic dream, and like most such dreams it was too good to be true. Some steps were taken toward realizing it. The city bought an old steel mill for the testing facility, and First Energy Corporation, a large utilities company with headquarters in Akron, promised funding. But the building in its present state was unsuitable and would have required a huge investment to convert it, and the funding from First Energy never came through. Ovshinsky's dream of Akron as the capital of a new hydrogen-based economy faded like a mirage.

10 Information: Displays and Memory Devices (1981–2007)

Ovshinsky's most important energy technologies, thin-film solar cells and NiMH batteries, were major commercial successes. But his information technologies—which were more radically innovative and based on his most original discoveries, the switching effects he first observed in the early 1960s—failed to realize their full commercial potential for ECD. The flat panel displays that Ovshinsky had envisioned in 1968, and which ECD's subsidiary OIS (Optical Imaging Systems) contributed greatly to developing, ended up enriching other companies. Ovonic optical memories, such as rewritable CDs and DVDs, enjoyed a period of commercial success but again mostly profited others. And while many in the semiconductor industry recognized the enormous promise of Ovshinsky's electrical phase-change memory, it lay dormant for years because it was not considered commercially viable. Finally, his innovative cognitive computer, based on a further extension of his phase-change technology, never advanced beyond its research phase.

This chapter about ECD's information research between 1981 and 2007 thus appears at first to be mainly a story of missed opportunities and unrealized possibilities. From a later vantage point however, the story looks quite different. As of this writing in 2016, it seems that the time for Ovonic phase-change memory has finally come (a story we briefly outline in the epilogue) and that Ovshinsky's information technologies, based on his crucial discovery of the Ovshinsky effect, may end up having the most impact of all his inventions.[1] Those technologies thus both grow out of his early efforts as an independent inventor working on his own and also depend on his later collaborative approach, inventing with and through others as he did in ECD's energy technologies.

Flat Panel Displays

Despite Ovshinsky's prediction in 1968 of flat TVs that could hang on the wall, it took some time for him to begin developing them, partly because, as in the case of solar cells, he hoped they could be made from his Ovonic chalcogenide materials.[2] That was a reasonable hope, because liquid crystal displays depend on a grid of thin-film switches that allow an electric field to rotate the crystals and let light through. But as with the solar cells, chalcogenides proved unsuitable.

In 1981 Robert Johnson, the former Burroughs executive who had served in the 1970s as a consultant (see chapter 6), became ECD's senior vice president in charge of developing thin-film technologies other than solar cells. Johnson recognized that the amorphous silicon material ECD was now starting to produce for its solar cells could also be used to make either diodes or transistors as switches for displays. He convinced Ovshinsky to set up a program of building active-matrix liquid crystal displays (LCDs) using diodes, which he saw as a unique opportunity because other researchers were using transistors.[3]

Johnson had hoped that Dick Flask would lead ECD's development of the LCDs, but "the first disaster," from Johnson's point of view, was that in early 1982 Flasck left ECD.[4] Hoping to learn more about the state of the art in active matrix displays, Johnson then contacted Professor J. William Doane of the Liquid Crystal Institute at Kent State University in Ohio, who suggested that ECD invite Zvi Yaniv to give a talk. A bright and ambitious Romanian-born physicist educated in Israel, Yaniv was writing his PhD thesis on order parameters in liquid crystals. He planned to return to Israel to become president of the newly established Practical Engineering College at Hebrew University in Jerusalem and so was surprised when the day after his talk Johnson brought him to see Ovshinsky, who offered him a well-paying research position at ECD.[5] Yaniv started working at ECD early in 1983.

He began by studying ECD's amorphous silicon solar cells with Vin Canella. Yaniv recalled, "I said to myself my God these guys know how to make these diodes very, very well." At Johnson's suggestion, Yaniv and Canella started developing the amorphous silicon diodes for LCDs. Both physicists were "emotionally built," which made them compatible. "There were times when we'd be in an office talking," Canella recalled, "but we'd be shouting and screaming and laughing, and people would knock on the door and say, 'Is everything okay?'"

After about a year, Yaniv and Canella, working with consultants Marvin Silver and Mel Shaw, had developed a 32 × 32 pixel prototype LCD using diodes. Johnson then convinced Ovshinsky that it was time to create an ECD subsidiary to commercialize diode-driven LCDs, and in May 1984, Ovshinsky created Ovonic Display Systems (ODS), with Johnson as president, Yaniv as vice president, and Ovshinsky serving as chairman of the board. Not long after that, Yaniv secured $300,000 to support the work of ODS from the large Israeli defense company Elbit. The growing staff of ODS also included Canella, John McGill, and Meera Vijan.[6] They started making rudimentary prototype displays based on diodes, and "Zvi had ideas for mass-producing them," Johnson recalled.

The new company proved fairly successful in attracting government research contracts because the military was interested in flat panel displays for aircraft cockpits, and ODS had the advantage of a unique approach. Many companies, including GE, Sharp, Canon, Toshiba, Philips, IBM Japan, Seiko-Epson, Hitachi, and AT&T, were developing amorphous silicon transistors for driving liquid crystal displays, but only ODS's displays used diodes.

In 1985 Ovshinsky renamed the company Ovonic Imaging Systems (OIS), because it had begun to explore using the diodes for other imaging devices as well as for displays.[7] In the summer of 1986, a new 50/50 joint venture called Quartet Ovonics was formed with the Chicago-based Quartet Manufacturing Co. The venture produced and marketed OIS's first product using the amorphous silicon diode, the electronic whiteboard. This highly successful technology, still on the market today, digitized writing or drawings made during presentations. About a year later, OIS followed up with the "wand," a small handheld scanner.[8]

At this point, however, losses of funding from Sohio in the solar program and from ANR in the battery program led to a steep decline in ECD's stock and drastic layoffs (see chapters 8 and 9). Ovshinsky needed to pull another rabbit out of his hat. To raise new funds, he followed Nancy Bacon's suggestion to take OIS public, and in December 1986, Ovshinsky and Yaniv traveled to New York to meet with potential investors.[9]

In the Park Lane Hotel overlooking Central Park, OIS demonstrated its imaging and display technology, including prototypes of a small liquid crystal display, image digitizers, and fax machines. "I remember every half hour we had another group on another floor and we were explaining to them how great we are," Yaniv said. In a room with a large window commanding a view of Central Park, a potential investor asked how large they thought the future TV displays could become. Ovshinsky turned to the big window and said, "I think we can make it as large as this window—correct Zvi?" Yaniv gasped and whispered, "Probably." It was another of Ovshinsky's visionary claims that seemed wildly improbable at the time. (Today, flat screens based on transistors rather than diodes can be even larger.) His typical exuberance and confidence in the technology helped raise substantial funding for OIS.

One of the investors at the meeting was William Manning of the Manning and Napier Investment Company of Rochester, a firm that managed several billion dollars in assets. Starting in the late 1970s, Manning had taken a liking to ECD's technologies and had many of his clients invest heavily in the company. "Bill Manning was always fascinated by technology," Canella said. "Stan could spin a story, throw out the hook, and Bill Manning would bite." Now, impressed with OIS's imaging and display possibilities, Manning invested roughly $15 million in OIS, which made the public offering successful, and about the same amount in another ECD subsidiary, OSMC (Ovonic Synthetic Materials Company), whose work included magnets and x-ray mirrors.[10] Yaniv remembered that at the end of the last talk he gave in New York, Manning put his hand on his shoulder and promised, "You'll have your money." Manning thus became ECD's largest shareholder.

Losing Control of OIS

As a condition for his investment, however, Manning required Ovshinsky to sign an agreement that soon became the focus of an intense legal dispute. Manning claimed to be concerned with Ovshinsky's management of ECD, which had repeatedly lost money and, to reduce his control, he sought to take away Ovshinsky's loaded vote, which at that point stood at 25 votes per Class A share.[11] The claim seemed plausible; others also complained about ECD's unprofitability. "Why do companies keep giving money to Stan Ovshinsky," *Forbes* magazine later asked, "the inventor who can create anything but profits?"[12] Yaniv put it more admiringly: "Stan was the Robin Hood of scientists. He was taking money from the rich people and hiring three hundred to five hundred scientists. No one else in the world did this." Investors like Manning would complain that Ovshinsky chose to plow all of ECD's profits back into research instead of paying dividends to shareholders. But ECD didn't actually have the profits with which to pay dividends. Since it was not a mature existing business with regularly recurring revenue, there was no way to quantify the future benefits of its research. Conservative accounting rules therefore required treating the company's large research costs as expenses instead of additions to its capital. Ovshinsky's ambitious business model with its several concurrent and interdependent R&D programs meant that research expenses almost always exceeded revenue, so ECD seldom showed a profit.

It soon became clear, however, that Manning was actually attempting to take over the whole of ECD, and when Ovshinsky refused to comply with the agreement, Manning sued to enforce it and gain control of the loaded vote. ECD filed a countersuit for violation of the 1934 Securities Act. The ensuing arbitration struggle ended with a "divorce" settlement in which Manning agreed to sell all his ECD stock and not buy more for ten years, but in return he gained all of OSMC and a controlling share of OIS.

Ovshinsky had managed to prevent Manning's attempted takeover, but at a steep cost. Instead of gaining new funding through OIS, ECD was left with even less resources and had to make painful sacrifices. The company downsized from close to five hundred employees to about a hundred, reducing the staff so severely that it was difficult to function. In the midst of this terrible time in 1987, Ovshinsky had heart surgery in New York City at St. Luke's hospital, where his daughter Robin then worked.[13]

While in the hospital Ovshinsky handled some work from his bed, mostly by phone. One piece of business was negotiating a license with the Korean company Samsung for OIS's hand-held displays, thus initiating Samsung's entry into the television display market.[14] Stan bitterly recalled licensing it "for nothing because we were under a lawsuit."

Meanwhile, under Yaniv's direction, OIS had been working on making larger flat panel displays with amorphous silicon diodes, but when the displays grew larger than six inches they suffered serious problems. Moreover, because competitors like IBM, Matsushita, and Mitsubishi were all using transistors, Yaniv explained, ECD would have had to develop its own production capability for diode-driven displays, an extremely expensive proposition. At the time, Yaniv was also negotiating for support for OIS from Sharp, which required them to work on transistors as well as diodes. Quietly, while most of the OIS staff worked on diodes, Yaniv had one researcher, Mohshi Yang, experiment with thin-film transistors (TFTs); he demonstrated "a superb three-inch color TV," Yaniv recalled.

When Samsung and the avionics industry (the market OIS was mainly aimed at) offered significant support to develop TFTs, that tipped the balance. On his own, Yaniv decided to change from diodes to transistors, announcing his decision in 1988 at a crucial management meeting with OIS officers. As he expected, they were dismayed. "These people developed the diode with me from time minus ten, and to come to a point where I gave up for commercial reasons, they couldn't understand it. I remember their faces." The change from diodes to transistors may have been necessary and inevitable, but it destroyed Ovshinsky's patent advantage. (OIS held the patents for switching with diodes, while using transistors was in the public domain.) Ovshinsky never forgave Yaniv for making the change.

Even though OIS had given up the advantage of using diodes, it still held patents for the design of active-matrix LCDs that would soon prove to be of value.[15] In July 1989 Yaniv negotiated a second, far more lucrative licensing agreement with Samsung, this one for $2.5 million, ten times larger than the initial license that Ovshinsky had approved from his hospital bed in 1987.[16] By this time OIS was virtually independent of ECD. Aiming to making a prototype three-inch color TV, Samsung sent personnel to Michigan to be trained by OIS staff. But the Samsung team behaved strangely, Vin Canella recalled. "They believed that we were trying to cheat them. We'd find people rummaging through the dumpster, trying to find the secret papers. We were very open and honest with them." Marv Siskind also remembered seeing the Samsung visitors searching the dumpster "for any of our paper that we threw out." The explanation for this strange behavior eventually became clear: the team from Samsung were television, not semiconductor, people. They only knew about cathode ray tubes, and when they returned to Korea and tried to use what they had learned at OIS, the devices they built failed. Only after an independent review by a Japanese TFT display expert did Samsung recognize that the failure did not result from OIS's withholding information but rather from their team's inappropriate background. Once Samsung replaced them with semiconductor people, they began making displays successfully.

From that point, Samsung went on to make ever-larger displays, eventually becoming the world's largest manufacturer of TVs and LCDs. Yet something of Samsung's collaboration with ECD and OIS remained embedded in its display design; much later, when the electronics systems designer Guy Wicker looked inside a Samsung display, he found that it retained the same pattern of connecting the transistors and other components that OIS had developed. And when the computer scientist and entrepreneur Tyler Lowrey visited Samsung and asked his hosts how they were able to develop their huge liquid crystal displays, the reply, as Lowrey later told Ovshinsky, was "You wouldn't know a man named Stan Ovshinsky, in Detroit? We got our license from him."[17]

By early 1991 OIS was seriously underfunded again and sought a partner with deep pockets. "By this point OIS is not doing Stan any good," Canella explained, and he was "looking to dump the company," for while ECD still owned a substantial share of OIS, the settlement with Manning had deprived Ovshinsky of so much control that he had no reason to continue. The successful business leader, William (Bill) Morse Davidson, owner of the Detroit Pistons and then among the richest men in Michigan, was an interested buyer. His company, Guardian Industries, was a large architectural and automotive glass manufacturer, and as Yaniv said, he believed it could make "anything built on glass." Davidson purchased OIS and immediately butted heads with Yaniv. "Displays are not just pieces of glass, just as a microelectronic chip is not just a piece of silicon," insisted Yaniv, who also let Davidson know that he believed he was underestimating the competition.[18] "He thought that because Guardian Industries competed with Asia in making glass, they could also compete with Asia in making displays." Yaniv also felt that Guardian had an exaggerated view of the size of the market and disagreed with Davidson's plan to build a huge factory to make avionic and military displays.

The result of this unwelcome advice was that Yaniv was asked to step down the day after Guardian took over, though he remained a paid consultant for the next two years.[19] Davidson went on to build his state of the art facility in nearby Northville, the first large-volume LCD plant in the United States. But when the Japanese and Koreans invested tens of billions in their TFT active matrix industries, Guardian's displays were too expensive to compete. (Indeed, no American display maker could compete.) Guardian OIS was bankrupt by September 18, 1998.

There was one more chance for ECD to play a role in the display industry. Roughly a decade after ECD had licensed its technology for LCDs to Samsung, Tatung, the largest Taiwanese electronics company, was having trouble scaling up its thin-film transistors. Guy Wicker and Rosa Young convinced them to try using Ovshinsky's threshold switches instead of TFTs. Representatives from Tatung who visited ECD were ready to offer funding to develop prototypes, but Ovshinsky disagreed with their plan to start small and work within a two- or three-year timeline. "He thought the timeline should be compressed to a much shorter time," Young recalled, and nothing came of the discussion. "I was really very unhappy with Stan's decision," she said. It was yet another missed opportunity.

Phase-Change Memory

The creation of phase-change memory in the 1960s is arguably the most important invention of Ovshinsky's career (see chapter 5). But as Marv Siskind said, "The world wasn't ready for it."[20] During the 1970s, after the failure of the West Coast memory company OMI that Keith Cunningham had started (see chapter 6), ECD's phase-change memory research languished because of inadequate funding. In the late 1970s, however, funding from IBM for imaging technologies had supported a small program in optical memory that allowed ECD to continue work on the materials.

In the mid-1980s the optical program got a boost from a lawsuit settlement. Matsushita, the largest Japanese semiconductor company, had introduced a rewritable optical memory, which they claimed was their invention. Angered by the infringement of his patents from the 1960s and 1970s, Ovshinsky sued the giant Osaka-based corporation in May 1983. With Momoko Ito handling the negotiations, the suit was settled out of court. She also persuaded Matsushita to collaborate with ECD on optical memory. ECD got $1.5 million for a two-year development program; Matsushita got a license to use the technology developed both before and during the program. It was not a large settlement, but Ovshinsky was satisfied to be acknowledged as the inventor of the technology.

Ovshinsky divided the optical memory work between groups headed by Dave Strand, who had been leading the program since 1980, and Rosa Young, who had recently joined ECD. Strand's group focused on the basic physics of the materials, building and operating test equipment, while Young's group focused on developing new materials and tailoring the properties of the phase-change alloys using Ovshinsky's principle of chemical modification, which had been developed in the early stages of the photovoltaic program (see chapter 6). She was assisted by Eugenia (Genie) Mytilineou, a professor of physics at the University of Patras in Greece, who often worked at ECD in summers and during her sabbaticals, and who would become one of Young's closest friends.

By systematically changing the recipe for the chalcogenide materials, Young's group managed to increase the speed of the optical memories and improve the cycle life. In Strand's group, work on recording and erasing led by Mike Hennessey significantly improved the process and resulted in a patent that was licensed to Matsushita and others.[21] Both Strand and Young recalled how smoothly the collaboration with Matsushita went, and as always, Ovshinsky kept closely in touch with their progress and kept up the pressure.

When the two-year Matsushita program ended in 1986, Ovshinsky continued to fund the optical memory research. By 1988 the work had paid off with the development of the much-faster 225 alloy (consisting of germanium, antimony, and tellurium, $Ge_2Sb_2Te_5$). With its repeatable switching time of 50 nanoseconds or less, it was around a thousand times faster than the earlier alloys, and required less energy for switching.[22] Additional work done primarily by Japanese companies greatly increased the storage density.[23] As a result, rewritable optical memory discs (CD-RW and DVD-RW) based on technology from ECD were widely used in the 1980s and 1990s and are still in use today. But while ECD received about $1 million in royalties from several Japanese licensees, when production later moved to China, the relatively low return on the discs made it impractical to enforce ECD's patents.

Meanwhile, in 1985, while the Matsushita collaboration was still going on, Ovshinsky took steps to revive ECD's electrical phase-change memory program.[24] Named NGEN (Next GENeration of computers), the new program received funding from two Japanese companies: $1.2 million from NTT (Nippon Telegraph Technology) for 3D phase-change memory development and $4.5 million from NSC (Nippon Steel Corporation) for threshold switching logic. Both companies discontinued their support when the team's effort to make a 3D memory failed, but ECD now had a clean room with deposition and lithography equipment that would be used continuously for further development of the threshold and memory switches.

In the mid-1980s, however, electrical phase-change memory had two fundamental shortcomings: it required too much current to switch, and it wasn't fast enough. It also faced the formidable competition of flash memory, which had been introduced in 1984.[25] The solution to increasing its speed came from the fast 225 alloy developed by the optical memory group. Ovshinsky was excited when it showed 50-nanosecond optical switching speeds and believed it would work even faster in an electrical memory. Many of his advisers disagreed, maintaining that the 225 alloy was a different class of materials and would not have suitable electrical behavior. But Ovshinsky insisted on trying it, so Wally Czubatyj, who by 1988 had become manager of the electronics group, assigned Pat Klersy and Dave Beglau to make the electrical devices using the 225 alloy. Guy Wicker, assigned to test them, was amazed to find he could set them with a 10-nanosecond pulse. Wicker emphasized, "It was Stan who motivated the use of fast optical alloys for electronic memory. Everyone thought he was crazy for insisting on it, but it was the biggest single improvement in the memory in more than twenty years."[26] It was another instance—and one of the most important—of Ovshinsky's strategy of cross-fertilization among research programs. By 1989 the electronics group had developed a good working model of a three-dimensional electrical phase-change memory.[27]

Now Ovshinsky was ready to commit more resources to the program. On a sunny New Year's Day, January 1, 1990, as Ben Chao recalled, Ovshinsky held a special meeting for roughly twenty of ECD's scientific staff in his home to announce the new effort. He told them that he wanted to make electrical phase-change the next memory device and predicted a time when all computers would use it. At this point, the optical memory group and the electronics group, which had been working primarily on threshold switches, merged. Initially Dave Strand and Czubatyj led the new group. Strand brought familiarity with the materials, and Czubatyj brought familiarity with fabrication.[28]

For about three years the group worked on improving the electrical phase-change memory. The working model was fast, but it was too large and still required too much current. Klersy and Beglau processed wafers in the clean room, trying to make the devices smaller. While developing an insulator etching process, they inadvertently left a residue of carbon and fluorine polymer that broke down after a single pulse. The resulting memory bits needed two orders of magnitude less current. While they had managed to produce a device that could compete in speed and current with existing ones, it took some time to learn how to produce it reliably. By 1993, ECD had developed a consistent process that clearly showed the potential for making a competitive memory device.

Ovonyx (Ovonic Unified Memory)

At this point, the commercialization of electrical phase-change memory required more funding and a larger network of associates. In 1994, during Strand's continuing efforts to raise money for the program, he made a cold call to Micron Semiconductors in Boise, Idaho, then the largest DRAM (dynamic random-access memory) manufacturer in the United States.[29] Strand was surprised when Micron's chief scientist, Tyler Lowrey, answered the phone himself, and he was absolutely elated when Lowrey expressed real interest in ECD's electrical phase-change memory.

Well-known in semiconductor circles for developing more efficient and lower cost DRAM fabrication processes without lowering the pay for labor, Lowrey was "a visionary in his own right," remarked Steve Hudgens. He had "the ear of the semiconductor industry," said Guy Wicker, and could easily get an audience with senior people at leading firms. Because of his deep knowledge of semiconductors, Lowrey understood the importance of nonvolatile memory. As he later explained, "When you turn off the power on your PC you've got to reboot it every time because it's got to bring in all the programs from the hard drive. You want an instant on. You want to be able ten years from now to turn it on and have it be right back where it was. Plus if you crash you want the data to be stored, not gone." Lowrey also understood that flash memory, the computer industry standard, would eventually run into problems because it could only be scaled down to roughly fifty or a hundred stored electrons. Chalcogenide phase-change memory does not have that limitation; indeed, since it needs less power it functions better as it scales down.[30] Moreover, it can be cycled many more times than flash memory, which wears out after some fifty thousand rewrite cycles. The many advantages of phase-change over flash memory also include its roughly a hundred times faster speed, greater efficiency, lower power, hardness to radiation (allowing the memories to function in space and in military applications) and greater potential for use with processors.

Lowrey recalled the original demonstration of the ECD devices at Micron. "We marched on down to the lab and we ran them, and we said, 'Man, these things work.' It was about a million [write/erase] cycles in a second, which was more than the lifetime of a flash memory. We signed a deal right on the spot. It was like a one-page agreement." Micron subsequently finalized it as a joint venture at a meeting that included Micron's CEO Joe Parkinson, Stan and Iris Ovshinsky, Marv Siskind, and Dave Strand.

Figure 10.1
The Ovonyx team in 1999. Left to right: Sergey Kostylev, Wally Czubatyj, Tyler Lowrey, Steve Hudgens, Pat Klersy, Boil Pashmakov, Guy Wicker.

Once the joint venture began, Micron's advanced fabrication lines were used to produce devices a hundred times smaller than could be made with ECD's simpler equipment.[31] But after two years of making progress, Micron faced a crisis when the price of DRAM, its main business, plummeted. It could no longer support the joint venture and pulled out.

Lowrey, however, continued to be enthusiastic about phase-change memory. When he left Micron in 1997, Ovshinsky invited him to join ECD. For eighteen months he was restrained by a noncompetition clause, but "the day after the eighteenth month was over, I was back in Michigan," he recalled. By the end of 1999, he and Ovshinsky had formed a separate company to develop and commercialize electrical phase-change memory. They called the new company Ovonyx, short for Ovonic Unified Memory (OUM), a name chosen because of the memory's versatility: "It can be used optically, electronically, thermally, whatever," Steve Hudgens explained. The new company was created as a joint venture between Lowrey and ECD.[32] Lowrey hired all the people at ECD who had been working in phase-change memory and were willing to join him, including Steve Hudgens, Guy Wicker, Boil Pashmakov, Sergey Kostylev, Patrick Klersy, and Wally Czubatyj, "Everybody was very happy," said Pashmakov. "Tyler was the biggest name in flash memory worldwide," and he knew what ECD didn't know yet, "how you actually make a memory chip."[33]

Ovonyx initially aimed to survive on the fees for licensing the technology to as many semiconductor manufacturers as possible. In these arrangements, it would be Ovonyx's job to do R&D, while the licensees were to commercialize the technology. In November 1999, Ovonyx formed a funding agreement with Lockheed Martin Space Electronics and Communications (now British Aerospace, BAE), and in February 2000 with Intel.[34] Part of the team began work in a Silicon Valley Intel facility in Santa Clara.[35] "Intel gave us a laboratory and all the money and resources we needed to develop this," recalled Wicker. And once Intel got involved in Ovonyx, other electronics firms wanted to follow suit. Among those who took licenses was the major Italian flash memory supplier, ST Microelectronics. When the dot-com bubble burst in March 2000, however, many companies were "scrapping to survive," Lowrey recalled, and could not afford to invest in new projects. Ovonyx spent this time improving the technology, working with its partners.

In 2001, ECD closed its Troy headquarters and moved into much larger facilities in nearby Rochester Hills, but as a result the Michigan-based Ovonyx team no longer had a clean room. Ovshinsky took responsibility for providing a state-of-the-art fabrication facility, which he had Klersy design with a $7 million budget. Instead of charging Ovonyx for the use of the facility, Ovshinsky used it to renegotiate his agreement with Lowrey, regaining some of the intellectual property that had been transferred to Ovonyx involving neural network applications of the threshold and memory switches, property that became the basis of the cognitive computer (discussed next). Upon completion of the new clean room, most of the Santa Clara Ovonyx staff moved back to Rochester Hills, leaving Hudgens to manage the California program.[36] The new clean room became Ovonyx's main asset, enabling the company to build its patent portfolio.

When the market started to recover after 2002, Ovonyx signed up other licensees, many joining as a protective move. In 2005 Ovonyx added the Japanese memory company ELPIDA (created by Hitachi and NEC). About a year later, Samsung, "the biggest memory supplier in the world," also signed up, and by 2007 Ovonyx had between ten and fifteen different licensees, "sort of all but Toshiba," Hudgens said. The hope was not only to replace flash memory, then a $10–$12 billion yearly market, but also to displace DRAM, whose market was estimated at $25 billion a year.

Excited by the development and prospective commercialization of his phase-change technology, Ovshinsky interacted with both Lowrey and Hudgens as much as possible.

He was impressed that Lowrey "follows me, where other people I lose." In turn, Lowrey continued to bring questions to Ovshinsky, finding him "still sharp as a tack." But as Joi Ito noted, only "the really smart people" were able to understand Ovshinsky and work with him.[37]

A continuing problem was that the corporate giants were pouring billions of dollars into research to improve flash memory, which was also getting cheaper. It was "like trying to jump on a moving train," Hudgens remarked. Ovonyx researchers knew that phase-change is a superior nonvolatile memory, and they believed that it would eventually replace silicon-based flash memory. Cell phone manufacturers especially liked phase-change memory because battery-operated cell phones need nonvolatile memory with low power requirements.[38] But the fact that phase-change memory was better than flash in many ways made no difference because, as Hudgens observed, "the marketplace wants cheap and good enough," and for more than two decades flash had been cheap and good enough. Only the few who were committed to phase-change memory opted to wait in frustration, and more or less in limbo, for the time when MOSFETs (metal-oxide semiconductor field-effect transistors) had scaled down to their limit, a time they thought might not come for three or four decades.

Envisioning a Cognitive Computer

While Ovonyx was struggling to commercialize electrical phase-change memory, Ovshinsky worked with a small team to develop an even more ambitious information technology based on chalcogenide switching. Exploring new possibilities, it also drew on the whole history of his information work: his efforts to probe the nature of human and machine intelligence, his nerve cell studies, his Ovitron, threshold, and memory switches. All culminated in his attempt between 2003 and 2007 to build what he called his cognitive computer.

Ovshinsky had thought for years about expanding his nerve cell studies to model a synapse, and perhaps eventually a human brain. The possibility of actually doing that emerged when Boil Pashmakov discovered that the energy of the set pulse triggering the phase-change memory device could be composed of a number of lower energy pulses. That enabled cumulative memory storage. Just as the neurons in the brain collect pulses from many inputs and fire when their sum reaches a threshold, the cognitive devices in the computer would accumulate information from different inputs, and when these added up to the threshold for switching from amorphous to crystalline, the device would "fire." Ovshinsky saw in the cognitive computer yet another physical confirmation of his belief that energy and information are two sides of the same coin.

As he put it, "You're adding energy, adding energy, adding energy, adding energy, and you're making little crystallized regions. And when these regions connect to form a percolation path, then it fires just like a nerve cell."[39] Combinations of these devices would offer a capability for parallel processing like the processing of the human brain, which similarly sums up the information it receives.

The key to modeling this threshold feature was to make the Ovonic memory such that it didn't have to be in either an "on" (conducting) or "off" (nonconducting) state, corresponding to the "one" or "zero" states in an ordinary binary computer.[40] Instead, the memory could have many intermediate states. For example, if the pulse packages each consisted of 10% of the energy needed to switch from the amorphous to crystalline state, one would need ten pulses to reach the threshold, or if the packages were twice as big, one would need only five pulses to reach the percolation limit. This opened up the exciting possibility of performing arithmetic calculations within a single nanostructure. It also allowed encryption, for if you had a device requiring ten successive pulses of a particular width and height to produce a percolation path and you wanted to store a three, you would put three of these pulses on the device, and to read the three would take seven pulses. But if you found that it took six pulses then you would know that a four had been stored there. Only a person who knew how the encryption worked in a particular device could read the encrypted information, because having the wrong pulse width and height would destroy the information already there. The cumulative memory of the cognitive device also enabled storing different intermediate resistance states by applying different intermediate size pulses. That offered the possibility of a multi-bit memory in a single device, which would not only greatly increase memory storage density but also allow the development of new computer architectures: instead of binary logic, they could use decimal, hexadecimal, or other bases.

Another device that Ovshinsky and Pashmakov developed to provide enhanced capabilities for a cognitive computer was the three-terminal threshold switch, sometimes called the Ovonic Quantum Control Device. In this device a signal applied to the third terminal, analogous in function to the third terminal of a transistor, could change and control the threshold voltages at which the cognitive device would fire. Invented during the summer of 2006, the three-terminal device was designed to work in the same circuit with the cognitive device as part of a chain in which the output of one device affects the input of the next. Pashmakov led the development but, Dave Strand emphasized, "the inspiration and the direction on how to make it came from Stan." As Pashmakov explained, in a normal computer "the processor has to interact with the memory all the time to retrieve data or send data to memory, and that's what slows down computers." Having everything in the same location eliminates such delay, and the greatly increased speed allows for many more sophisticated applications, like pattern or voice recognition, that require very fast processing of large amounts of data. "We've found that the device actually has the functionality of an artificial neuron," Pashmakov said.

As Ovshinsky envisioned it, the cognitive computer would first perform relatively simple cognitive functions, such as pattern recognition, before progressing to more-sophisticated functions like inference. He looked forward to integrating groups of circuits to create an analog of human intelligence. He would emphasize that all his information inventions were only models, not actual synapses or brains, but he believed that "evolution is going to have to happen," and that his cognitive computer could evolve into a device "that could make decisions and could learn" based on its interactions with external devices. The most advanced version of the cognitive computer that Ovshinsky achieved had sixteen linked devices, which he liked to call "synapses," but he remained certain that in time he would be able to build a machine with a thousand or more synapses, approaching the cognitive power of the human brain. "Will it have built-in consciousness? No, but it will have a great deal of intelligence," Ovshinsky said.

Once again, Ovshinsky's ability to see the potential of something that no one else would consider remarkable was the key to developing the notion of the cognitive computer. Pashmakov recalled Ovshinsky's excitement when he showed him that the device took two or more pulses to crystallize, recognizing that this "was how a neuron works." Ovshinsky remarked, "That is what I always thought it was going to do." He added, "We have to make a patent."[41]

Progress was slow, however, because the small ECD cognitive computer team working between 2003 and 2007 was badly underfunded and understaffed, consisting mainly of Ovshinsky, Strand (who acted as the group's manager), and Pashmakov (who carried out much of the experimental work).[42] The biggest problem, however, was that designing such a computer on a large scale required developing a new architecture different from the von Neumann paradigm for binary computers. While Ovshinsky and his team took some steps toward this end in the early 2000s, they did not solve the problem. Strand later summarized the work on the cognitive computer as providing "really good bricks to build a house. But if you didn't have a plan to build a house, you couldn't get there with just a pile of bricks."

Despite its innovative concepts and the promising steps taken toward realizing the cognitive computer, the research did not progress beyond its early stages. Perhaps because Ovshinsky was now eighty-three, he had trouble interesting others in the invention. When he called a press conference on June 6, 2006, to announce the new Ovonic Quantum Control Device, "one guy showed up." Less than a year later, ECD's new board would abruptly terminate the cognitive computer program (see chapter 11). Ovshinsky had hoped to continue the project when he set up Ovshinsky Innovation in 2007 (see chapter 12), but he never had the time or funding to do so.

Interlude: Science, Art, and Creativity

In chapters 8, 9, and 10, Ovshinsky has figured as just one among many actors in our accounts of the most important technologies he developed from his amorphous and disordered materials. The following three chapters will bring him back to center stage again as we trace his story to the end. But before resuming that story we offer a brief supplementary interlude that expands on some of the themes from the introduction, showing how Ovshinsky's creative imagination extended beyond his work as an inventor.

In his own reflections on his work, Ovshinsky emphasized the way he refused to recognize conventional separations between areas and kinds of thought. "In all the fields I've ever worked in, I have felt that man made disciplines, that nature did not, and therefore I was free if I could make a contribution to work in whatever area I found exciting. For me, it was all science, all art, and all creativity."[1] This attitude, affirming the unity of all spheres of human thought, imagination, and activity, informed not only his work as an inventor but also his recreations, which ranged from making thousands of drawings to developing theories of cosmology. Indeed, if we take him at his word, we should also question the conventional separation between work and recreation. His restless energy, imagination, and curiosity fueled everything he did.

Drawings and Paintings

We begin with his drawings, the activity apparently farthest from his scientific work. He turned them out constantly, using whatever paper was at hand—hotel notepads, company letterhead, graph paper, and the like—usually while participating in a meeting or talking on the phone (or during the many hours of interviews for this book). Yet, as can be seen from the few reproduced here, they were not simple doodles but deft, shrewd caricatures, sometimes made with a witty and economical use of lines, sometimes with heavy saturation. They turn up already in his high school notebooks, including a page covered with thumbnail sketches of fellow workers at Akron Standard Mold, and they continue through every stage of his long career. Ovshinsky sometimes struggled to convey his thoughts in speech, but his drawings communicate effortlessly.

Stan also produced a few oil paintings, whose vigorous brush strokes and bold colors give a sense of passion and urgency. In particular, his double portrait of the doomed anarchists Sacco and Vanzetti conveys the intensity of his sympathy with these emblematic figures.

Figure 10a.1
Shop workers, c. 1940 (found with Ovshinsky's high school notebooks).

Science, Art, Creativity

Figure 10a.2
Drawing collection. Photos by Genie Mytilineou.

Figure 10a.3
Grumpy "suits."

Figure 10a.4
Isaac Stern (sketched on the back of a concert program, November 10, 1968—the night before the *New York Times* story about Ovshinsky appeared; see chapter 6).

Figure 10a.5
Drawing on notepaper

Science, Art, Creativity

Figure 10a.6
Drawing on memo pad.

Figure 10a.7
"Iris: do you want these?"

Figure 10a.9
Indignation

Figure 10a.8
Top hat under scrutiny.

Figure 10a.10
Sacco and Vanzetti.

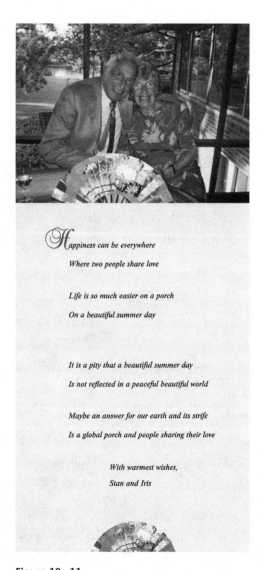

Figure 10a.11
Image and text from holiday card, "A summer evening on our porch / Photo by our good friend, Dr. Takeo Ohta / Poem by Stan."

Toys and Models

The drawings and paintings show the artistic ability his early teachers recognized, but their continuous playful creation is also part of a mentality that contributed to his inventions. Discussing the topic "Creativity and Innovation" in one of the fragments of his autobiography, Ovshinsky follows a brief account of inventing his switches with this comment: "Some of the work was playful even though it had fine results. I always had on my desk several simple toys—permanent magnets, photosensors, particles in a small plastic container which could be rubbed to electrostatically separate and recombine, etc. To this day, I have magnets and molecular models on my desk. All of this is to indicate that thinking by visualization and analogy and playful models helps me as I do the deeper thinking of physics and chemistry and the incredible amount of reading that goes along with it."[2]

Playing with such toys not only helped his work but was also part of the larger realm of activities pursued for their own sake that he called "civilization."

Writing

In addition to constantly drawing, Ovshinsky also wrote poetry and fiction. His files include several examples, including an accomplished comic short story, "You Should Have Heard Her Scream!" Drawing on, and probably dating from his early days of working as a machinist, it unfolds as an earthy first-person vernacular narrative of masculine sexual boasting and anxiety. It focuses on the boisterous welder Jonesy, whose tales have gained him a reputation as an authority on women. "Now this is no mean feat in a shop where everybody bullshits everybody else about last night's happenings, what's going to happen tonight, and with dim voice and moist eyes, tells the beautiful, unbelievable stories of what happened once, purely by accident, when a big, beautiful babe came up, etc. To be considered an oracle of ass in this atmosphere is quite a compliment." From this preamble about storytelling, the story builds to its "tragic" denouement, in which a shattered Jonesy confesses his secret tormenting sense of inadequacy ("I'm hung real small") and his disastrous attempt to substitute a dildo. Told with comic but sympathetic detachment, this modern fabliau shows a side of Ovshinsky unsuspected by those who knew him only as a visionary entrepreneur in three-piece suits.

Fiction writing may have stopped after Ovshinsky's early years, but he continued writing poetry, as is fondly remembered by the many who received his and Iris's holiday greetings each year. Each card had pictures, usually nature photos taken by Iris from their house, and a poem by Stan. The poems typically take off from the beautiful images, contrast the absence of peace, beauty, and social justice in the larger world, and end with hope for a better future.[3]

The goal of "a better and more beautiful world" was, of course, the aim of the Workmen's Circle, whose values were so influential in Ovshinsky's development. Inseparable from his and Iris's commitment to social justice was the belief that beauty and the enrichments of culture should be available to all. Along with generous economic benefits, they often gave theater and concert tickets to ECD staff to help them enjoy the pleasures of civilization. Ovshinsky himself greatly enjoyed music, with particular enthusiasm for rousing anthems like the old union songs. A sheaf of sheet music for several, from "Solidarity Forever" to "The Internationale," was filed along with his own creative writing; his particular favorite, "Hold the Fort," was a watchword with him, as his son Steven recalled. "We used to sing the song, the union song. I must have heard it a million times."

Hold the fort, for we are coming. Union men, be strong!

Side by side we battle onward; Victory will come.

Science and Art: An Analogy

As Steven developed into a successful musician, music became not only a pleasure shared between him and his father but also an analogy. Stan would compare the way he thought about science with his appreciation of Steven's music: both science and music dealt with deep, universal values, and both offered an experience of beauty. Ovshinsky also drew an analogy between musical and scientific gifts in explaining his intuitions. "You have to have what I call perfect pitch. You have to know yourself whether or not an idea is really right."

Like all art, he told Steven, music is "part of what makes life worth living," and it should be democratically available to all. "Music is not just for the people who know or just for the rich, but it is for everybody." That was part of what it meant to work for not just a better but also a more beautiful world, as Ovshinsky had learned from his own father; Steven especially loved to hear Stan tell how on Saturday afternoons Ben would come home in his work clothes "and just plop himself down in front of the radio and listen to the opera."

Science, like art, was something Ovshinsky valued for its own sake, not just as a means to practical ends, and it gave him strength to endure in the face of financial problems, lawsuits, and the rejection of his work. "I was always able to experience the happiness of creativity," he insisted. "I had Iris, and I had the joy of science, that could overcome any rejection and any bullshit that I had to go through. To me science was a manifestation of civilization, just like art, literature, poetry, music."

Ovshinsky found support for this view in the writings of outstanding scientists. One was the Japanese theoretical physicist and Nobel laureate Hideki Yukawa, whose book, *Creativity and Intuition: A Physicist Looks at East and West,* he reviewed in 1991.[4] Yukawa says he took his inspiration from early Chinese philosophers, but Ovshinsky found more general principles that echoed the terms he used to explain his own creativity: "The precepts implicit in Yukawa's thinking show intuition to be dependent on the ability to draw analogies." Besides considering Yukawa as an example of independent, adventurous innovation like his own, Ovshinsky also honors him as both a "humanist" and a scientist who "transcends … specialties and national barriers. … The universal intuitive process, whether expressed in art, literature, music or science, is the ability to 'see' connections between facts or concepts which to others are unrelated."

Another prestigious model was Einstein, from whom several quotations and references appear in Ovshinsky's files.[5] In addition to sounding the themes of visualization, intuition, and imagination, Einstein also provided an eloquent affirmation of the connection between science and art.

Where the world ceases to be the stage for personal hopes and desires, where we, as free human beings, behold it in wonder, to question and to contemplate, there we enter the realm of art and of science. If we trace out what we behold and experience through the language of logic, we are doing science; if we show it in forms whose interrelationships are not accessible to our conscious thought but are intuitively recognized as meaningful, we are doing art. Common to both is the devotion to something beyond the personal, removed from the arbitrary.[6]

Cosmology

Ovshinsky took real pleasure in such wonder, questioning, and contemplation, and he made time for it amid his more practical concerns. Scientific visitors to ECD recall how, after one of Iris's fine dinners, he would invite them down to his study. "Let us relax and discuss science," Hellmut Fritzsche recalled him saying. And of all the topics he pursued with them, none better showed the spirit of pure intellectual inquiry than cosmology. Pondering cosmological questions was a release: "It was his way of relaxing from work-related stress," Rosa Young recalled. "He would pick up a cosmology book to read, think, and talk about when he needed to switch off his mind from his work."

Ovshinsky had been fascinated with cosmology since childhood. Probing the mysteries of the origin and ultimate fate of the universe, it appealed to his imagination as it does to many readers. But Ovshinsky not only pursued the subject with extensive reading in astrophysics, he also made sustained efforts to contribute to it by developing his own cosmological theory. He worked on this in close collaboration with Fritzsche, going together through Einstein's general relativity theory and reading other papers on the expansion of the universe. The two friends also kept up a voluminous correspondence on cosmology conducted over many years.[7] The typical pattern of these exchanges had Ovshinsky proposing his own solution to some cosmological problem, which Fritzsche would restate or criticize by drawing on his training as a physicist, particularly his command of the mathematical formulations that Ovshinsky himself could not produce. They would argue and exchange drafts through many iterations and refinements of their emerging theory, greatly entertaining themselves but also hoping to make a recognized contribution.[8]

The main problem they addressed was that of "dark matter." (They themselves often put the phrase in quotes to indicate their skepticism about the prevailing theory.) Cosmologists postulate a large volume of invisible matter to account for gravitational effects that could not be caused by all the visible matter in galaxies. Ovshinsky and Fritzsche instead proposed that the observed effects result from wrinkles in space-time curvature caused by the uneven expansion of the universe. In later extensions of the theory, Ovshinsky claimed that it also accounts for the recently observed acceleration in the expansion of the universe, supposedly driven by an even more mysterious "dark energy." The details and merits of this model matter less than the enthusiasm and pleasure with which the two friends constructed it. "To work with you on cosmology," Ovshinsky wrote, "is my idea of fun." And in a later letter, written when he was in the hospital awaiting surgery (September 13, 1998), he used the time to respond to theoretical points Fritzsche had recently raised and closed by saying, "In any case, the fact that I am writing just before my operation gives you the idea that I am enjoying this very much—and I don't mean the operation."

In presenting his theories, Ovshinsky insisted on playing the game on his own terms. Rather than apologize for not being able to formulate his theories mathematically, he invoked Henri Poincaré—"It is by logic that we prove, but by intuition that we discover"—adding, "Sure, the mathematics is going to be absolutely necessary, but it is sterile as much of cosmological mathematical speculation seems to be without having physical meaning." Ovshinsky sought to find "physical meaning" in the cosmos just as he had sought it on the atomic scale, applying his educated intuition by visualizing and exploring analogies. Indeed, a recurring feature of his letters to Fritzsche is his

Figure 10a.12
Fritzsche and Ovshinsky working on cosmology.

use of analogies drawn from the condensed matter physics of his inventions. "I find it helpful," he wrote early in the correspondence, "to view the vacuum as I have the world of amorphous materials." Later he brought in such terms as "lone pairs," "band gap," and "mobility edge," all of which figure as well in explanations of the Ovshinsky effect. As for the exponential cosmic expansion just after the Big Bang, he commented, "I like inflation because it allows us to use phase change theory in disordered materials which you and I have been more than pioneers in." In all these analogies, Ovshinsky is highly aware that he is working in the same way he does as an inventor. Fritzsche warns him against "inventing new particles," but he explains that "what I have tried to do is to play my role as inventor where I take seemingly unrelated physical facts and put them together intuitively in new and unexpected ways to solve problems."

Ovshinsky felt that he and Fritzsche had indeed helped to solve some important problems of cosmology, and he was eager to publish their ideas. They did succeed in publishing a short piece in *The American Journal of Physics* (September 1997) as a comment on an earlier article, but when they later submitted a more ambitious account of their theory, "The Origin of Dark Matter in the Universe," it was rejected by *Physics Letters A*. The referee brusquely describes it as "unacceptable for publication in any journal," and adds that the authors "should study the Einstein field equations."[9] Undaunted, Fritzsche passed on the referee's report to Ovshinsky, adding, "It is clear that Einstein's field equations have to be amended to include the possibility we are talking about." Their long friendship began when Ovshinsky's threshold switch forced Fritzsche, and later other physicists, to revise their understanding of amorphous materials. Similarly, Fritzsche remarked that although "our idea contradicts the present theories, the present theories don't explain dark matter either." This nicely catches the independent spirit of all Ovshinsky's work, part of the creativity that he sensed running through both art and science. The time and energy he devoted to cosmology may seem misplaced, part of a quixotic effort to gain greater scientific recognition, but as he would explain to Iris, for him it was a way of affirming and advancing the values of civilization (see chapter 11).

III Later Years

11 Losing Iris, Losing ECD

In many ways, things were looking rosy for Stan and Iris in the summer of 2006. Popular appreciation of alternative energy had reached what appeared to be a tipping point. United Solar was selling so many solar panels that production could not keep up with demand, and orders were backlogged. Important politicians, even President George W. Bush, were publicly acknowledging man-made climate change and the need to address it with renewables. No longer considered a suspect outsider as in 1968, Ovshinsky had even been celebrated as a "hero for the planet."[1]

But there were also signs of impending loss. Control of ECD was shifting from Ovshinsky and his allies to an increasingly hostile board of directors. "It was a civil war," Ben Ovshinsky said, "between Dad and his loyalists and everyone else." Ovshinsky had no intention of giving up his company without a fight, and even if he lost this battle, he had already begun envisioning a new future for him and Iris in a new company.

The Last Vacation

In making such plans, Stan had apparently not noticed the gradual weakening in Iris that other family members saw over the last several years. Robin observed that "she was cooking less and getting very tired" and for the past five years or so, Iris had sometimes said she did not want to travel or work as hard, or wanted to start working at home. For years, Stan had promised to work less, but he never did. Even in visits to their beloved "resting place," their cottage near Montreal, he often spent hours taking meetings on the phone or responding to ECD faxes. "At first, Iris would cringe when the phone rang," Harvey recalled, "but I think eventually she just stopped hearing it." Stan had told Iris on her birthday, on July 13, 2006, that they would work less. "That's my birthday present to you. We'll work at home one day a week now." But, Robin noted, "It never happened."

Living nearby in Ann Arbor, Harvey and his wife Cathie had been more aware than the other family members of Iris's decline. Over the past year they had seen her becoming emotionally frail and physically exhausted, always attempting, but more often failing, to keep up with Stan. As he had since childhood, he still wanted and needed "more." But now, Iris couldn't keep up. She wanted and needed less. Besides struggling with the demands of work, she also suffered as a result of Stan's own frequent health crises and trips to the hospital. "There were times she thought he was going to die any minute," Harvey recalled, "and it worried her sick."[2]

Harvey and Cathie saw signs of Iris's mental decline when "she would play with her food or interrupt Dad in the middle of a story to show him a cut on her finger. And she forgot core recipes that we all knew and loved. Sometimes she would be fine; other days she would be somewhere else." As a nurse psychologist, Cathie wanted to intervene. With their children Noah and Natasha, Harvey and Cathie began to help Iris with tasks like cooking and cleaning up at family holidays, and Natasha began to take her shopping and out to lunch.

But when Harvey asked Stan, "Wouldn't this be a good time for you and Iris to slow down and just take care of each other and write your memoirs?" Stan retorted angrily, "That's not who I am. I'm not interested in the past. I need to move forward. We're going to start another company and pick up where we left off." "But what if Iris can't do that?" Harvey asked. "Iris and I are a team," Stan said. "She will do whatever I tell her we need to do."

In this troubled context early in August 2006, Stan and Iris enjoyed the first family vacation they had taken in some fifteen years. They went to Santa Fe with Steven, who had been invited to perform in a chamber music concert. Steven later recalled that "incredible trip." Stan and Iris not only attended the concert, but also went to the opera. "We took amazing walks," he said. "We had great meals. We went to art galleries. It was the kind of experience that one would plan with one's parents, if one knew that one of them was about to go away forever."

The relaxed setting also made it easier to discuss relieving Iris from the stress of work. When Steven suggested a plan for her to retire in the next year, when she reached eighty, Iris's excitement brought her close to tears. Still, she continued to deflect attention from her health. At one point she complained of a dizzy spell, but when Stan suggested they go to a hospital she had resisted. "Oh, no no. I'll be fine."

During the vacation in Santa Fe, Steven recalled Stan talking about his work in cosmology. Surprisingly, Iris objected. "I don't get it. I don't think this is as important as your other kinds of work." What struck Steven was that she "really stood up for herself in a way that felt sort of rare and new." Although Robin felt that Iris did not look well

Figure 11.1
Stan and Iris in Santa Fe, August 8, 2006. Photo by Steven Dibner.

in the photos that Steven took in New Mexico, her energetic challenge to Stan on his cosmology work made Steven feel "she was doing very well."

On the flight back to Detroit, Stan took the opportunity to explain to her more completely than he had before why cosmology was important to him. He began with a broad historical account that set the advance of scientific reason against the forces of ignorance and superstition that had long engendered conflicts and persecution. He appealed to Iris's French background, invoking the heritage of the Enlightenment, and spoke of how the progress of science had reduced religious fanaticism, how "science has always brought along civilization." Stan remembered being "on fire trying to explain all this," as he correlated "the rise of science with everything rational, including the abolition movement," because he "could see that she was paying attention intently." This sense of science as the agent of civilization, Stan explained, was the reason he cared so much about cosmology. Grappling with unsolved problems like dark matter and dark energy was a way of enlarging human possibility, just as Steven's music was also a way of "creating a greater civilization." At that point, Stan recalled, "She turned to me and said, 'Why didn't you talk to me like this before? I would have understood immediately.' And then we just kissed." As their plane descended, he told her, "You've made me very happy tonight. We've had wonderful times, but tonight you've made me especially happy."

Death by Water

This section is written in the first person by Lillian Hoddeson because she took part in the events it recounts.

On Wednesday, August 16, 2006, I traveled to Detroit to conduct oral history interviews with Stan and Iris Ovshinsky. As noted in the preface, I had been invited to write a scientific biography of Stan, but I was not yet committed to the project. Impressed with both Stan and Iris when I met them about a year earlier, I had planned to do a series of interviews and perhaps work with a younger colleague who would write a book using that material. The events of that day made me decide to write the book myself.

When I arrived at ECD shortly after the lunch hour, Stan and Iris were still at a local restaurant continuing the discussion about cosmology and civilization they had begun on the trip back from Santa Fe. Both were radiant when they showed up and greeted me at ECD. I complimented Iris on her beautiful purple dress. "It's my favorite color," she replied. "It's my daughter's favorite color. My granddaughter's favorite color. My other granddaughter's favorite color." While Stan and I proceeded with our interview, Iris popped in frequently from her adjoining office to check on Stan and address his spur-of-the-moment needs. It was clear that she wanted to stay in constant touch with him, and it was clear that he wanted that too. During one of her stops in, Stan took a call from Dave Strand to talk about progress in the work on the cognitive computer, and I spoke briefly with Iris. She asked whether I'd like to join them for dinner at a new Thai restaurant nearby, and also to swim with them before dinner in the lake behind their house, explaining that they were trying to swim in both the morning and evening, to stay fit. "Sounds wonderful," I responded.

The swim was immediately stressful however, because the water was icy.[3] I had to swim as quickly as I could to maintain my body temperature. Swimming even faster, Stan headed toward the center of the small lake, and I tried to follow. But within minutes I reached my aerobic limit and had to turn back. I hoped that I would have enough energy to make it to shore.

Neither Stan nor I realized at first that Iris had been lagging. As I swam back, it suddenly occurred to me that I couldn't see her. Then I noticed what to my blurred vision first looked like a small rock poking up out of the water. I hadn't remembered any rock. As I approached, I realized that it was actually Iris's shoulder and that her head had dropped

Figure 11.2
Lake behind the Ovshinsky home.

below the surface. When I reached her, I scooped up her slim body with one arm and holding her tightly swam sidestroke to shore. Once in shallow water, a flood of adrenaline allowed me to carry her limp body to the water's edge. I remember thinking how light and delicate she was as I laid her in the sand and proceeded with mouth-to-mouth CPR. She remained blue-lipped and motionless as I breathed and pumped for what seemed like a very long time. Optimistically, I expected her to come back to life at any moment.

When Stan emerged from the lake, he rushed over screaming, "Iris, Iris. I want to talk to you." She remained silent. "Go to the house and call 911!" he ordered, and announced that he'd take over the mouth-to-mouth. When I reached the unfamiliar house, I could not at first find a phone. After I did, I had trouble answering the very first question. I soon found the address on a piece of mail. But then there were other questions that only Stan could answer, and I ran back to the lake carrying the phone. Soon a paramedical crew arrived and after working on Iris for a long time took her to St. Joseph Mercy Hospital in Pontiac. It was a hospital that neither Stan nor Iris would ordinarily have gone to, but the emergency workers were legally required to take her to the nearest one.

I asked the friendly cop on the scene whether he could drive Stan and me to the hospital. He agreed. My next problem was to convince Stan to ignore his sandiness and dress quickly without taking a shower. He put on his three-piece suit and, before leaving for the hospital, grabbed his address book, so that as we drove in the police car I could make calls for him on my cell phone. He wanted me to let family members and Bob Stempel know "the news was very bad," and to ask them to please come to the hospital right away. Mostly I just left messages.

At the hospital, Stan and I were led into a small dimly lit room, where we sat and waited. After a long time, one of the physicians appeared and confirmed what Stan already knew, that Iris had died. Stan later told me that he had known immediately. "Her eyes were not looking at me," he said, "they were looking at eternity." We were then led to the bed where her slight body lay. Stan sobbed gently and spoke to her in a low, broken voice. After a while members of the family arrived: Herb and Selma, Harvey and Cathie, Noah and Natasha. We all sat around the bed. Suddenly, Stan realized that it was time for his nighttime medicine. Until that evening, Iris had administered all Stan's medicines, and he was not completely sure what he needed.[4] With pad in hand, I pushed him to try to recall what medicines he thought he needed, and where I might find them. "Look behind the turtle," he said in his raspy voice. What turtle, I thought. Herb offered to drive me to the house and help me look for the medicines. Selma and Natasha came along, too. Miraculously, Natasha found both the asthma inhaler, which looked like a turtle, and the medicines.

Meanwhile, working from home, Stan's administrative assistant Freya Saito had rebooked my itinerary so that I could fly back to Illinois the next morning. Herb told me that I didn't need to return to the hospital; the family could handle things now. I don't remember how we managed to retrieve my suitcase from the trunk of Stan's car, now locked in his garage. But we did, and late in the evening I was dropped off at my motel.

Several days later, Stan phoned me to talk about what we had both seen and felt the evening Iris died. He was in deep mourning. He asked me whether Iris had had any last words. I could not tell him. Then he asked, "What should I do?" I suggested that he continue the work that he and Iris had started. He told me that was just what he planned to do. As Robin explained, "He decided that it was his duty to carry the flag after Iris's death and continue their mission to improve the world using science and technology." In a later call he thanked me warmly for helping him on that fatal night. "I was out of my mind," he said. He also told me that he had learned that the medical team had worked on Iris for an unusually long time and "didn't find a lot of water in her lungs," suggesting that she had died of a heart attack. He and the children had opted not to have an autopsy. "I didn't want to see her body mutilated," Stan said.

Afterward

Iris's funeral was on Sunday, August 20. As she had requested some months earlier when she and Stan were in Akron discussing hydrogen plans (see chapter 9), she was buried in a peaceful and beautiful spot on the hillside at the Akron Workmen's Circle cemetery where Ben and Bertha rested. "We knew saying goodbye to Iris was going to be emotional," Harvey said. But nothing prepared the family for the drive from the cemetery entrance to the gravesite. Dozens of current and former ECD colleagues, some who flew in from all over the country, lined up along the dirt road to silently pay their respects and say their last good-byes.

At the funeral, Stan explained why Iris was being buried there, rather than at Royal Oak, near Detroit, where the two had purchased plots. He added he had always felt a little uncomfortable about being buried in Royal Oak, since that had been the territory of the radio priest, Father Coughlin, who "was always a true, true fascist." But, he explained, Iris had insisted that she wanted to be able to go there every day. "She thought for sure I would die sooner."[5]

Family and friends grieved the loss of Iris in their own ways. Stan received over two thousand cards and letters from people all over the world who had been touched by Iris's generosity, warmth, and intelligence, and many honored her memory at the moving afternoon memorial service on September 25. Stan's 2006 holiday card was another poignant tribute.

The cause of Iris's death haunted Stan for the rest of his life. He was overcome with guilt for not making sure that she had been properly checked by physicians. After her dizzy spell in Santa Fe, Stan insisted on making an appointment for her to take a stress test when they returned home. She was to have had it three days after the day she died.[6] There had been other warning signs that Stan didn't know about. Iris had told one of the secretaries (who later told Harvey) that she was afraid to go swimming; she was having chest pains, but kept that to herself.[7] Stan deeply regretted that she never told him.

In the months after Iris's death, Stan mourned a great deal. She had been his best friend for more than half a century, and now he felt alone and lost. He remarked, "We were so intensely in love that not one day, not one hour [passed] that she didn't know how I felt about her or that she didn't let me know." He remained emotionally broken for months, but he could still work, which he did day and night because sleeping was so difficult and work distracted him. Both Hellmut Fritzsche and Morrel Cohen stayed with Stan in his house for periods of time. And when they did, they worked with him on the cosmology project, offering the severe criticism he requested. According to Stan,

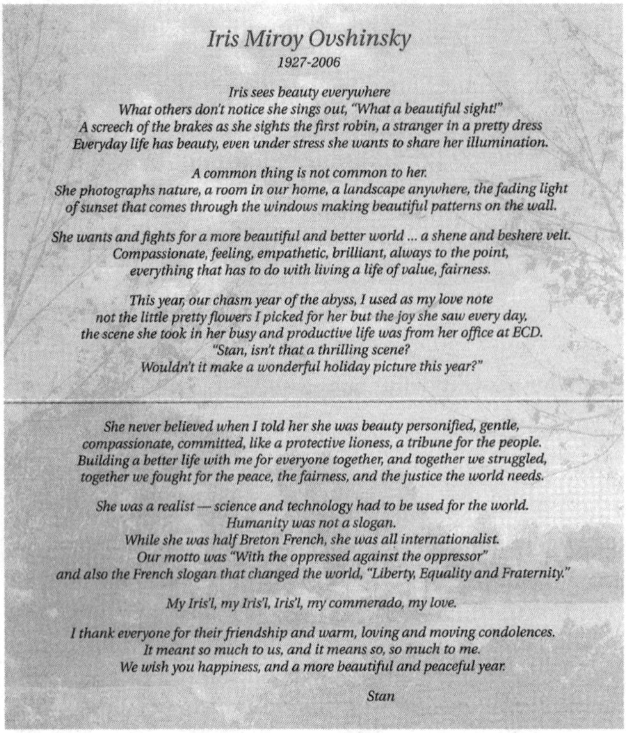

> **Iris Miroy Ovshinsky**
> 1927-2006
>
> Iris sees beauty everywhere
> What others don't notice she sings out, "What a beautiful sight!"
> A screech of the brakes as she sights the first robin, a stranger in a pretty dress
> Everyday life has beauty, even under stress she wants to share her illumination.
>
> A common thing is not common to her.
> She photographs nature, a room in our home, a landscape anywhere, the fading light
> of sunset that comes through the windows making beautiful patterns on the wall.
>
> She wants and fights for a more beautiful and better world ... a shene and beshere velt.
> Compassionate, feeling, empathetic, brilliant, always to the point,
> everything that has to do with living a life of value, fairness.
>
> This year, our chasm year of the abyss, I used as my love note
> not the little pretty flowers I picked for her but the joy she saw every day,
> the scene she took in her busy and productive life was from her office at ECD.
> "Stan, isn't that a thrilling scene?
> Wouldn't it make a wonderful holiday picture this year?"
>
> ---
>
> She never believed when I told her she was beauty personified, gentle,
> compassionate, committed, like a protective lioness, a tribune for the people.
> Building a better life with me for everyone together, and together we struggled,
> together we fought for the peace, the fairness, and the justice the world needs.
>
> She was a realist — science and technology had to be used for the world.
> Humanity was not a slogan.
> While she was half Breton French, she was all internationalist.
> Our motto was "With the oppressed against the oppressor"
> and also the French slogan that changed the world, "Liberty, Equality and Fraternity."
>
> My Iris'l, my Iris'l, Iris'l, my commerado, my love.
>
> I thank everyone for their friendship and warm, loving and moving condolences.
> It meant so much to us, and it means so, so much to me.
> We wish you happiness, and a more beautiful and peaceful year.
>
> Stan

Figure 11.3
Stan's 2006 holiday card.

"They tried to tear it apart. And finally they gave up and said, well, you may have something here." Members of the broader physics community who worked in the field disagreed, however (see the interlude, which precedes this chapter).

Dismissal

Iris's death came as a sudden and devastating blow for Ovshinsky. His forced "retirement" from ECD a year later was in contrast the final stage of a long and painful process of losing power in the company he and Iris had created. It had begun in the early 1990s with the Canon joint venture, when his dominant Japanese partner had excluded him from the management of United Solar (see chapter 8). The tension between manufacturing and research and development that began at that time reached a breaking point in 2007.

Some of the factors contributing to Ovshinsky's loss of power arose from his own strengths and flaws. His passionate commitment to his goals could make him demanding, impatient, and unyielding, and as he grew older some found him harder to deal with. He refused to accept the limitations of age, and, aware that his time was growing short, he was more and more intent on realizing his goal of replacing fossil fuels. Yet as he advanced into his ninth decade, he was less and less able to persuade others to provide the support he needed.

The main cause of Ovshinsky's departure, however, was the Sarbanes-Oxley Act of 2002, enacted in the aftermath of the Enron scandal in order to combat corporate accounting fraud. Among the law's provisions was the requirement that a majority of a corporation's board of directors be independent, with no other connection to the company. Bringing ECD into compliance entailed replacing several insiders, including Hellmut Fritzsche, Nancy Bacon, and the COO Jim Metzger, who stepped down in the fall of 2003, with outside directors who had no understanding of the company's technology and no sympathy with Ovshinsky's social goals. Narrowly focused on maximizing profits, the new directors were opposed to supporting work in areas that were not already profitable, and once they were in the majority they moved quickly to eliminate them. Ovshinsky argued against their shortsightedness, explaining that new technologies take time before they start to make money and that even currently successful ones require continuing research to stay ahead of the competition. But he failed to persuade the new board, and his control over the company had already been reduced when he lost his loaded vote in September 2005.[8]

From the perspective of some shareholders who had for decades seen no return on their investments in ECD, the changes may have seemed not only warranted but also long overdue. Yet the new board's decisions were not in anyone's long-term interests. Of all the divisions and research programs, only United Solar and the Ovonic Battery Company brought in significant, recurring earnings; all other activities were sacrificed. Beginning in April 2007, a series of reorganizations decimated the company. Projects like the cognitive computer and the hydrogen car were abruptly ended, and the lengthy list of layoffs ranged from scientific researchers to the machinists in the model shop. Even programs that were not completely eliminated were drastically reduced. Ben Chao, who was then responsible for hydrogen storage, had his staff cut from more than thirty to seventeen, then nine, and then six. "You can still call it a hydrogen group," he said, "but it won't move technology anywhere. It just doesn't have enough resources."[9]

Some of those who were laid off were treated with brutal insensitivity. Boil Pashmakov was summoned to an office, where two people he'd never seen before told him he was fired. As Hellmut Fritzsche, who had recruited Pashmakov, reported, "Two security guards took him out. He was not allowed in any more. It took seven minutes." For those who remained, the effect on morale was devastating. Instead of feeling like members of a supportive community, employees felt isolated and insecure. "I'm very lonely there," Chao said; Dave Strand observed that the board, disregarding individual contributions and expertise, considered everyone replaceable: "People are treated just like widgets."

For Ovshinsky, it was especially painful to see the machine shop closed. He considered it "the basis of the company, because everything I wanted to build we built there." The staff was "unionized by the International Association of Machinists, who made me an honorary member. So when they fired everybody, escorted them out no matter if they worked there thirty years, and didn't give them any time to get their stuff together," he saw the ECD culture being destroyed in front of his eyes. He was moved to tears when the men tried to comfort him. "I couldn't stand it, and I just left." He later wrote the machinists a letter (May 18, 2007) expressing his grief and reminding them he had "come from your ranks many years ago, but I never left them." He closed the letter by paraphrasing Eugene Debs: "I never wanted to rise from you but with you. Please excuse me for not being able to do more."[10]

Ovshinsky could do nothing because in March 2007, before the layoffs began, the new board had asked him to resign as president.[11] He was given the honorific title of Chief Scientist, and in place of his ECD office, at the insistence of Bob Stempel, who was still chairman, a new one, almost a complete replica, was built for him in the Institute for Amorphous Studies, where his and Iris's papers, books, pictures, and artifacts were moved. In addition to the new office, ECD provided staff support, but he was completely excluded from the company, "in exile," as he said. In July, the board asked him to take his retirement on August 31.[12] The negotiated separation agreement specified that ECD would transfer the institute property to him, would provide a company secretary for life, and would continue his health insurance.[13] Forty-seven years after he and Iris had founded the company, Ovshinsky's involvement with ECD was over.

Ovshinsky had lost his company, but he was not defeated. At age eighty-four, he planned to start over. "I have the inventions that could impact the climate change catastrophe," he would assert. By this time, having studied all the alternatives to fossil fuel, "including wind, tidal, hydro, bio, geo-thermal, and nuclear," he was convinced that the only alternative that could save the planet was solar energy. He added, "If we build enough nuclear plants (and what a great world it would be in terms of terrorism!), what would you do with the waste?" On the other hand, he said, "In one hour we get more energy from sunshine on the globe than you would ever need in one year. We know how to capture it, and we know how to store it, so that is what we should be doing." And that is what he devoted himself to for the rest of his life (see chapter 12). "I want to continue. I want to continue with the same objective as I had when I formed the company with Iris, to use science and technology to change the world and answer its problems. And I know I can do it."

12 New Love, New Company

In the months after Iris's death, Stan deteriorated physically, losing so much weight and appearing so frail that friends feared he would also die soon. He succumbed to a series of illnesses and worried about everything. To John Ross, the physical chemist who had joined ECD as a consultant in 1976, Stan appeared "a broken man." His friends, family, and colleagues tried to comfort him, but only work could distract him. "I couldn't sleep. And yet I could work," he recalled. "That's how I knew I was alive, that I was still me."

But work was not enough. His enormous personal needs, which Iris had once satisfied unconditionally, remained unfulfilled and unabated. He told himself, "I really should try to have a relationship." At first friends tried to fix him up, but he did not want to date strangers. He told his assistant Georgina Fontana that he needed to find someone he already knew and was comfortable with. "I can't do blind dates," he confided to Harvey. "I need to see what I'm getting into with my eyes wide open."

Rosa and Stan

Rosa Young, Stan's colleague for over twenty years, was shocked to hear of Iris's death when she arrived at work around 9 a.m. on August 17, 2006. Like many others, she could not at first believe it. Over the past two decades, Rosa and Iris had become close friends, and just two days earlier, Stan and Iris had visited the hydrogen lab.

Without thinking, Rosa drove over to Stan's house, where Harvey greeted her at the door and confirmed the news. Rosa asked whether she could see Stan. "Normally I would not want too many people coming in today," Harvey said, "but since you have been very close to Stan and Iris, I'll let you come in." Rosa waited in the kitchen for Stan, who was downstairs in the shower. When he appeared, "Stan just hugged me and cried and cried and cried. There's nothing we can say," Rosa recalled. Then he walked to his bedroom to dress. Rosa remembered saying to Harvey, "Poor Stan, how is he going to live without Iris?" Then she asked what would help at this point. Harvey said that during the next few weeks, family members would give Stan whatever support he needed. Steven was planning to live there for a couple of weeks, and Harvey also planned to take time off from his work to help Stan begin to manage his life. After that, help from friends would be much appreciated, and he asked her to spread the word. He suggested people could invite Stan to their homes, or bring food to eat with him in the house.

Rosa passed on Harvey's suggestions, but she continued to worry about Stan's survival. She remembered that her Greek colleague and friend Genie once remarked that when a couple is very close and one of them dies, the other often can't survive. Over the next few months, when Rosa and her team went out for dinner, she would often say, "Why don't we invite Stan?" Some wondered whether he would be comfortable going out with them, especially on short notice, but she'd say, "Let's try." And Stan almost always said, "Sure." He was moved by Rosa's outreach. Soon he decided that he wanted to see much more of her.

About a week after Thanksgiving, Rosa was surprised by Stan's invitation to join him for dinner at his home. He explained that Harvey and Robin had hired someone from Grand Rapids to cook for him. Rosa recalled, "This guy put on chef's clothes and fixed a very fancy dinner for us." Also serving as a personal assistant and valet, he was supposed to live in the house and be available full-time. But Stan felt uncomfortable sharing his home with a stranger and moved him to a hotel. After about a month the "chef" quit.

Some days afterward, Stan asked Rosa whether he could take her out for her birthday on December 15. It had become a tradition for Stan and Iris to take Rosa to a good restaurant on her birthday. At that point, Rosa sensed "a type of personal affection," but she was not yet aware of how deeply he felt about her. She got a stronger sense of that when she tried to fix him up with one of her friends. Stan told her irritably that he didn't want to be fixed up with anybody.

Closer to Christmas, Stan asked Rosa what she was doing for the holidays. When she told him that she had booked a two-week trip to Egypt, Stan said, "This is not a good time for you to go to Egypt. It's not safe." But Rosa ignored his caution and left for Egypt. In the meantime, to prevent Stan's being alone in the house on New Year's Eve, a day always associated with the time when he and Iris fell in love, Robin planned a family trip to Hawaii between Christmas and New Year's. Besides herself and Stan, the trip included Robin's daughter Sylvie, Steven, and Steven's recently adopted one-year-old son from Guatemala, Pablo. Stan, Iris, Robin, and Steven had been "the nuclear family that grew up together," Robin said, and they were "probably the most comfortable with each other of any combinations. I think Mom would have been so thrilled to be on that trip, but I don't think it would have happened," she added, because as she aged Iris lost her desire to travel.

While Rosa was in Egypt, she turned off her cell phone to avoid the high roaming charge. When she turned it back on in January after arriving in New York, there were "ten or fifteen messages from Stan," she said. "So I knew that he was very serious about me, but I tried to tell him, no." She reminded him that she had resigned from ECD a few weeks before Iris's death and was planning to stay as a consultant only until June, when she planned to move to San Diego, where she had bought a house near her sister's. Meanwhile she had sold her Michigan house, rented an apartment in Birmingham, and crammed it with all her furniture. She had also agreed to start serving in July as science and technology adviser to the city of Chongqing, one of the largest in China.

Rosa's objections didn't deter Stan from spending as much time as he could with her. He would often drive to her apartment, take her out for dinner, and stay with her afterward as long as possible. It was becoming increasingly clear that Stan wanted a permanent relationship, but Rosa doubted that they could make a happy couple. She told him, "I'm not the type of person who will always agree with you. I speak my mind. And secondly, I don't cook. I live a single life and a simple life.[1] I just don't think this will work." But Stan said, "I like a woman with her own mind, and I wouldn't marry a woman to cook for me."

Stan knew what he wanted, but Rosa did not. She struggled to decide what to do. "At this stage of life," she explained (she was sixty-three; Stan was eighty-four), "it is not like when you were young and fall in love, when you were just in love with being in love. You do things more rationally." She kept going over her choices—go to China or stay with Stan—discussing them with her sister and two daughters, but not yet telling her mother. She recognized that Stan needed her, and she wanted to help, but as she told him, "If I use my brain to analyze, this won't work." Stan responded, "Don't use your brain to analyze. Listen to your heart."

Listening to her heart, she was at least willing to consider their living together, but she felt that her two-bedroom apartment, filled with all her furniture, was too cramped for the two of them. Stan, however, loved Rosa's Birmingham place. After she moved out, he would insist that he missed her apartment, where he had felt comfortable and safe.[2] On the other hand, the prospect of moving into Stan's house on Squirrel Road made Rosa uncomfortable, for Iris seemed to be everywhere there, in the many pictures and objects, and especially in the kitchen. Stan would say, "If you don't feel comfortable, we can build another house, or we can stay in Birmingham and rent a bigger apartment, if that's what you want." Appreciating that Stan was ready to "do everything," she decided to try living in the Squirrel Road house while still keeping her Birmingham apartment. This was right at the time when Stan was asked to resign as president of ECD and sent into "exile" in the Institute for Amorphous Studies (see chapter 11), a time when his need for Rosa was greater than ever. In late March 2007, she moved in.

This point, just seven months after Iris's death, was also when family members had to deal with Stan and Rosa's relationship.[3] For many, it was a painful adjustment. Cathie Ovshinsky remembered that Robin had called in tears. Stan had asked her to go through Iris's things and make room for Rosa. Harvey, with his children Noah and Natasha, met Robin at the house to help. Cathie stayed home. "It wasn't just my grief over losing Iris," she explained, "or even my feelings toward Rosa." It was also her anger. Despite over forty years as an integral member of the family, she had long felt increasing resentment toward Stan for what she perceived as his selfishness and lack of consideration for Iris as her health declined. After Iris died, Cathie began getting physically ill just from being in Stan's presence; she announced to Robin and Steven that she could no longer have anything to do with him. Other family members came to terms with the situation in their own ways, and to the end Stan kept hoping Cathie would also be reconciled.

Steven accepted the change most easily. In early April, he recalled, "Stan came to California specifically to talk to me about Rosa even though I already knew all about it." Steven's view was, "My mother is dead. He and my mother were absolutely joined at the hip for fifty-one years, but now he was clearly miserable. And I think that any kind of happiness that he can find at age eighty-four—this amazing man—he should have it." Steven acknowledged, "It's been harder for other people in the family," especially so soon after Iris's death. "The nice thing for me and maybe why I accepted it so easily is that I don't believe there's been any conflict related to my mother. It's not like she's trying to replace my mother in any way. She loved my mother. That's not what it's about at all."

For the next couple of months, Rosa and Stan lived happily in the house, and "from then on, we talked about marriage," Rosa said. For Stan it made no difference whether they were formally married. "He has a very liberal mind," said Rosa, "but with my Chinese upbringing, I am more conservative." She told Stan, "I know that marriage is only a piece of paper, but if I'm going to stay with you, we should get married." He agreed.

But Rosa had not yet actually committed to marrying Stan. She was instead still committed to starting her new job in China in July. At one point, she told Stan that she was going to leave. "I never saw him so upset." He told her, "You should stay here.

I want you to have a life with me." She recognized how much he needed her. "He not only lost his wife. He lost the company. So you feel that with any human being, you would want to give him a helping hand. He was fighting on so many fronts." Thinking over her choices, she realized that she also had doubts about whether she could fit into Chinese Communist society. Faced with the approaching deadline, she recognized, "I have developed a profound feeling about him. If I just drop him and leave and go to China and something happens to him, I will feel regret for the rest of my life."

In early June, Rosa traveled to San Diego to visit her mother, who was in a hospital. She discussed her dilemma with her younger sister Marietta, who knew Stan. After listening carefully, Marietta encouraged Rosa to marry Stan. "How can you miss a chance to be loved by a man like that?" She also urged Rosa to tell their mother. When she heard that Stan and Rosa had a twenty-one-year age difference her mother's first reaction was negative. Marietta said, "Stan is young at heart, and where can Rosa find a man who appreciates and loves her as much as Stan?" Rosa recalled, "My mom smiled and said, 'Well, Rosa, you will make the decision.' And so we reached the conclusion that I should stay with Stan. That was in June. I didn't really make a decision until the last moment." A few days later, Marietta and Rosa learned that their mother had terminal cancer with only ten days to live. Rosa stayed in San Diego until her mother died on June 24. She told Stan not to come to the funeral because it meant traveling alone, but he insisted, and attended.

When Stan told Robin that he and Rosa were planning to get married, she said, "if this makes you happy, you do it, but I want you to wait one year after Mom's death." Stan and Rosa waited even longer, until October. Of Stan's five children, Steven accepted the marriage most easily. His response was, "It's fabulous." And he added that it was "exactly what my mother would have wanted." Robin agreed, "My mother would have wanted Stan to be happy." At the same time, she "would have wanted to tear the eyes out of anyone who tried to do that."

It was harder for Harvey and his family to accept what was happening. "Part of the problem with accepting the relationship was that it was shoved down our throats, especially so soon after Iris died." Stan would say to him, "Look, this is what I need. Do you want a dead father or a live father?" "Okay, well," Harvey said, "I want a live father, but I don't have to like it." Ben spoke with Stan and "made it very clear to him that I felt it was totally his life to lead." Ben also sensed that he and Stan had "an enhanced mutual understanding—much more than at any other time in my adult life—as my wife was dying and his had just died." After the wedding he sent an email to his siblings: "I think Rosa is good for our dad, because Rosa always speaks the truth to him."

Some friends were initially negative about the marriage because they had loved Iris, but most were positive or gradually became so. Jeff Yang said that he told Rosa, "Thank you so much for taking care of Stan," and added, "I think her heart opened." Rosa insisted she wasn't concerned by the mixed reactions to her marriage from family and friends. "Once I decided to marry him, I don't care what the other people think. I care what I think and Stan thinks."

One problem to deal with before the wedding was hiring a new housekeeper. After firing Harvey and Robin's second choice for a personal assistant, Rosa found Irina Youdina, a former schoolteacher with a college degree, from Bishkek, Kyrgyzstan. When she began in September she immediately got along well with Rosa, but her relationship with Stan took time to develop. At first, Stan would get upset when Irina did not instantly understand what he wanted, and in the early months, she recalled, "He's always sick." In time, both the communication and Stan's health improved, and when Rosa was away on trips with her daughters, Irina would stay in the house with Stan and they would spend time together watching the news, listening to music, or just talking. She became a valued member of the household and stayed to help even after Stan's death.

This was also the time when the house was being renovated to make it feel more like home to Rosa. The renovation was stressful for the whole family. One part that particularly distressed Robin and Harvey was the removal of the gallery of family photos that had filled a long hallway with generations of memories. When Stan had told Robin of this plan, she recalled, "I screamed at him for the only time in my life." Harvey called it "the neutralization" and found it deeply alienating. "All the pictures of me and my brothers were removed. After that I never felt welcome in that house. I felt excluded from Dad's new life." And Cathie resented that she and her children could no longer spend time in the kitchen cooking—"all these family traditions just out the window."

Rosa noted that the other family members managed to adjust. "Natasha and Noah would often come to have dinner with us. Robin, Steven, Ben, and Dale would come and stay with us."[4] But "Cathie never came to visit, although Harvey would come to the Institute and he and Stan would go for lunch."

On October 4, 2007, Stan and Rosa were married quietly at the courthouse with only Rosa's sister and brother-in-law as witnesses. The next day Stan called Lillian for a phone interview and to tell her about his marriage. "I don't pretend, nor does she pretend, that she's Iris, or that I have the same feelings about her as I did Iris. But there's love there, and there's understanding there, there's a devotion there. That is a basis for building something."

A month after their wedding, Stan and Rosa went to New York so that he could have a complicated surgery. He had been in and out of the hospital every two or three months because of recurrent blood infections, and the surgery was to correct the source of bacteria spreading from his colon. They found a cancerous tumor that had to be removed; fortunately it had not spread. There was also severe diverticulitis to be corrected. It was a long and risky surgery, but it was successful, and Stan recovered well.

During the surgery, Rosa stayed with Robin, whose apartment was in walking distance from the hospital. It was an opportunity for them to get to know each other better. Rosa also learned at this time that she would have to move her things out of her Birmingham apartment immediately because the building had been sold. While still in New York, she asked Irina, just recently hired, to work with the moving company to pack her things and ship them to San Diego. "It was a very difficult month," Rosa recalled.

The Christmas holidays brought another medical crisis. Stan had an esophagus tear and was vomiting blood. He had to be rushed to the emergency room, where he received transfusions while continuing to lose blood.[5] Robin said, "We were prepared: we're going to lose him today. He was losing probably more than half of his body's

Figure 12.1
Rosa and Stan in New York, 2007.

blood." But after receiving six pints, Stan stopped bleeding and healed. Soon after, Stan's health improved dramatically. Rosa joked that the young blood he had received rejuvenated him. "He is a fighter," she said. "Nothing could defeat him."

Everyone in the family could see that Stan was happy in his marriage. Robin noticed that Stan improved physically, exerting himself to meet Rosa's hopes and expectations. She and Steven could always tell when Rosa came into the room while he was on the phone with them. "If by any chance he had been complaining about anything, saying he had a bad day or that he had an ache or a pain or something, all of a sudden it would be, 'We've been having a great time.' He would light up and minimize anything negative. He did not want to be a complainer in front of her. And he always wanted her to feel that he was young and vibrant and strong, and capable, and sharp. He wanted to be his very best for her."

Robin contrasted this with the way he had often presented himself to Iris in the last years of her life, where any physical pain he was experiencing radiated to her, and she absorbed it. "She took it on like a sponge, my mother." Robin added that Stan also knew that Rosa "was signing on for a very tough task, and that he had to make it as positive as he could, because he really needed her at the beginning of the marriage. I don't think she had any illusion. She knew that he needed her, and that she was the only one who could really fill this hole."

Stan never got over missing Iris, but he also loved Rosa. And Rosa loved him back. Robin saw their relationship develop over several years, recognizing that though set at a later stage of life, "it was physical, too." She could see that "he wanted someone to take care of him because they cared about him. So it was important that Rosa care about him, but better yet that she love him. He was a very seductive guy, and he was on a campaign. I saw it in the early days, because it was so culturally different for her. She was not someone for public touch—she's Chinese, for goodness sake. And when he would take her hand in public, I would see her kind of wince or pull back, while he spoke in effusive language about how beautiful she looked, or called her by her nickname, Tingela—he kind of Yiddishized it," Robin said.[6] "But there was a turning point, which I could not see until it was over. I saw that they had fallen in love, and that it was mutual, and that it was deep, and it was very, very sincere and sustaining." As Robin observed, "For a self-sufficient and a strong, independent woman to even have the constitution, at sixty-three, to find those feelings, to me it's just a fairy tale."

Stan and Rosa both had to adjust. "We used to have some terrible fights when she worked for me," he recalled, and there were many points where they differed. For one thing, her political views were much more conservative than his. Stan didn't try to argue but instead involved her in discussing books about social and political issues. "She came in as a Republican; now she's a Social Democrat," he bragged. For Rosa, this was just part of their growing rapport. "After two years, I really know him much better on a personal level. He's so well read; we read books together, he reads poems to me, we discuss things, and we read the paper, we express our opinions about all this, and I think it's very stimulating. To share a life with him has really opened my eyes and enhanced my life." Stan summed up, "She's a marvelous person and I have genuine real affection and respect and admiration for her. Hell of a person to get into an argument with."

For her part, Rosa found Stan to be a "very good husband, very understanding, very accommodating, and very loving and caring." She was also struck by his generosity. For example, when they went to a restaurant he would always leave a huge tip. At first she questioned this, but he explained, "These waiters and waitresses don't make much money. They have to raise their families and send their kids to college. This is our way to help them out." This, Rosa said, made a big impact on her. She added, "He is a high-maintenance man, but he makes up for that."

It was harder to coordinate their professional lives. Stan very much wanted Rosa to work with him, but she decided otherwise. "If I want to keep this marriage, it is better for me not to get involved directly with his work, because we often had different opinions and different approaches. I didn't want to have fights. And I'm not like Iris, who wanted to stay with him all the time. I need my space. And so he agreed, and we gave each other space."

Stan understood that he had to make concessions and avoid conflicts with Rosa. In his interviews, he said over and over again, "I really want to make and keep her happy, so it is important to keep her free of stress." Since quitting her work and marrying Stan, Rosa became more relaxed. "Before that she had high blood-pressure," he said. Instead of trying to replicate his relationship with Iris, Stan consciously developed a new part of himself for Rosa. "When you build a new life, you have to build a new appendage to be able to function," he said. Rosa, in turn, encouraged and supported Stan's working on his life-long goal to make solar energy cheaper than fossil fuel.

Ovshinsky Innovation

The renewed energy and hope that Ovshinsky found in his marriage to Rosa fed his work as an inventor. He wanted to revive all the research programs that were cut off when he was pushed out of ECD, such as the cognitive computer and hydrogen storage. Most of all, he wanted to pursue his ideas for dramatically increasing the production rate of solar panels, significantly lowering their cost. He believed he could at last make solar energy "cheaper than coal" by building a gigawatt machine.[7] Ovshinsky had already conceived this new invention before losing his position at ECD, but when he had tried to discuss it with the new board, "they just laughed." Now he set out to achieve his vision on his own.

Early in 2008, Stan and Rosa set up a new corporation called Ovshinsky Innovation, with a subsidiary, Ovshinsky Solar, dedicated to research on the gigawatt machine. Both were housed in the offices of the Institute for Amorphous Studies across the lake from their home. In February, Ovshinsky began fundraising and asked Dave Strand to put together a formal business plan and presentation for potential investors. Interest came from a group in France, to whom Ovshinsky and Strand made a presentation in May. In Japan, contacts at both Canon and Sharp tried to help, and there were extended discussions with a Chinese group, but no money came. In the midst of these unsuccessful efforts, Rosa suggested that Stan fund the work with his own money, arguing that once he had achieved proof of principle he would be better able to attract investors.[8] So, in October 2008, at the age of eighty-five, he invested $3 million of his savings in Ovshinsky Innovation. Bob Stempel also invested $500,000, saying, as Strand recalled, that it was his "civic duty."

Although Rosa encouraged Stan in his new effort, she was firm in her refusal to be part of it and instead took a teaching position at Wayne State University. Ovshinsky organized his research team around Strand, who joined the new company full-time in October. While he recruited people for the team, an Ovshinsky Solar lab was set up in the roughly 1,000 square feet of two rented rooms in an industrial building in nearby Troy. About fifty small companies and other users were renting space there. "We had Unit S," Strand recalled. "Right next to us, they were making pickles; behind us was a music studio, and on the other side there was a dental business making crowns. The previous occupant before us was storing a Porsche there." The research began its work in January 2009 and Ovshinsky Solar remained in this space, reminiscent of Ovshinsky's original storefront, until his death. "I come from the storefront and go out in the storefront," he would say.

The new company initially consisted of Ovshinsky, his brother Herb, Strand, and Ovshinsky's long-time assistant Freya Saito, who was paid by ECD, the last in a series of indispensable, and sometimes much put-upon, administrative assistants. Besides these core members, five others—Boil Pashmakov, Mike Hennessy, Pat Klersy, Paul Gasiorowski, and Tim Barnard—made up the actual working team.[9] Hellmut Fritzsche, who had resumed his role as scientific consultant, compared the team to a string quintet, where "each player was just the right one for his part of the music." With Strand in charge, Pashmakov focused on the physics; Hennessy used his electronics expertise to do measurements, while Klersy and Gasiorowski, as vacuum technologists and plasma experts, produced the thin films with the new method Ovshinsky had designed. Barnard did the computer programming, and Herb, at the drawing board, designed equipment. The technology for the project included a high-power microwave generator and an ultra-high vacuum system with turbo pumps. Rosa described the state-of-the-art lab as "very sophisticated and very impressive."

To achieve a gigawatt production volume, Ovshinsky planned to speed up the deposition rate, but the challenge was to do that without degrading the quality of the material at the same time. The system Ovshinsky had his team build in the Troy laboratory addressed this challenge by incorporating several novel ideas. In creating the plasma he used microwave excitation, which he located at a distance from the cells so it was not in contact with the film growth surface. He also added fluorine to the reactive gas, believing that in the right proportion it would promote a superior film structure: stable, nanocrystalline (to improve current conduction), and with many fewer defects. (As seen in chapters 6 and 8, Ovshinsky had long been enamored with fluorine because it forms a stronger bond with silicon than hydrogen does, so he worked to balance the proportions of reactive fluorine and hydrogen.[10]) The microwave excitation made the material cheaper per square foot by increasing the deposition rate; the fluorine made it cheaper per watt by increasing efficiency.

Many trials and adjustments followed, as the team deposited and tested the material produced by the new system. In evaluating the results, Strand recalled, "Hellmut agreed with Stan that we first had to measure the density of states because the density of states is really critical to having a good film.[11] And so we worked, and worked, and worked to get the density of states low, and then Hellmut said, 'Now we have to measure the photoconductivity, because if the photoconductivity isn't good then the cell won't be good.' I'm thinking, 'Hellmut why didn't you tell us this at the beginning? Why did we spend all this time just focusing on the density of states, because we could easily have been measuring photoconductivity all along?'" But Strand came to see this as Fritzsche's way of "managing Stan. He was helping him to have that first victory of the low density of states and then establish a second goal. It really buoyed Stan up to have that success, and it was a necessary but not a sufficient result."

The next tests were also encouraging. Fritzsche had been skeptical whether they could get adequate photoconductivity at such a high deposition rate, but he said, "It turned out to my great surprise that the material was very, very close to photovoltaic quality. I was amazed. Our measurements showed that this really could be done."

After these positive results, the next step was to make actual solar cells with PIN junctions, which would be more likely to impress potential investors. Working with the experimental microwave system made this difficult, but the team managed to produce cells with a very respectable 4 or 5% efficiency. "So," Fritzsche recalled, "we were very happy, and we opened a bottle of champagne."

Ovshinsky believed that adding fluorine could also prevent the Staebler-Wronski degradation (see chapter 8), but here the results were less successful. For almost two years, he kept pushing his team to get more and more fluorine into the film, but that just reduced the photovoltaic quality. Finally, Fritzsche persuaded him to go back to zero fluorine, where they knew the photovoltaic properties were very good, and add fluorine in small steps. They found that a small concentration increased the deposition rate tremendously without harming the photovoltaic properties, but adding more lowered the film quality. "We realized," Fritzsche said, "that our attempts to keep increasing the fluorine content were absolutely wrong. That futile effort had lasted almost two years and cost a lot of money. So how did Stan react to this? Amazingly, he was able to accept his failure. Stan surrendered and considered it an important learning experience. Of course he has his intuitions, but when nature tells him something, he listens and accepts it." "Materials I can control," Ovshinsky once said. "Nature, not so easily."

Over a period of three years, Ovshinsky had basically accomplished his initial goal for Ovshinsky Innovation, achieving proof of principle for his audacious plan to make solar panels much faster and cheaper.[12] To go further, however, would require significant outside funding. The experimental samples, which were only one square centimeter, would have to be enlarged a hundredfold, requiring a much larger deposition machine. Pashmakov recalled, "It was estimated that initially we would need like $20 million, and a production line would be in the hundreds of millions." Companies in Switzerland, Italy, Germany, and China expressed interest, but none made commitments. In the meantime, the effects of the 2008 recession and Chinese price-cutting on their polycrystalline solar cells had drastically reduced the demand for thin-film solar panels. The prospects for outside funding were thus poor, and Ovshinsky's own investment, which had grown from $3 million to $6 million, was nearly exhausted. In March 2012, Ovshinsky Solar began to lay off staff and wind down, though some research continued. Had Ovshinsky lived longer, the story of the gigawatt machine might have ended differently, but for now it remains yet another unrealized possibility.

Closest to the Sun

Yet the significance of Ovshinsky Innovation cannot be gauged only by the gap between Ovshinsky's daring inventive vision and his incomplete achievement. The new company also embodied his continuing efforts, pursued with vigor late into his ninth decade, to make the world better. A fine example is the trip he and Rosa took to Chile in October 2009. Earlier that year, Harley Shaiken had brought former Chilean president Ricardo Lagos to Detroit to visit ECD.[13] Lagos was already interested in developing sustainable energy programs for Latin America and, strongly impressed by what he saw at United Solar, invited Ovshinsky to visit Chile.

The trip was a triumph. "I think in many ways the trip to Chile was the experience of a lifetime for Stan," Shaiken said. "It seemed to fulfill so much of what he had struggled so long and so hard for." Ovshinsky was warmly received by numerous public and private leaders, whom he inspired with his vision of an independent energy future for Chile and of Chile as a model for all of Latin America. From a conference on renewable energy, where his keynote speech received a standing ovation from five hundred participants, to a private dinner hosted by President Michelle Bachelet, where the two

Figure 12.2
Rosa and Stan with Michelle Bachelet.

socialists immediately formed a close bond, he delivered his message with urgent conviction. Standing with Rosa on a high and windy mountain in the Atacama Desert, he gestured energetically at the scene while speaking extemporaneously for a Chilean television crew. "The beautiful part of Chile is it has all the energy possibilities and potential, that is being wasted really." Tapping into this potential, he said, "You'd have a showcase of how to have energy without pollution, without climate change, without war over oil. And build new industries in Chile from your own natural resources." "Being here is so moving," he added, "because I am now closest to the sun."[14]

Figure 12.3
Ovshinsky speaking in Chile, October 2009.

13   Last Days

Stan's birthday on November 24th had always been an important family event. To honor the tradition, and also highlight Stan's ninetieth birthday, Rosa planned a huge celebration. Rather than wait until late November, when many of those to be invited would want to be with their families for Thanksgiving, she decided to hold the party over Labor Day weekend, when the warmer weather would also allow having the party outdoors. In May, Freya Saito and Georgina Fontana began emailing guests to save the date of Sunday, September 2, 2012. Simply finding the addresses of Stan's many friends and colleagues was a huge job. Full invitations went out in July.[1]

The Birthday Party, September 2, 2012

The evening before the party, Rosa organized a dinner at home for about twenty, including close family—Harvey, Dale, Robin and her family, Herb and his family, and Rosa's family. (Steven and his son Pablo flew in later; Ben couldn't attend the event because of knee injuries.) Also at the dinner were Harley and his wife Bicky, and the Russian scientist Alex (Sasha) Kolobov, who made a special trip from Japan. Although Stan had been suffering for some time, racked with back pain and struggling to walk unassisted, he seemed well and happy that evening. "He actually looked great," recalled Robin. "He and Sasha sat together on the couch and talked and talked."[2] But on the day of the party, Stan was not in good shape. "I thought, he looks like hell," Robin recalled. Irina, whom Rosa had asked to come to the house and help Stan dress, assumed that the dinner party the night before had exhausted him.

But Stan still enjoyed the party, which was a spectacular tribute. Attended by approximately three hundred family members, friends, and work associates, former and present, as well as a number of honored guests (including Senator Carl Levin and UAW president Bob King), the party started early in the afternoon with drinks and hors d'oeuvres. The feast that followed included salmon, filet mignon, and at the end, a gigantic birthday cake. During and after the dinner, the participants, and especially Stan, enjoyed a program expertly hosted by master of ceremonies Harvey Ovshinsky at the microphone.³

Speeches and tributes celebrated Stan's life and achievements.⁴ Senator Levin emphasized the impact of Stan's vision and passion on a world "in which science lights the way to a brighter future, in which justice and fairness prevail." Hellmut Fritzsche recalled how, at the time he first met Stan and examined his threshold switch, he had been "flabbergasted, astonished, puzzled, and curious about the materials covering the two crossing wires which formed his device." After mentioning many of the distinguished scientists who regularly visited ECD, he recounted a dream that he said he had had of Stan talking with Einstein, who expressed his admiration and insisted that his difficult work on relativity had actually been much easier than Stan's work. Harley Shaiken recalled first meeting Stan and Iris at age fifteen when he had attended a meeting to organize a Detroit chapter of CORE, and he also told how Stan had introduced him to books and ideas that had changed his life. Joi Ito extended the tribute by telling how Stan had shaped his values and future career. Shorter tributes from family and friends reflected on other aspects of Stan's life and work, many of them touched on in this book.⁵ There was laughter and music, and a parade of six grandchildren holding ninety balloons, accompanied by a bagpiper.

Figure 13.1
Birthday party: Harley Shaiken.

Figure 13.2
Birthday party: Hellmut Fritzsche.

Figure 13.3
Birthday party: Joi Ito.

Figure 13.4
Birthday party: bagpiper and grandchildren.

Stan looked happy as he sat in his chair and listened for most of the evening. Occasionally he rose to thank people, and he was cordial even to those with whom he had had strained relationships. In retrospect, many signs were apparent at the party that Stan was in extreme physical pain. His own speech, which he delivered in a weak and raspy voice, was mainly limited to thanking Rosa and a few others. True to his style, he promised to keep working to make the world a better place. Later, with Harvey's son Noah helping him to walk, Stan left the party.

Within days he would have to be taken to the hospital because of his agonizing back pain, and it would be downhill from there. He never went back to work afterward. Still, for many days, as Irina recalled, Stan would relive his memories of the many people who shook his hand and congratulated him at his wonderful early ninetieth birthday party.

Last Trips

Stan's health had already suffered in the months before his birthday party, when a sequence of illnesses and mishaps had occurred, foreshadowing his decline in the weeks afterwards. He had seemed strong and happy when, in late winter of 2012, Rosa brought him to New York for a urological procedure at Lenox Hill Hospital, where Robin worked. "He did very well. He was in for a day or two, and then he recovered at my house," Robin recalled. She viewed Stan's agreeing to stay with her as a sign that he was in good health, and she was pleased to see that he could go out for a couple of hours with Rosa, even with a leg catheter. "They walked from Fifth Avenue to Lexington Avenue, which was kind of a lot for him." It was a sharp contrast to his visits with Iris in her last years, when they would stay in a hotel and take cabs. But Rosa "wasn't that into cabs just for short rides," so they walked when possible, not only to the doctor's office but also to several restaurants.

In May, however, Robin got a much less hopeful view of Stan's health when she accompanied him and Rosa on a trip to the Pacific Northwest. Junior high school students at the Louis Riel School in Calgary, Alberta, had invited Stan to talk about his work; he told Rosa, "This is my civic duty. I have to go." Knowing that Stan loved trains, she suggested that they follow the school visit with sightseeing, taking the spectacular two-day Rocky Mountaineer train ride through the Canadian Rockies. She thought taking this luxurious train with its glass-topped observation cars would be an exciting and easy way for him to see the mountains and glaciers. They would take meals on the train, stay over in a nice hotel on the way, and afterward spend a few days in Vancouver. Robin had never been in the Canadian Rockies, so she eagerly accepted when Rosa invited her to join them on the adventure.

The trip began badly however, when Stan fell in the Detroit airport. Fortunately, Irina managed to grab him and he did not get hurt. He fell again the first night in Canada, hitting his head on the bathtub of their hotel in Calgary. "Luckily," Rosa commented, "that bathtub was fiberglass." But Stan's talk the next day was a high point. The school children had turned his visit into a school-wide study of energy, and to express their appreciation, their hosts presented Stan and Rosa with white cowboy hats, a Calgary tradition.

When Stan and Rosa met up with Robin the next day she was alarmed to see Stan "leaning on Rosa, huffing and puffing, barely able to walk." So they could see the mountain lakes and glaciers, Rosa had planned for them all to drive on the Icefields Parkway to Jasper and stay overnight before boarding the Rocky Mountaineer. But when they reached Lake Louise, which is about a mile above sea level, Robin realized that Stan's weak lungs couldn't handle the altitude. With some difficulty, they managed to rent a portable oxygen concentrator, and with that Stan could walk with a cane. Robin and Rosa were still anxious, but Stan insisted, "I want to continue. We're going to be on the train, it will be fine."

On the way to Jasper the next day they saw bighorn sheep on the side of the road, Robin recalled, and Stan was able to enjoy the drive. In beautiful Jasper, they stayed at a hotel on a small lake "and elk were coming right up to our cabin." When they finally boarded the train, "We were like, phew, we made it, because it was just one day of train ride, and then you stay in a hotel, and then another day of train ride, and you're in Vancouver. And I figured, okay, by then we've got good medical care, if we need it."

As it turned out, they hadn't made it. After settling for the first day on the upper level of the train to enjoy the beautiful scenery, Rosa and Robin took the stairs to the lower level for breakfast while Stan used a small handicapped-accessible elevator. But during breakfast, "all of a sudden, I hear this commotion outside," Robin recalled. Stan had gotten back in the elevator, while Rosa used the stairs. Missing a sign warning people not to put their hands on the edge, Stan caught his finger and sliced it open. "He was bleeding like a stuck pig, he was in agony, screaming," and there was blood everywhere. "Amazingly, the steward for our car was an EMT-type guy, and he did an amazing job," said Robin. By this point they'd been on the train for about two hours, and Robin knew that Stan's hand had to be taken care of immediately. The train made a special stop "in a little nowhere town, but they had a clinic with a very well-trained doctor from South Africa, and he sutured Stan up." Then they hired the only cab in town (for $600) to meet the train at its evening stop.

The next morning, they had to decide whether to board the train again and go to Vancouver by train or fly from a local airport. Taking the flight, however, required waiting, and Robin thought that staying with the train, which had first-aid staff, was a better choice. Stan also preferred to stay with the train because he did not want to ruin the trip for Rosa. So they did. On the train he and Rosa wore the hats they were given in Calgary, and this time the ride went well. Robin recalled, "We saw bald eagles," and Stan was treated as a minor celebrity, which seemed to please him. "But the minute we hit Vancouver, we had to go to the emergency room, because the stitches weren't really holding." "That's it," Robin insisted. "We're going home." When Robin told Stan that he had to see a hand surgeon, he "was really depressed," she recalled, but "I didn't care at that point, I just wanted him out of this degree of vulnerability, and Rosa did too." She told Rosa, no more trips for Stan.

During June, Stan saw a hand surgeon, and with physical therapy for his finger he recovered enough to forget Robin's travel ban. Well before the trip to Calgary they had purchased tickets for a trip to Scandinavia. Stan had been nominated for a European Inventor Award based on his work on improving the NiMH battery for use in cars

Figure 13.5
Stan and Rosa on the Rocky Mountaineer wearing their Calgary hats.

(see chapter 9). He was eager to attend the award ceremony in Copenhagen and also wanted to attend the annual meeting of E\PCOS, the European Phase Change and Ovonics Symposium, to be held that year in Tampere, Finland, two weeks later. It was a meeting at which Stan, the acknowledged father of the technology, had often delivered the keynote speech. To cut down on the amount of airplane travel and also to enjoy some sightseeing, Rosa had suggested spending the intervening weeks on a cruise seeing the Norwegian fiords. They invited along a number of family and friends—Rosa's daughter Angela, with her husband Jim and their children Lolo and Norah, Steven and Pablo, and Rosa's friend Genie. (Robin could not come because of work commitments.)

When the time drew close, Stan finally admitted that he was not well enough to attend the Copenhagen meeting because the stairs in the auditorium would have been too much to manage. Mike Fetcenko agreed to accept the award in his place. Stan insisted, however, that he wanted to attend the meeting in Finland, so they decided to take the cruise and then the flight to Finland. The cruise was a special occasion for the

Figure 13.6
Rosa and Stan on the Norway cruise.

Figure 13.7
Family on the Norway cruise. Back: Lolo and Steven; front, left to right: Pablo, Genie, Stan, Rosa, Norah, Angela, Jim.

family, and everyone got to have time alone with Stan. Steven, who had brought along his bassoon, would play for him in his room. Stan enjoyed the spectacular scenery and being with his family, but he was not well and had spells of extreme back pain.[6] Rosa phoned Robin, who said it was probably an osteoporosis compression fracture, and Rosa's son-in-law Jim, a radiologist, thought so too. Stan and Rosa decided that he was not well enough to attend the Finland meeting; they left the cruise at Bergen, the last port, and flew home directly.

These aborted trips before Stan's birthday party were signs of more serious health troubles ahead. After they returned from Norway, Stan had continuing excruciating back pain that forced him to use a walker. Still, he went to work every day and managed to keep up his normal activities until about a week before the September 2 birthday party.

After the Party (September 3 to October 17)

The party was a high point for Stan, but afterward things went from bad to worse. The next day, Labor Day, he stayed home with Robin, Steven, and Natasha, while Rosa went to lunch with her family. At one point, Robin recalled, "He started to scream in pain. Howl." He told her, "I've been in pain since the cruise!" "We've got to get you back to the orthopedists. They're missing something," she said. He was diagnosed as having an osteoporosis compression fracture, as Robin and Jim had thought, and he continued to be treated for pain.

Over the next two weeks, Stan suffered persistent, terrible back pain. He was in and out of the hospital, but test results were inconclusive and he was sent home with more pain medicines. Finally, on September 16, a CAT scan yielded the correct diagnosis of metastatic prostate cancer. "It was everywhere," Robin said, "all over his bones." That explained both the pain and the debilitating weakness that had caused him to fall so often. A cure was impossible, but the doctors felt that they might be able to give him six months to two years. Meanwhile, because they were failing to control the pain, Robin found a palliative care nurse. Stan loved her, and "she adored him; they had profound conversations," Robin said. Stan's doctors decided to use radiation just to alleviate the most painful areas of his spine. Robin came back for three-day weekends for the next several weeks.

The cancer diagnosis came as a shock to Rosa, and all the more so when she learned that for years other family members had been aware that Stan had prostate cancer. He had decided against being treated twenty-five years earlier, refusing surgery as well as radiation or chemical therapy. Yet he continued to have his PSA (prostate specific antigen) checked and would panic to learn each time that it was rising higher. "Then remind me why you keep checking this result," Robin said, "since you've decided that none of those options are going to be done? And finally he stopped."[7]

Rosa was angry with Stan for not telling her about his history of prostate cancer. Had she known, she would have insisted on treatment. At one point, he admitted that he had made a mistake. "I'm paying the price," he said. "Are you going to punish me for it too?" That made Rosa very sad, said Robin, "and she never said anything about it again, to my knowledge." Once it was clear that Stan would not recover, they arranged hospice care, as Harvey and Cathie (who taught hospice and end-of-life care) had suggested. Hospice helped Stan and Rosa come to terms with the situation. Robin observed Rosa passing through the standard Kübler-Ross stages, including denial and anger, "roughly one a day." Stan also passed through them, but "not as logically as she did." When Rosa reached the stage of bargaining, "she started to do literature searches about experimental therapies. She was calling me; I said, 'Rosa, the diagnosis is extremely clear, and it's extremely advanced, and there are very, very few, really if any, choices right now.'" Rosa soon reached the stage of acceptance, coming to peace with Stan's death before he died. "And that was amazing," Robin said. "And he did too."

Saturday, October 13 was an important day. The hospice nurse told Rosa that Stan had no more than a week to live. The prognosis was another shock to Rosa because only two weeks earlier the oncologist had said Stan might live several more months. It was up to Rosa to tell Stan that he needed to prepare for his final days. "It's not good news," she told him. "It's time to talk about what's most important to you." "I want to live," Stan said. "Yes," Rosa said, "I want you to live with me for many years to come, but we need to prepare for the worst." He spoke then about Dale, concerned that his son would have enough money and be watched over.[8] Even this close to the end, Stan was thinking about the future.

Stan had never been much concerned with his own death. "I'd rather talk about life," he would say. "The end of life is not nearly as important as being in life. The process of living means expressing yourself to the fullest. I think it's the process of living, of putting meaning in life, that is the important part." Stan always said he wanted to die with his boots on, and even this close to death he struggled to press on with his work. He wanted desperately to bring his gigawatt project to fruition; anxious to keep this "setback" from interrupting the research, he wanted to see Dave Strand and talk about the progress being made at Ovshinsky Solar. During his last visits with his children, however, Stan apologized for spending so much of his own money on Ovshinsky Innovation. "I wanted to leave you more," he said. They tried to reassure him: "You left us plenty," Harvey said. "More than you'll ever know."

Late that afternoon, Stan suddenly sat up and said to Rosa, "I want a date with you." The two would often go out to dinner on Saturday evenings, sometimes to a Buddy's Pizza place a few miles away. "I want Buddy's Pizza," Stan insisted, and when Rosa told him it was impossible to go out, he said, "Let's order it in." Rosa asked Harvey to pick up Stan's usual: double sliced tomato, double pepperoni, double anchovies. Ben, who had flown in a week earlier, joined them. Harvey recalled, "It was a wonderful meal. He ate at the kitchen table with us, and we talked, and we listened, and we were all together. It was lovely."

Then Stan "went to bed, and never got out of it. And he never ate again," said Robin, who was sad to have had to miss the pizza party because of an important meeting in New York that day. She remarked about something she had read in the hospice brochure they received: "They talk about a kind of an arousing, an awakening, a last time. And that was it. And I missed it." From then on, Stan was "weak, incontinent, and bed-bound, but pain-free. He slept a lot, but between, we could talk. Dale flew in from Florida on Monday, and Herb came to visit. Stan was awake, and I remember we were all just sitting around the bed. It was nice. We would stroke him or talk to him, and someone was with him all the time." In the middle of one of the last conversations with his children, he struggled to raise his arm. "He could barely talk by then, so we didn't know what he wanted," Harvey said, "until we realized he was pointing at the

Figure 13.8
Photocopy of the drawing of Stan's father Ben.

framed drawing of his father on his bedroom dresser. My brother Ben understood and brought it over for Dad to hold. He just reached out, touched the image of his father, and quietly, tenderly kissed it."

Stan wasn't fully conscious most of the time from then on, but there were some significant exchanges. At one point, Rosa asked Stan, "Do you see me?" And Stan answered, "I see you everywhere. I look for love." He asked Harvey, "Is Cathie okay?" "Does Cathie understand?" Harvey tried to assure him, "Dad, she does. We all understand." There were also several mumbled fragments that suggest what was in his "drugged, almost delusional mind during the days when he was actively dying," said Robin, who wrote down as much as she could make out. "Department of Energy," "All those Nobelists. I never got the chance." "What is all this, sounds like the VOIG3." And "such a beautiful spot." "I got somewhere. A journal published what I said." "Freya, get me the summary!"

At some point on Monday, Robin said, "I think we need some Schnapps." The idea was to toast to Stan's health. Stan said, "Great idea," and they had some of the good Russian vodka from the freezer. Stan toasted in Russian, "Nostrovia!" "It was a wonderful, wonderful moment," said Robin. "And then he and Herb started singing Yiddish songs together. It was wonderful." Dave Strand came by on that Monday too. Stan smiled and seemed happy, even though he was hardly conscious, Robin recalled, when Dave said the experiments were going very well, "which was hardly true, I think."

By Tuesday Stan was mostly mumbling, but at one point he woke up and started to talk clearly. He looked at Rosa, who was lying on the bed with him. "She would crouch facing him," Robin said. "She wanted to see his face more, and he wanted to see hers. He asked her, 'Can I talk to you?' And she said, 'Of course.' And he looked around at the rest of us in a slightly hallucinatory way and said, 'With all these people around? I don't want to talk to you with all these people around.' So we all left. And she told me later, he said, 'I really love you. Do you love me?' And she said, 'How can I not love you?' Those were his last words."[9]

On Wednesday, October 17, Stan's last day, they played calming music that Steven had selected for him. (Steven, who couldn't be there because of his performance schedule, called often and had asked Rosa to download music from iTunes.) With the music in the background, "it was very beautiful," said Robin. "It was fall: the windows were open, it was very sunny, and there were yellow leaves all around, falling. It was like a metaphor. Stan could see the scene from his bedroom windows." By late Wednesday afternoon when Herb and Selma brought Chinese food over for dinner for the family, Stan "was just barely hanging on. His breathing was slow, his blood pressure was so low, everything was low." With her hand on his chest, Robin could feel him cooling down. Feeling his pulse through his thin chest, she could tell it was getting irregular. Herb compared what came next to "a scene out of a movie." When Robin told Stan that Herb had come, "it was as if he was waiting for him to arrive." Stan flinched as the door opened. "I said, 'It's your brother Herb, he's here to see you.' And I started to feel the pulse getting a little more irregular. She called in the others. "Herb kissed him and held his hand, and within two minutes, he was gone." Herb recalled the first thing Rosa said: "It was the best five years of my life."

Then Rosa, Herb, Robin, Stan's three boys, and Irina "each kissed him goodbye, and let Rosa have some time alone with him," recalled Robin. "Irina was the first to leave the room. She was very choked up. And, you know, we all were, and we touched him or held onto him, or hugged each other, or whatever." When they eventually left the room, at the head of the dining room table was a glass of vodka, with a piece of dark bread on top. Irina told them that it was a Russian tradition, a tribute, when the father or head of the household died. It had to be left there for nine days, the period when the spirit of the one who died remained in the house. Robin found it very moving.

"And then," Robin continued, "we did sit down and eat the Chinese food that Herb and Selma had brought. It was a little difficult, but we knew we needed to, and we were very grateful to all eat together." Later that night, Stan's body was taken to the same Jewish funeral home they had used when Iris died. Rosa "was crying so much," Irina recalled. Robin said, "I feel good about the way he died. I feel good that we were able to help him to have that comfort, and that he knew how much he was loved."

After Stan's Death

The family had already made most of the decisions about what to do after Stan's death in the hours of sitting around talking during his last days. They decided not to have a memorial, partly because everyone was exhausted and also because they'd already had the commemorative speeches at his birthday party. Robin remembered Rosa saying on Monday or Tuesday night, "What else are we going to do or say?" The others agreed. "Like Tom Sawyer, he got to be at his own funeral," Robin remarked. "He got to hear his eulogies. What could be better?"

As Stan and Iris had planned, he was to be buried next to her in the Akron Workmen's Circle cemetery the next Sunday. Before then, Rosa started hearing from many ECD people who wanted a way to say goodbye to Stan without having to travel to Akron for the burial. At the last minute, she decided to have a viewing on Friday. It was advertised with less than twenty-four hours' notice, mainly by word of mouth plus a single email from Dave Strand. Rosa and Irina went to the funeral home beforehand to decide whether to have an open or closed casket viewing. As Irina recalled, Rosa worried that Stan would not look good if the casket were open, and she began to cry when she saw him because he did not look like himself. His curly hair had been made to lie straight. "Just give me water; I will make his hair curl," said Irina, who then fixed his hair so that he looked as though he had just fallen asleep. Irina had also brought along the Swiss army pocketknife that he liked to carry and one of the coins that said "Energy" on one side and "Information" on the other. She put these into his pockets, adding a few other items: his handkerchief, the little pin he never forgot to wear in his lapel, plus some notepaper and a pen.[10] "He looked totally normal, in his three-piece suit. He looked great, no makeup, no nothing," Robin said. At least a hundred people came to see Stan one last time.

Meanwhile, Harvey heard from a friend about online virtual memorials. He showed Robin a few samples. She didn't like any of them, but the suggestion led her to find a beautiful site called "Forever Missed," with pages for tributes, stories, photos, and documents. Irina recalled that in the first week after Stan's death, Rosa "was looking every night." On Saturday, the family commemorated Stan by eating at Buddy's Pizza. Vicki, Rosa's daughter, arrived from San Diego on Saturday, as did Steven from San Francisco, so that they could come along on Sunday to attend the burial. Robin recalled that in one of the very first conversations that she had had with Stan about Rosa, after it was clear that he was going to try to make a life with her, he had asked, "How am I going to tell Rosa that I want to be buried next to Iris?" Robin told him that she didn't think Rosa expected anything else, "because I don't think that you're going to be together fifty years like you were with Mom."

On Sunday, they all drove behind the hearse to Akron, where the burial took place about 1 o'clock, without any prepared speeches or ceremony. It was a sunny fall day, and the cemetery was beautiful. They each threw in a few of the red and white roses they had brought from the viewing and placed stones from near the lake on Iris's and on Stan's parents' graves. "It was hard to see them lower him next to Mom, and to see her stone again," said Robin. "It was really hard."

Robin later designed Stan's stone according to some handwritten notes she found after his death, asking that he be buried with the amorphous materials symbol from the Institute of Amorphous Studies flag. She found other instructions, most of which had already been followed: "No religious anything, no service, no Jewish stars, no prayers, everything he had done for Mom."[11] Robin decided to use the amorphous materials symbol at the top of Stan's gravestone, just as Iris's has an iris at the top. Steven suggested adding a rose to symbolize Rosa.

While his father was dying, Harvey wrote a press release and arranged for the obituaries. Dozens of newspapers, social media sites, and even NPR celebrated his life and mourned his passing. Most of the obituaries were thoughtfully written, especially those in the *Wall Street Journal* and *Crain's Detroit Business*. But the *New York Times* notice appeared to have been written years earlier and included a number of errors.[12]

No matter what their relationship had been, those who had interacted with Stan a great deal felt the loss. Irina, for example, felt empty. She remembered her conversations with Stan "about everything, like in the springtime, if we walk to institute, and some tree is blooming, and talking about how beautiful it is. He is a gentleman, a very wonderful man." Irina especially missed the times when he asked her to sing Russian socialist songs, "and we're singing together, and sometimes he's singing in English, I'm singing in Russian, the same song."

The burden of clearing the house and the Institute of all the books, papers, furniture, and other belongings, so that the properties could be put up for sale, fell on Rosa. Her friend Genie came from Greece for an extended stay so Rosa would not be alone in her grief, and she helped Rosa go through the papers and household things. "We all owe her a great debt," Robin said. Among the many objects in the house that

Figure 13.9
Stan's gravestone.

Figure 13.10
Inscription on Stan's gravestone.

needed to find another home after Stan's death was the oscilloscope in the basement attached to one of the first threshold switches, still displaying the familiar cross pattern that he first observed in the storefront. The device went to the Detroit Historical Society.[13] As Hellmut noted, it was "not the original oscilloscope, but the switch had been switching on and off for all these forty-seven years, 120 times per second. And you walk past it, and you see the switching cycles absolutely stable and steady after all these years."

When the house was finally sold, the family was happy that the buyer was planning to live there with his children, for they had feared that the house would be torn down and the property subdivided. It turned out that the buyer had just won the state lottery and could well afford to keep the property intact.

In the months that followed, Rosa continued to find surprises from Stan. Irina turned up many love notes he had written to Rosa on the cardboards from his laundered shirts. Every time she returned from a trip she would find one at the front door. "He would write some special message for her," Irina remembered, "like 'Tingela, welcome home, from your man, Simcha.'" One surprise that Rosa found a few months after Stan's death was part of a birthday gift he had given her on December 15 the year before. He had worked with Robin to select the present, a lovely piece of modern cloisonné Chinese jewelry with an image of his favorite bird, the heron. The printed card read, on one side, "Who stole my heart?" On the other, "Oh, it was you." And Stan wrote on it, "Happy Birthday Dear Heart. My Izon. Your lover, Stanford" (*izon* means "wife" in Chinese). He had placed the card inside the box behind the jewelry, but when Rosa first opened the gift, she didn't read the message inside. She found it shortly before her next birthday.

Figure 13.11
Oscilloscope still displaying the "cross" after Stan's death.

Last Days

Figure 13.12
Packing up at the Institute for Amorphous Studies.

Figure 13.13
Stan's birthday gift to Rosa.

Figure 13.14
Stan's card.

Epilogue: Deaths, Survivals, and Revivals

What became of ECD after Ovshinsky left? What remains now of his other enterprises, and how much of his legacy is likely to continue? Here we summarize the recent histories of the most important companies and technologies he created.

ECD and United Solar

After Ovshinsky's "retirement," when ECD's new managers discontinued all its programs except batteries and solar (see chapter 11), they focused on maximizing returns from the solar program. United Solar had become the largest US manufacturer of amorphous silicon solar panels.[1] ECD's CEO and its board of directors now attempted to greatly increase production capacity and lower unit costs by raising over $400 million in convertible debentures.

That proved to be a disastrous strategy. The effects of the 2008 recession, which led to the loss of government subsidies for solar energy, combined with drastic price cuts by Chinese makers of polycrystalline silicon solar cells, which their government continued to subsidize, crippled the whole US solar industry. With its huge debt load, ECD could not survive; the company filed for bankruptcy on February 14, 2012.[2] United Solar had to sell its assets to pay its creditors. By the following summer the company's machinery, equipment, inventory, and real estate holdings, as well as its intellectual property, had been auctioned off. It was a heartbreaking experience for those who had invested large portions of their careers in developing the solar technology. Arun Kumar recalled "dismantling stuff for sale as scrap, and that really hurt me. I saw those machines being scrapped, and I wept."[3]

Ovonic Battery Company

Led by Mike Fetcenko, the battery company (now Ovonic Materials Division) continued its work for almost four years after Ovshinsky's departure from ECD.[4] On February 13, 2012, the day before ECD filed for bankruptcy, the division was sold for $58 million to the world's largest chemical company, BASF (originally Badische Anilin und Soda Fabrik). All the employees received job offers from BASF, and all accepted.

The nickel metal hydride battery continues to be commercially important; more than thirty-five BASF licensees pay millions of dollars each year in royalties from their manufacture and sale of Ovonic batteries.[5] Five hundred million cells are sold each year, and the NiMH batteries are used to power hybrid electric vehicles, such as the more than five million Toyota Priuses sold over the last nearly twenty years.[6] As the leader of all BASF battery activities in North America, including the Battery Materials–Ovonic Division, Fetcenko remarked that BASF is "proud to carry on some of Stan's legacy. It is very rare at BASF for an acquired company to keep its former name. BASF recognized Stan's reputation by retaining the name Ovonic."

Ovshinsky Innovation

Ovshinsky Innovation and its subsidiary Ovshinsky Solar (jointly referred to as OI/OS), ended after Ovshinsky's death in October 2012. Despite significant progress toward achieving the gigawatt machine, for which Ovshinsky claimed proof of principle, he had been unable to secure funding for the next phase. The project had been supported only by his personal savings, and already in March 2012 he had agreed to start winding it down. In the months after his death, further efforts to find support were also fruitless.

The only real possibility for funding was a Chinese group that had been very interested earlier but withdrew when Ovshinsky demanded more control than they were willing to grant. Now Rosa wrote to them, explaining that with Ovshinsky no longer involved there would be no restrictions. They proposed supporting the research for two or three years, bringing some of their scientists to Michigan to work with the OI/OS researchers and equipment, but only if Rosa, with her physics background and knowledge of Chinese, agreed to participate. This created a painful dilemma for her. Rosa had been initially skeptical about the gigawatt project, but she recognized it had achieved results that warranted further research. She also felt it was important to give Ovshinsky's last and most ambitious vision a chance to be realized. At the same time, she was still in deep mourning and struggling with the demands of settling the estate.

Deaths, Survivals, and Revivals

She felt emotionally and practically unable to take on another big commitment and so told the Chinese she could not get involved. For years she continued to regret the loss of this last chance.

The OI/OS assets were liquidated; the same firm that had disposed of ECD's physical assets and intellectual property auctioned off the expensive, specialized laboratory equipment. "It took us until about February 2013 to dismantle everything," Dave Strand recalled. Like the bankruptcy of ECD, it was another sad ending, but unlike that earlier debacle, the termination of Ovshinsky's unrealized gigawatt vision was not a case of total failure but an interrupted story of partial success.

Ovonyx

For two years after Ovshinsky left ECD, Ovonyx continued to develop phase-change memory. In 2009, for unknown reasons, this work stopped abruptly and the company became dormant, staffed only by its leader Tyler Lowrey and a skeleton administrative crew. In 2012, after ECD's bankruptcy, its 38.6% share, including the phase-change memory patents, was sold to Micron.

Little more was heard for some time, but in July 2015 it became clear that the development of phase-change memory had been continuing. That was when Intel and Micron announced their new 3D Xpoint memory chip, "a major breakthrough in memory process technology and the first new memory category since the introduction of NAND flash in 1989."[7] Intense speculation and debate followed about the composition of the new device, which was not initially disclosed.[8] Nearly six months later, Intel and Micron confirmed what experienced ECD veterans had already deduced, that this revolutionary new device was essentially the same as the one they had created for Ovshinsky in 1989 (see chapter 10). "Chalcogenide material and an Ovonyx switch are magic parts of this technology with the original work starting back in the 1960s," said an Intel-Micron executive.[9] While not directly acknowledging Ovshinsky as the inventor, this clearly indicates the origins of the 3D Xpoint, which uses Ovonic phase-change memory as its storage element and an Ovonic threshold switch as the access device. Those who helped to develop phase-change memory at ECD and Ovonyx had thought it might be decades before flash memory reached its limits and phase-change came into its own, but that may already be happening. It seems likely that Ovshinsky's breakthrough discovery in his storefront will play an important role in the information technology of the twenty-first century.[10]

The Cognitive Computer

A further development of phase-change memory, the cognitive computer never went beyond the research stage at ECD (see chapter 10), but it too now seems to be coming closer to realization. In August 2016, IBM announced its development of an "artificial neuron," a device that will be able to "handle huge volumes of data at a fraction of the energy cost of conventional chips."[11] From the description of this device, it appears to be essentially the same as the cumulative phase-change switches Ovshinsky reported in 2008.[12] It is "made from a chalcogenide-based crystal," and "fires when it reaches a certain threshold," changing "from an ordered crystalline structure to a more glass-like amorphous state."[13] Again, there is no acknowledgment of Ovshinsky as the inventor of phase-change memory, much less any recognition that his cognitive computer had anticipated IBM's work. But this "artificial neuron" further confirms the increasing importance of his fundamental discovery.[14]

Hope in a New Barn

Others are also working to revive and extend Ovshinsky's work. One of the people who bid at the Ovshinsky Solar auction was the former ECD electrical engineer Guy Wicker. Spending down his savings, he bought a considerable amount of the equipment, which he used to set up a high-tech lab in the barn behind his house in Southfield, Michigan.[15] Ovshinsky would have smiled, remembering the old barn he rented in 1946 to start Stanford Roberts. Wicker planned to conduct experiments following in Ovshinsky's footsteps with a small team of other former ECD scientists, including Boil Pashmakov and Marshall Muller. Their hope was to resume the effort to make solar energy cheap enough to replace fossil fuels. In addition to the OS equipment, they received the OI patents and the right to use the name Ovshinsky Innovation from the Ovshinsky Foundation.[16] An article in *Crain's Detroit Business* from 2014 describes their solar program, proclaiming, "Though Ovshinsky and his companies may have died, the dream of affordable solar didn't die with them."[17] Wicker and his colleagues hope to build on Ovshinsky's later work to produce a more efficient and lower cost thin-film, flexible solar cell that can compete with the Chinese.

When Intel and Micron announced their 3D Xpoint memory chip in July 2015, Wicker and his colleagues recognized the structure they had built over twenty-five years earlier and knew, as was later acknowledged, that it must be based on Ovonic chalcogenide phase-change and threshold switches. They began to explore possibilities of developing the technology further in ways that would not be covered by the existing patents. (Micron acquired all the Ovonyx intellectual property.) Competing with the two technology giants is a formidable challenge, but at least one other chipmaker has been willing to fund their research. No matter how the story of the revived Ovshinsky Innovation develops, it is already clear that the important role Ovshinsky long ago envisioned for his amorphous devices will continue to grow in the information economy of the future.

Conclusion

Despite his scientific breakthroughs and technological achievements, despite the dozens of national and international honors and awards he received (listed in Appendix II), Ovshinsky would often worry that in the end, he and his life's work would end up as only a footnote in the chronicles of science and invention. The growing importance of his phase-change memory should give him much more prominence than that, and already he has gained the kind of recognition that eluded him in his lifetime.

In May 2015 an event occurred that Ovshinsky had always hoped for but eventually stopped believing would ever happen. For his more than four hundred patents, for "dramatic improvements in battery technology, electronics and solar power, with special recognition for the invention of the first working nickel-metal hydride battery," he was posthumously inducted into the National Inventors Hall of Fame. Ovshinsky would have been pleased for many reasons. In the induction ceremony he was described as "a prolific self-taught inventor and physicist whose pioneering work in multiple fields had an impact on many aspects of modern life." Equally important, the citation added, "Ovshinsky was known for his passion to use science and technology to solve social problems with the goal of bettering the world and the quality of life for humanity."[1] This recognition concisely notes several important features of Ovshinsky's work. In concluding, we want to enlarge that sketch and propose some further ways of understanding his achievements and distinctive qualities.

We have called Ovshinsky "the man who saw tomorrow" to emphasize the way he recognized possibilities and grasped consequences more quickly and surely than others. This ability is most apparent in his frequently accurate predictions, predictions that initially provoked scornful disbelief, whether he was foreseeing flat screen televisions or telling General Motors engineers that their electric car would go 200 miles on his batteries. But it was already present in his early days of learning to be a machinist, when he saw that grinding his tool bits differently would reduce friction and heat, the beginning of the process that led to his first invention, the Benjamin Lathe.

Figure 15.1
National Inventors Hall of Fame banner.

Once he and Iris had started ECL, his inventive work became firmly linked with his social ideals, and his anticipations of the future joined scientific insight with the vision of a better and more beautiful world. In a small laboratory beaker he could see the possibility not only of a new kind of battery but also of using it in electric cars because he was already seeking ways to replace fossil fuels, just as from tiny experimental solar cells he looked forward to producing them by the mile, part of the effort to make solar energy cheaper than coal that continued to the end of his life. His grandest visions, like the hydrogen economy, have not yet been realized and may never be, but their scope and ambition were typical products of a mind that was always eagerly reaching into the future.

Ovshinsky's most important scientific achievements also arose from his ability to see possibilities that others could not imagine: the enormous potential of amorphous and disordered materials. Where others had seen only defects in their irregularity, he realized that, unlike rigidly structured crystals, they offered the flexibility of compositional freedom, allowing for what he called "atomic engineering" in the design of materials with the properties he needed, whether for semiconductors or battery terminals. And because amorphous semiconductors can expand to cover large areas, devices like thin-film solar panels or flat panel displays became imaginable and achievable.

Beyond individual inventions and new materials, Ovshinsky always aimed for making connections and creating larger organizations. For him, every technological innovation started with the material and ended as a system, with the manufacture of affordable products that could change the world for the better. He often proudly told visitors to ECD, "We invented the materials. We invented the systems. We invented the manufacturing technologies."

Pervading all of Ovshinsky's work, whether as an independent inventor or as the leader of ECD, are his insistence on intellectual freedom and his resistance to arbitrary divisions and constraints. All his insights, and the technological innovations they led to, blend advanced physical science with commercializable technology, blurring the distinction between them. Just as his creation of new materials depended on the freedom to mix many elements, Ovshinsky claimed the freedom to cross or ignore disciplinary boundaries. To his more conventional critics, his way of mingling disciplines and his intuitive approach (not to mention his self-promotion and exaggerated claims) discredited him as a scientist, while many eminent scientists admired and were glad to work with him. Rather than rehearse the controversies that surrounded his work, the recognition or rejection he received, we find it more useful to consider a remark by the physicist Richard Zallen, an expert on amorphous solids and the author of the leading text in the field, *The Physics of Amorphous Solids*: "What Stan does isn't science. What Stan does is more interesting than science."

Both parts of Zallen's statement are significant. To say that Ovshinsky's work isn't science is clearly not a criticism here but a description: what he did was not confined to the methods and goals of an established discipline (though his scientific publications show that some of his results were recognized as disciplinary contributions). To sense how what he did could seem more interesting than science, we can recall the invention of Ovshinsky's first amorphous device, the Ovitron. He conceived it as his "nerve cell analogy," a model for testing his theory of how neurons work, and yet the result was not a neurological discovery but a new kind of switch that led to his later threshold and memory devices. Scientific and technological strands were so closely interwoven in this discovery process that trying to separate them seems pointless. Ovshinsky was deeply engaged in neuroscience, writing papers, giving talks, and doing laboratory experiments, yet that interest had first arisen from his work in automating the Benjamin Lathe. And while the Ovitron did not in fact work quite the way he thought and so did not precisely correspond to his conception of the nerve cell, his focus on the role of the cell membrane came from an intuition that led to his use of thin films in many of his most important technologies. This mix of elements and their unexpected results can indeed make what Ovshinsky did seem more interesting than the disciplined work of normal science.[2]

Demonstrating the possibilities of amorphous materials itself involved a rejection of what Ovshinsky called "the tyranny of periodic constraints," the exclusive focus on crystals. The political resonance of his terms indicates the way his inventive work is linked to his social values, not only in its aims but also in its methods. The intellectual freedom he claimed to make all kinds of connections, unconstrained by disciplinary boundaries, was inseparable from political freedom, from resisting social pressures to conform.

Finally, returning to the historical perspective we proposed at the end of the introduction, we consider once more how Ovshinsky's work was a part of the changing world he lived in, how his career spanned and contributed to the transition from the industrial to the information age. There we noted (and in chapter 5 developed further) the way his path from the shop floor to the research laboratory suggests an alternative genealogy, in which important new information technologies emerge from the old industrial world. Here we want to stress how Ovshinsky's unchanging social values kept him connected with his industrial roots. Had Ovshinsky moved to Silicon Valley, Joi Ito remarked, he would have been a billionaire. That was not Ovshinsky's aim: "I never had any intention of becoming a billionaire." He said he preferred to work "in the belly of the beast. Where else do you struggle? And without struggle we can't change the world."

Choosing to remain in Detroit, the declining capital of the industrial age, and struggling to transform it set Ovshinsky somewhat aside from the ascendant information economy. He insisted on energy and information as "the twin pillars of the global economy," and he felt his energy technologies held the most promise for changing the world—indeed for saving it from the environmental consequences of industrialization. And unlike those who simply celebrate the economic and cultural shift from manufacturing to information work, Ovshinsky stayed true to his industrial roots. He always kept ECD involved in manufacturing and promoted his solar and hydrogen technologies as the means for creating new manufacturing industries.[3] While he pioneered those high technologies, Ovshinsky never lost his love for the world of shops and factories that first drew him to become a machinist and toolmaker. "To me," he said, "manufacturing has always had glamour to it," and he saw his inventions as a way to help restore the social and economic benefits of that world.

Harley Shaiken summed up the career of his old mentor and comrade by locating him in relation to both the industrial past and the future he envisioned. "He was the last of his kind. Henry Ford transformed the 20th century with a moving assembly line and a car that was suited to mass production. Stan Ovshinsky did what Ford did but he really went beyond him in that he also developed the science that allowed new materials and new approaches that laid the basis for a global transformation in energy and information."[4] Ovshinsky himself had suggested such a comparison when he juxtaposed pictures of ECD's first roll-to-roll solar machine with ones of Ford's Model T and assembly line (see chapter 8), proclaiming the hope that his mass production of thin-film solar panels with its economies of scale would have a transformative effect like Ford's.

Looking back at the comparison, however, we can see not only the parallels but also an important divergence, for Ovshinsky's whole career was dedicated to aims and values opposed to Fordism, with its standardized mass production that subordinated workers to the demands of technological rationality.[5] He advocated automation as an alternative to the repetitive routine of the assembly line, and the community he and Iris created at ECD was dedicated to nurturing individuals and helping them to realize their potential. And of course the alternative energy technologies he developed were aimed at undoing the effects of rising fossil fuel consumption for which Ford's Model T could also be an emblem. If there is to be the kind of better future he envisioned, his inventions and his example will have helped make it happen.

Appendix I: Interviews

We have greatly benefitted from oral history interviews with Stanford Ovshinsky, as well as with his family, friends, colleagues, and former staff. LH denotes Lillian Hoddeson, and PG denotes Peter Garrett. Transcripts and voice files of the interviews listed below are in the possession of Hoddeson and will be added to the Ovshinsky papers.

Interviews by the Authors

Nancy Bacon (LH and PG): Feb. 15, 2013.

Lee Bailey (LH and PG): Jan. 21, 2016.

Arthur Bienenstock (LH): Apr. 17, 2007.

Dick Blieden (LH): May 13, 2008, Oct. 31, 2008.

Vin Canella (LH and PG): Apr. 22, 2013.

Ben Chao (LH): Dec. 18, 2009; (LH and PG): Apr. 23, 2013.

Morrell Cohen (LH): Jan. 15, 2007.

Dennis Corrigan (LH): May 9, 2008; and (LH and PG): Apr. 25, 2013.

Wally Czubatyj (LH): Feb. 16, 2007, Mar. 11, 2010.

John de Neufville (LH): June 1, 2009.

Subhash Dhar (LH): Oct. 31, 2008; and (LH and PG): Apr. 25, 2013, Nov. 4, 2016.

Robin Dibner (LH): Jan. 16, 2007, Jan. 11–12, 2013; and (LH and PG): Nov. 6, 2014.

Steven Dibner (LH): Apr. 18, 2007; and (LH and PG): Sept. 29, 2013, Nov. 21, 2015.

Joe Doehler (LH): Dec. 13, 2006.

Elif Ertekin (LH and PG): May 11–12, 2016.

Ed Fagen (LH): Oct. 11, 2010.

Bruce Falls (LH): Dec. 27, 2008.

Julius and Mete Feinleib (LH): May 1, 2007.

Mike Fetcenko (LH and PG): Feb. 14, 2013.

Richard (Dick) Flasck (LH and PG): Sept. 27, 2013.

Hellmut Fritzsche (LH): Dec. 23, 2006, Feb. 15, 2007, Mar. 6, 2007, Mar. 8, 2007, Sept. 9, 2007, Feb. 21, 2010, Sept. 5, 2012, May 3, 2013; and (LH and PG): Nov. 12–14, 2014, May 15–20, 2015, Oct. 17, 2016.

Subhendu Guha (LH and PG): Apr. 24, 2013.

Steve Heckeroth (LH): June 15, 2010.

Eric Hintz (LH): Oct. 18, 2009.

Steve Hudgens (LH): Feb. 15, 2007, Jan. 2, 2009.

Steve and Coleen Hudgens (LH and PG): Nov. 22, 2015.

Joi Ito (LH): Dec. 14, 2008.

Mimi Ito (LH): Dec. 26, 2008.

Alice and Sato Iwasa (LH): Apr. 30, 2007.

Masat Izu (LH and PG): Jan. 12, 2013.

Robert Johnson (LH): June 29–30, 2009.

Jim Kakalios (LH): Nov. 16, 2007.

Chet Kamin (LH): Mar. 25, 2009; and (LH and PG): Jan. 8, 2016.

Gordon Kane (LH): Apr. 1, 2012.

Mark Kastner (LH): May 1, 2007.

Ghazaleh Koefod (LH): Mar. 11, 2011.

Alex Kolobov (LH and PG): June 30, 2016.

Arun Kumar (LH and PG): Apr. 23, 2013.

Tyler Lowrey (LH): Feb. 16, 2007.

William Lipscomb (LH): May 2, 2007.

John Marine (LH): Mar. 7, 2007.

Eugenia Mytilineou (LH): Aug. 7, 2007.

Interviews

Larry and Barbara Norris (LH and PG): Dec. 27, 2008.

Ben Ovshinsky (LH): Apr. 19, 2007, Sept. 26–29, 2013, Jan. 20, 2016; and (LH and PG): Nov. 23, 2015, Oct. 27, 2016.

Harvey Ovshinsky (LH): Apr. 30, 2009; and (LH and PG): Jan. 12, 2013.

Harvey and Cathie Ovshinsky (LH and PG): Apr. 21, 2013.

Herb Ovshinsky (LH): Dec. 13, 2006, Feb. 16, 2007, Mar. 5, 2007, May 9, 2008, Apr. 24, 2009, Mar. 11, 2011, Jan. 14, 2013, July 19–22, 2016, Sept. 23 and 29, 2016.

Herb Ovshinsky and Rosa Young Ovshinsky (LH and PG): Sept. 8, 2015.

Iris Ovshinsky (LH): Jan. 6, 2006, July 20, 2006.

Iris and Stan Ovshinsky (LH): Jan. 4–6, 2006, July 19–20, 2006, Aug. 16, 2006.

Iris and Stan Ovshinsky, Dick and Nancy Blieden (LH): July 19, 2006.

Rosa Young Ovshinsky (LH): Dec. 13, 2006, Apr. 23, 2009, Dec. 18, 2009; and (LH and PG): Apr. 24, 2013, Feb. 7–10, 2014, Aug. 17, 2016.

Rosa Young Ovshinsky and Eugenia Mytilineou (LH): Aug. 7, 2007.

Stan Ovshinsky (LH): July 19, 2006, Aug. 16, 2006, Dec. 11, 2006, Dec. 13, 2006, Mar. 6, 2007, Aug. 6, 2007, Oct. 5, 2007, Dec. 13, 2007, May 6, 2008, Oct. 30–31, 2008, Apr. 22, 2009, May 21, 2009, July 13, 2009, Sept. 2, 2009, Dec. 16–18, 2009, Mar. 10–12, 2010, Oct. 8, 2010, Mar. 9–11, 2011, Sept. 28, 2011, Oct. 27–28, 2011, Dec. 11–12, 2011, Feb. 9, 2012.

Stan Ovshinsky and Hellmut Fritzsche (LH): Feb. 15, 2007, Mar. 5, 2007, Mar. 6, 2007, May 7, 2008, Dec. 18, 2009.

Stan Ovshinsky, Hellmut Fritzsche, and Brian Schwartz (LH): Feb. 15, 2007.

Stan Ovshinsky, Helmut Fritzsche, and Jeffrey Wilhite (LH): Mar. 7, 2007.

Stan Ovshinsky and Freya Saito (LH): Oct. 31, 2008.

Stan Ovshinsky and Dave Strand (LH): May 6, 2008.

Boil Pashmakov (LH): Mar. 8, 2007; and (LH and PG): Feb. 11, 2013.

Max Powell (LH): Mar. 7, 2007.

Max Powell and Herb Ovshinsky (LH): Dec. 17, 2009.

Benny Reichman (LH and PG): Apr. 23, 2013.

Lionel and Delores Robbins (LH and PG): Sept. 9, 2015.

John Ross (LH): Apr. 17, 2007.

Freya Saito (LH): Aug. 8, 2007.

Krishna Sapru (LH): June 2, 2009.

Brian Schwartz (LH): Jan. 16, 2007, Feb. 17, 2007.

Mel Shaw (LH and PG): Apr. 22, 2013.

Harley Shaiken (LH): Apr. 16, 2007; and (LH and PG): Nov. 23, 2015.

Charlie Sie (LH): Dec. 28, 2008.

Marvin Siskind (LH and PG): Feb. 14, 2013.

Robert (Bob) Stempel (LH): Apr. 24, 2009.

Dave Strand (LH): May 8, 2008; (LH and PG): Jan. 10, 2013, Oct. 6, 2014.

Srinivasan Venkatesan (LH and PG): Apr. 25, 2013.

Meera Vijan (LH): Apr. 24, 2009.

Guy Wicker (LH and PG): Feb. 14, 2013, Sept. 9, 2015, July 22, 2016.

Guy Wicker and Boil Pashmakov (LH and PG): Feb. 11, 2013.

Guy Wicker, Rosa Young Ovshinsky, et al. (LH and PG): Apr. 26, 2013, Sept. 12, 2015.

B. J. and Barb Widick (LH): Mar. 8, 2007.

Jeff Yang (LH and PG): Jan. 14, 2013.

Zvi and Monica Yaniv (LH and PG): Aug. 2, 2015.

Irina Youdina (LH): Feb. 14, 2013; and (LH and PG): Feb. 13, 2013.

Richard and Doris Zallen (LH): July 30, 2009.

Interviews Audio- or Videotaped by Harvey K. Ovshinsky

Stan Ovshinsky: Dec. 21, 1993, Mar. 4, 1996, June 13, 1996, Dec. 28, 2001, April 25, 2002, June 27, 2002, Feb. 6, 2003, Oct. 12, 2010.

Iris Ovshinsky: June 13, 1996.

Steve Hudgens: Mar. 4, 1996, Mar. 26, 1996.

Subhendu Guha: June 27, 2002.

Appendix II: Ovshinsky's Major Honors and Distinctions

1968   Diesel Gold Medal, presented by the German Inventors Association (Deutscher Erfinderverband) for discovery of the semiconductor switching effect in disordered and amorphous materials

1983   Induction into the Michigan Chemical Engineering Hall of Fame

1985   Fellow, American Physical Society, presented for contributions to understanding and applications of amorphous electronic materials and devices

1987   Fellow, American Association for the Advancement of Science

1988   Coors American Ingenuity Award, presented for work in amorphous materials, particular photovoltaics

1991   Inducted into Coors American Ingenuity Hall of Fame

1991   Toyota Award for Advancement, presented for Ovonic nickel-metal hydride batteries for electric vehicles

1992   Honorary member, International Association of Machinists and Aerospace Workers, Local Lodge PM2848

1993   Corporate Detroiter of the Year, *Corporate Detroit* magazine

1999   Named as one of the "Heroes for the Planet" by *Time* magazine

1999   Karl W. Böer Solar Energy Medal of Merit, awarded jointly by the University of Delaware and the International Solar Energy Society

2000   International Association for Hydrogen Energy Sir William Grove Award

2000   (with Iris Ovshinsky) "Heroes of Chemistry," presented by the American Chemical Society

2004   Hoyt Clarke Hottel Award of the American Solar Energy Society

2005   Innovation Award for Energy and the Environment, presented by *The Economist*

2005 Induction into US Solar Hall of Fame

2006 Frederick Douglass / Eugene V. Debs Award

2007 Walston Chubb Award for Innovation, presented by Sigma Xi, the Research Society

2008 Engineering Society of Detroit Lifetime Achievement Award

2008 Environmental Hall of Fame Award, Solar Thin Film Category, Father of Thin-Film Solar Energy

2009 IEEE Vehicular Technology Society Presidential Citation, in recognition of a long and outstanding record of pioneering accomplishments and service to the profession

2009 Thomas Midgley Award, presented by the Detroit Section of the American Chemical Society

2012 Finalist for European Inventor Award, presented by the European Patent Office for development of NiMH batteries

2012 Honorary Calgarian award, presented by the Louis Riel School in Calgary, Canada

2015 Posthumous induction into National Inventors Hall of Fame, for invention of nickel metal hydride battery

Honorary Doctorates

1980 Science—Lawrence Technological University, Southfield, Michigan

1981 Engineering—Bowling Green State University, Bowling Green, Ohio

1989 Science—Jordan College, Cedar Springs, Michigan

2007 Science—Kean University, Union, New Jersey

2008 Science—New York Institute of Technology, Old Westbury, New York

2009 Science—Wayne State University, Detroit, Michigan

2009 Engineering—Illinois Institute of Technology, Chicago

2009 Engineering—Ovidius University, Constanţa, Romania

2010 Engineering—Kettering University, Flint, Michigan

2010 Science—University of Michigan, Ann Arbor

Notes

Preface

1. Nevill Mott (ed.), *The Beginnings of Solid State Physics* (London: Royal Society of London, 1980). Presently the more comprehensive name of "condensed matter physics" is used. For a careful historical discussion of the relationships between condensed matter physics and solid-state physics, see Joseph D. Martin, "What's in a Name Change? Solid State Physics, Condensed Matter Physics, and Materials Science," *Physics in Perspective* 17, no. 1 (2015): 3–32.

2. Lillian Hoddeson, Ernest Braun, Jürgen Teichmann, and Spencer Weart (eds.), *Out of the Crystal Maze: Chapters from the History of Solid State Physics, 1900–1960* (New York: Oxford University Press, 1992). Reference to additional work on this history can be found in Joseph D. Martin, "Resource Letter HCMP-1: History of Condensed Matter Physics," *American Journal of Physics* 85, 87 (2017); doi:10.1119/1.4967844 http://dx.doi.org/10.1119/1.4967844.

3. Lillian Hoddeson and Vicki Daitch, *True Genius: The Life and Science of John Bardeen* (Washington, DC: Joseph Henry Press, 2002).

4. To sense the cathedral-like space of the plant, see the pictures of the machine at the end of chapter 8.

5. For an account of the "heroic age" of independent inventors in the later nineteenth and early twentieth centuries and their eclipse by the new industrial laboratories after World War I, see Thomas P. Hughes, *American Genesis: A Century of Invention and Technological Enthusiasm 1870–1970* (New York: Viking, 1989). More recent studies have qualified this narrative, showing that the change was more gradual. See Peter Whalley, "The Social Practice of Inventing," *Science, Technology, and Human Values* 16, no. 2 (Spring 1991): 200–232; and Eric S. Hintz, "The Post-Heroic Generation: American Independent Inventors, 1900–1950," *Enterprise and Society* 12, no. 4 (December 2011): 732–748.

6. For a discussion of normal human memory, see e.g., Eric R. Kandel, *In Search of Memory: The Emergence of a New Science of Mind* (New York: W. W. Norton, 2007). For an essay from the point of view of a historian who uses memories as sources, see Lillian Hoddeson, "The Conflicts of Memories and Documents: Dilemmas and Pragmatics of Oral History," in *The Historiography of Contemporary Science, Technology and Medicine: Writing Recent Science,* ed. Ron Doel and Thomas

Söderquist (London and New York: Routledge, 2006), 187–200. For a history of memory sciences, see Alison Winter, *Memory: Fragments of a Modern History* (Chicago: University of Chicago Press, 2012).

Introduction

1. Nearly everyone called him "Stan," as we do when telling about his personal life. When discussing his career as an inventor, however, we use "Ovshinsky."

2. The first desktop computer seems to have been the Olivetti Programma 101, which was marketed in 1965. It was basically a programmable calculator and very expensive. Prototype flat screen displays had also been produced in 1964 at the University of Illinois, but the plasma display was monochrome, expensive, and had poor resolution. Ovshinsky's materials offered the possibility of much better and cheaper devices.

3. Stanford R. Ovshinsky, "Reversible Electrical Switching Phenomena in Disordered Structures," *Physical Review Letters* 21, no. 20 (November 11, 1968): 1450–1453.

4. Unlike crystals, amorphous materials do not have a completely regular, repeated atomic structure, which was generally believed necessary for making semiconductor devices. As the term is usually used, "amorphous" is a subset of "disordered." There are degrees of disorder in materials, ranging from weak to strong. If the degree of disorder in a material is strong enough, it is considered amorphous. Materials composed of multiple elements, like alloys, are disordered whether or not they include amorphous elements. (The NiMH battery is an important instance of such disorder. See chapter 10.) Ovshinsky emphasized how "by using many different elements I was able to design many synthetic new materials. And by doing that I was able to get new mechanisms—electronic, chemical, structural. So disorder was a working tool to give me degrees of freedom to design materials that would not fit in any periodic structure." See also S. R. Ovshinsky, "Amorphous and Disordered Materials—The Basis of New Industries," *Materials Research Society Symposium Proceedings* 554 (1999): 399–412.

5. For a history of the rise of materials science as a discipline, see Bernadette Bensaude-Vincent, "The Construction of a Discipline: Materials Science in the United States," *Historical Studies in the Physical Sciences* 31, part 2 (2001): 223–248; for a comprehensive technical history as well as an institutional account, see R. W. Cahn, *The Coming of Materials Science* (Oxford: Pergamon, 2001).

6. "The Edison of Our Age?" *The Economist*, December 2, 2006, http://www.economist.com/node/8312367.

7. On the concept of technological systems, see Thomas P. Hughes, *Networks of Power: Electrification in Western Society, 1880–1930* (Baltimore: Johns Hopkins University Press, 1983). As Hughes and later historians of technology emphasize, for any invention to have a significant impact, it has to be integrated in a larger system. Ovshinsky clearly grasped this from the very beginning of his energy work, although he was never able to implement his systems fully.

8. Interview on October 27, 1987, for the NOVA documentary produced by Marian Marzynski, *Japan's American Genius* (Boston: WGBH, 1987). Edison's most authoritative biographer confirms

Wilson's point: "Edison ultimately was an inventor, not a scientist." Paul Israel, *Edison: A Life of Invention* (New York: John Wiley & Sons, 1998), 471. Another possible comparison might be made with Steve Jobs, who is much better known than Ovshinsky—and in contemporary popular culture probably better known than Edison. But although Jobs was brilliantly successful at designing innovative consumer products, he was not really an inventor. See Walter Isaacson, *Steve Jobs* (New York: Simon & Schuster, 2011). A more relevant contemporary comparison would be Elon Musk, who is, like Ovshinsky, a technological visionary who has developed a clean energy system linking solar power, batteries, and electric cars (not to mention his even more visionary projects of space exploration and colonizing Mars). But what Wilson says about the difference between Ovshinsky and Edison also applies here: Musk's projects do not depend on the kind of new science that Ovshinsky's discoveries produced.

9. Like Wilson's comparison of Ovshinsky and Edison, these words were spoken in an interview. Throughout, whenever no specific reference is given for any quotation it has been taken from one of the interviews listed in appendix I.

10. Norbert Wiener, *Cybernetics: Or Control and Communication in the Animal and the Machine* (Cambridge, MA: MIT Press, 1948).

11. "The effect by which certain kinds of glass films that are normally nonconductive become semiconductors upon the application of a small voltage." *The American Heritage Dictionary of the English Language* (Boston: Houghton Mifflin, 2000). Actually, they are already semiconductors. A more accurate definition would be "glass films that are normally nonconductive become conductive."

12. This kind of creativity, making novel connections by placing ideas in a new context, appears in the accounts other innovators have given of their thought processes. Linus Pauling describes a similar approach: "One thing that I do is to bring ideas from one field of knowledge into another field of knowledge." In Mihaly Csikszentmihaly, *Creativity: Flow and the Psychology of Discovery and Invention* (New York: HarperCollins, 1996), 118. For an account of how such connections also worked in ECD's concurrent research and development programs, see chapters 6 and 7.

13. Analogy has been studied by cognitive scientists as a motor of cognition both in everyday life and in scientific and technological innovation. See Dedre Gentner, Keith J. Holyoak, and Boicho N. Kokinov, eds., *The Analogical Mind: Perspectives from Cognitive Science* (Cambridge, MA: MIT Press, 2001), esp. 1–19 in chapter 1 (the introduction) and all of chapter 15, Douglas R. Hofstadter's "Epilogue: Analogy as the Core of Cognition," 499–538. Also see Lillian Hoddeson, "Analogy and Cognitive Style in the History of Invention: Inventor Independence and Closeness of Compared Domains," *Proceeding of EuroCogSci07*, the European Cognitive Science Conference, ed. Stella Vosniadou, Daniel Kayser, and Athanassios Protopapas (New York: Lawrence Erlbaum Associates, 2007), 413–418.

14. Other scientists have also gained insights from thinking in pictures, most notably Einstein in the "thought experiments" that led to the theory of special relativity. See Gerald Holton, *The Scientific Imagination: Case Studies* (Cambridge: Cambridge University Press, 1978).

15. In Yiddish, *a shenere un a besere velt*. In addition to furthering social justice, the Workmen's Circle also served to transmit Yiddish culture. It included workers from many trades—carpenters, peddlers, painters, toolmakers, and tailors. Influential in the labor movement in its early decades, the organization exists to this day, offering schools, camps, retreats, old age homes, health insurance, and a wide variety of lectures, concerts, and other events. In various ways, three generations of Ovshinskys were a part of it. See "A Brief History of the Workmen's Circle / Arbeter Ring," in the pamphlet, *Stand Up & Celebrate! The Workmen's Circle Arbeter Ring 2004 Gala Celebration*, Stanford R. Ovshinsky papers, Bentley Historical Library, University of Michigan (hereafter Ovshinsky papers).

16. Patent litigation was an occupational hazard and source of stress for most independent inventors, who could ill afford the legal expenses of conflicts with large corporations. A well-known case is radio pioneer Edwin Armstrong, whose struggles with giants like RCA cost him over a million dollars and eventually drove him to suicide. See Lawrence Lessing, *Man of High Fidelity: Edwin Howard Armstrong* (New York: Bantam Books, 1969); Tom Lewis, *Empire of the Air* (New York: Harper-Collins, 1991); and Thomas Hughes, *America Genesis: A Century of Invention and Technological Enthusiasm 1870–1970* (New York: Viking, 1989), 141–150.

17. See Michael Riordan and Lillian Hoddeson, *Crystal Fire: The Birth of the Information Age* (New York: W. W. Norton and Co., 1997).

18. Ovshinsky's lack of a proper academic pedigree and his use of press conferences to publicize his work were some of the reasons for his rejection by those who felt such discoveries "should" have come from their own, more prestigious institutions. The disturbing impurity of his materials themselves can be sensed in the astonishment of the physicist Hellmut Fritzsche, who became Ovshinsky's first scientific consultant, when he was told that their exact composition was not critically important and was shown that rough handling and contamination did not affect their performance (chapter 5). The initial resistance to Ovshinsky was a response to both social and physical impurity.

19. See S. R. Ovshinsky, "Amorphous and Disordered Materials: The Basis of New Industries," *Materials Research Society Symposium Proceedings* 54 (1999): 339–412.

Chapter 1: Young Years (1920s–1930s)

1. For a comparable event in 2002 featuring Ovshinsky as the visiting industrialist, see the last section of chapter 9.

2. See Hugh Allen, *Rubber's Hometown* (New York: Stratford House, 1949); John Tully, *The Devil's Milk: A Social History of Rubber* (New York: Monthly Review Press, 2011), 153–181.

3. On the predominance of East European Jews in the New York garment industry, see Irving Howe, *World of Our Fathers: The Journey of the East European Jews to America and the Life They Found and Made* [1976] (New York: Book-of the-Month Club, 1993), 154–159; for a social history of Jewish assimilation into American culture, see Neil M. Cowan and Ruth Schwartz Cowan, *Our*

Parents' Lives: Jewish Assimilation and Everyday Life (New Brunswick, NJ: Rutgers University Press, 1989).

4. Although he disliked factory work, Ben was impressed by the man who was hired by the workers to read them great literature on the job. This tradition, originating in the cigar industry, was one of the ways immigrant workers tried to acquire cultural literacy.

5. Harvey Leff, "Reminiscences and Appreciations of Stan Ovshinsky," in *Reminiscences and Appreciations: Presented to Stanford R. Ovshinsky on the Occasion of his 80th Birthday*, electronic manuscript, Ovshinsky papers.

6. Leff, "Reminiscences."

7. Stan recalled that to make room for them, he would give up his bed and sleep on two chairs.

8. There were striking parallels in Stan's first marriage (see chapter 2).

9. Herb explained that she took her second name because of her life-threatening health problems. "One of the superstitious Jewish things to do when you have a close-to-death experience is that you take another name to fool the angel of death."

10. When Stan was almost ten and Herb was four, the family moved from their apartment on Moon Street to a much nicer, big rented house on Euclid Avenue where they lived until 1937, when they bought their own house on Leonard Street.

11. On the role of unions, socialism, and the Workmen's Circle in Jewish immigrant life, see Howe, *World of Our Fathers*, 287–324, 357–359.

12. Stan's sister Mashie told of going to Cleveland with Ben to see a play starring Muni. They went backstage afterward, and when Muni, who was known to be very shy and had difficulty showing emotion, saw Ben, he shrieked "Schmieke!" and put his arms around him and picked him up.

13. When Stan was on his deathbed, he and Herb sang some of these songs. See chapter 13.

14. There is an interesting suggestion amid Stan's recollections of his frustrating school experience with mathematics. He recalls the teacher telling the class to hand in their math problems and then adding, "Stan, you can hand in your drawings." It may be that already at this early stage (fifth or sixth grade) he was trying to compensate for his deficiency in abstract mathematical thinking with visualization. "Before I solve a problem," he would later say, "I have to see it." Stan's characteristic learning style may not have helped him in grade school, but later his ability to visualize atomic structures made a vital contribution to his accomplishments as an inventor. In the years when they lived on Moon Street, Stan attended elementary school at the Samuel A. Lane School; while living on Euclid Avenue, he went to George W. Crouse, Sr., School.

15. The impact on young inventors of such popular science magazines as *Science and Invention*, *Popular Invention*, and *Radio News* (all published by Gernsback), is illustrated in Evan I. Schwartz, *The Last Lone Inventor: A Tale of Genius, Deceit, and the Birth of Television* (New York: HarperCollins, 2002), 8, 19–20, 129, 133, 145.

16. The term "sewer socialist" was coined to disparage the constructive programs of the Milwaukee socialists who focused their efforts on practical improvements like public sanitation. See the Wisconsin Historical Society discussion in "Turning Points of Wisconsin History," http://www.wisconsinhistory.org/turningpoints/tp-043/?action=more_essay.

17. Widick came to America from Serbia at age three, shortly before World War I. After graduating from the University of Akron, he became a journalist at the *Akron Beacon Journal* and was heavily involved in the 1930s labor movement, assisting in the United Auto Workers Flint-GM sit-down strike in 1936–1937. In the 1930s he also helped to organize the United Rubber Workers, and in the 1940s served on the national staff of the United Auto Workers. After spending time in Mexico with Leon Trotsky, Diego Rivera, and Frida Kahlo, he served in World War II and in the 1950s helped to clear many activists, including himself, before the House Un-American Activities Committee. He then studied and taught economics at Wayne State University, and moved on to become a professor in the School of Business at Columbia University, retiring in 1983. See obituaries in *Akron Beacon Journal*, July 2–6, 2008; and "B. J. Widick (1910–2008)," *Encyclopedia of Trotsky On-Line: Revolutionary History* 10, no. 1, https://www.marxists.org/history/etol/revhist/backiss/vol10/no1/wald.html. Also, "An American Idealist: An Activist's Life Adventures," unattributed copy, Ovshinsky papers.

18. Ben was also not observant, but he accepted Bertha's rules at home. (He would eat whatever he wanted elsewhere and worked on Saturdays, unlike Orthodox Jews.) Despite his decision not to observe Jewish rituals and to reject what he regarded as superstition, he nevertheless maintained a strong "cultural identity and an emotional relationship to the religion," Herb recalled, adding, "On my dad's side too, the rest of the family were deeply observant."

19. Years later, as an old man, this teacher apologized to Stan at a class reunion.

20. Stanford Robert is the English equivalent of Stan's Yiddish name, Simcha Rachmeil.

Chapter 2: Passion for Machines (1940–1944)

1. Lying about his date of birth caused him some trouble later with the Social Security Administration.

2. Shop heads in those days typically subscribed to some form of the notorious system of so-called scientific management developed by Charles Bedaux (1886–1944). Aimed at improving worker productivity, it built on the exploitative methods of Frederick Winslow Taylor. For an account of Taylorism, see Thomas P. Hughes, *American Genesis: A Century of Invention and Technological Enthusiasm* (New York: Viking, 1989), 184–248.

3. Among the practices Stan disliked was the check-off system of requiring employers to collect union dues, which suggests he valued individual freedom more than class solidarity. Consistent with his identity as a "sewer socialist" (see chapter 1), it further suggests he supported unionism as a way of improving life for workers, not as a general political commitment.

4. J. R. Williams, *The Bull of the Woods* [1944] (Almonte, ON: Algrove Publishing Ltd., 2002).

5. Stan also sent out a mimeographed notice describing the incident and announcing, "The plant will be shut down from 12:00 midnight, 10-6-41 until 12:00 midnight, 10-7-41, and your officers will appreciate your full cooperation." "Goodrich Workers," undated mimeographed notice filed with Stan's high school notes, Ovshinsky papers.

6. Thanks to Ben Ovshinsky, Stan and Norma's first son, for this information. Thanks also to Herb Ovshinsky.

7. Stan didn't recall the date, which appears here courtesy of the dossier the FBI compiled on him and which he later obtained through a FOIA request. These reports, though sometimes inaccurate, offer information unavailable elsewhere. A heavily redacted copy of the dossier is in the Ovshinsky papers.

8. About three weeks later they moved into a tiny house at 2019 North 10th Street; later, on June 6, 1943, they moved to a one-bedroom rented cottage at 920 West Adams Street. These details also come from the dossier of the watchful FBI.

9. Until Stan found something better, Norma worked nights in a drug store.

Chapter 3: Smarter Machines (1944–1952)

1. "My grandmother was generous, but she could also be very controlling," Stan's son Harvey recalled. "And I think, even then, my mother could only handle one Ovshinsky at a time."

2. "We were like relatives," Herb said. Frances's brother "Tiny" (Harold Wolinsky) would later be best man at Herb's wedding.

3. Barney continued the business until his death, and he and Stan stayed friends.

4. Ohio Department of Heath Death Certificate. Thanks to Harvey Ovshinsky for a copy. The average age to which men lived in Russia at that time was fifty-three, Stan noted.

5. Being unable to be with Ben on his deathbed was a wound that never healed for Stan. Many years later, Harvey said, "Dad was very aware that his children were with him while he was dying" (see chapter 13).

6. "He was loyal to her," Harvey recalled, "but I don't think he ever forgave 'Bubbe' for keeping him from seeing his father when he was so sick. One of the few times I ever saw Dad cry, was when he talked about how close he was with his father and how much he still missed him."

7. As we explain in the preface and introduction, we use the more formal "Ovshinsky" when narrating his career as an inventor but continue to use "Stan" in more personal contexts.

8. Schankler later worked for Ovshinsky at ECD on creating early electronic memories made from amorphous materials (see chapter 6).

9. Versions of the Benjamin Lathe include those protected by US Patents 2,619,709; 2,619,710; 2,656,588; 2,697,610; 2,699,083; and 2,699,084.

10. It is difficult to gauge the direct contribution of the Benjamin Center Drive Lathe to the development of later machine tools, but several of its features, such as increasing rigidity to eliminate vibration and performing multiple operations with the same machine, have now become common.

11. Stanford Ovshinsky, "How I Invent: Case Histories: The Benjamin Center Drive Lathe," April 6, 1997; June 23, 1997, Ovshinsky papers. This document is part of Ovshinsky's fragmentary autobiography, which he seems to have dictated to his second wife Iris.

12. "Proof of principle" or "proof of concept" are terms that seem to have come into use with the growth of information technologies in the 1970s and 1980s, but the basic idea was always important for Ovshinsky, particularly in the later years when, as head of ECD, he would assign the work of development to others.

13. M. Kronenberg, "A Report on the Benjamin Center Drive Automatic Made by the Stanford Roberts Manufacturing Company in Dover, Ohio," May 15, 1948, copy, Ovshinsky papers. These findings were later confirmed and extended in an elaborate inspection by master mechanics from General Electric. GE's report included micrograms, taken on March 8, 1949, of the structure of the white chips that came off during machining. Under the microscope the chips showed no signs of stress: the lathe's high efficiency in reducing friction had left them unchanged. GE expressed enthusiasm in several letters and promptly ordered a lathe. See, e.g., H. E. Kohler, letter to Stanford Roberts, January 11, 1950, and the report, "General Electric Data on Benjamin Center Drive Automatic Machine Efficiency," May 12, 1950, Ovshinsky papers.

14. During that year, Selma, his girlfriend and later his wife, was away at college at Ohio State.

15. Ben told us in an interview that his "first conscious memory as a child (two or three years old) is of driving up to the Dover shop, and Uncle Herb meeting the car and giving me half a salami sandwich to hold in one hand (a first!), and a micrometer to play with in the other."

16. The fact that their three young boys were born over a period of just thirty-three months amplified the growing marital problems; according to Herb, Norma suffered serious postpartum depression.

17. The FBI report states that Ovshinsky subleased quarters in the rear of 247 East Exchange Street from October 30, 1948, to September 30, 1949, and that "on or about the latter date the offices of the company were moved to Toledo, Ohio."

18. Ovshinsky had hired "the only black secretary in the city," and neighbors in other offices complained to the landlord that she used the restroom. When Ovshinsky was told to fire her or move out, he moved.

19. Only a couple of hours' drive from Akron, Toledo was a more convenient location for Herb because Selma was now living there.

20. Recent support had come from two wealthy local Jewish businessmen. As Herb recalled, one of them, Mr. Holub, ran a large scrapyard with which Ben Ovshinsky had successfully competed; the other, Mr. Nobil, owned several shoe stores.

21. This was an early form of binary programming, "the equivalent of a computer using old fashioned relays to do the logic," Herb explained. It was "literally an electrical diagram which was like a ladder with a relay card, and you hit yes/no contacts that allowed you to describe a sequence of events by controlling various components of the machine."

22. As Ovshinsky put it in an undeveloped autobiographical aside, "The question of servomechanistic controls ... led directly to my work on amorphous materials." See his autobiographical chapter, "How I Invent," Ovshinsky papers.

23. Norbert Wiener, *The Human Use of Human Beings: Cybernetics and Society* [1950] (New York: Avon, 1967), 47. For a discussion of Wiener's role in the emergence of "information" as a dominant concept in many areas of science and technology, see Lily E. Kay, *Who Wrote the Book of Life? A History of the Genetic Code* (Stanford, CA: Stanford University Press, 2000).

24. The book review, preserved (at least in part) in the Ovshinsky papers, and apparently never published, dwells less on the technical than the social aspects of cybernetics, probing the question of whether computers and automation pose a threat or can be a "liberating force for good." Drafts of the review indicate that it was written in Akron, while Ovshinsky was still at Stanford Roberts, before he left for Connecticut. According to Ovshinsky's recollections, his brief correspondence with Norbert Wiener also occurred at this time, but a letter from Wiener to Ovshinsky in May 1953, preserved in the Ovshinsky papers, suggests that either the correspondence lasted much longer than Ovshinsky remembered or that it did not take place until about two years later. Wiener wrote, "It is always gratifying to find that one's ideas have reached responsive ears." After sending "good wishes" to Ovshinsky, Wiener added, "It is my hope that these may encourage you in your endeavor to stimulate thinking about the great human problems of the future."

25. For discussions of Walter and other cyberneticians, see Andrew Pickering, *The Cybernetic Brain: Sketches of Another Future* (Chicago: University of Chicago Press, 2010).

26. The paper titled "Electro-Mechanical Motion," June 5, 1950, is apparently lost, but it is cited and quoted in Ovshinsky's later 1955 paper titled "Nerve Impulse," Ovshinsky papers. For another instance of Ovshinsky's envisioning the future, in this case sixty-six years in advance, see "Chip, Implanted in Brain, Helps Paralyzed Man Regain Control of Hand," *New York Times*, April 13, 2016, https://www.nytimes.com/2016/04/14/health/paralysis-limb-reanimation-brain-chip.html?_r=0.

27. See Jonathon D. Shea and Barbra Proko, *The Polish Community of New Britain* (Chicago: Arcadia Publishing, 2005), 7.

28. Stan may have been away from home often, and an increasingly estranged husband, but he was still a loving and engaged father. A letter he wrote from a Detroit hotel to Ben in fall 1951 offers advice and encouragement as his oldest son started school. "Listen to the teacher and learn all you can, and when you start studying we will both go into our study room and study together because I try to learn all the time too." Ovshinsky papers.

29. While Ovshinsky successfully demonstrated the superiority of the Benjamin Lathe, the superiority of welded steel bases to cast iron is still in dispute. As one current website notes, "The

debate over which platform is better still rages on." "Cast Iron or Steel Weldment," http://www.walkermachinery.net/fw/main/cast-iron-or-steel-weldments-1643.html. Cast iron may damp vibration better, and even the old-timers' belief in the need to season castings still has advocates.

30. The smaller ones were for machining artillery shells and the bigger ones for up to eight-inch US Navy shells.

31. A series of memorandums indicating where some 205 of the machines were shipped can be found with a letter from F. R. Downs, Assistant Sales Manager of the New Britain–Gridley Machine Division of the New Britain Machine Company to Stanford Ovshinsky, November 12, 1954, Ovshinsky papers. Stan and Herb also built a machine for Goodyear Aircraft that was intended for machining aluminum hubs but was soon mothballed. Also, working with a small machine tool company in Milford, they started to build a more advanced, fully automated machine that was never finished. These two lathes were later reworked and purchased by the Hupp Corporation around 1953 (see chapter 4).

32. US Patent 2,674,331: "Automatic Steering Control Apparatus for Self-Propelled Vehicles."

33. A closed loop system responds to feedback with adjustments.

34. See, e.g., Louis Bromfield, *Pleasant Valley* (1946) and Aldo Leopold, *Sand County Almanac* (1948). These were among the early books pointing toward the creation of an ecological conscience in modern agriculture.

35. Examples of performance evaluations by: Rheem Manufacturing Co., Burlington, NJ, "Outstanding"; Ekco Products Co., Chicago, "Excellent"; Chase Brass and Copper, Inc., Waterbury, CT, "By far the best cartridge case"; Ingersoll Products Div. of Borg-Warner Corp., Chicago, "Satisfactory"; Robertshaw-Fulton Controls Co., Knoxville, "Tops." A decade later, Ovshinsky wanted to have one of his machines at ECD but could not afford its cost of $10,000.

Chapter 4: Love Story (the 1950s)

1. Frank B. and Arthur M. Woodford, *All Our Yesterdays: A Brief History of Detroit* (Detroit: Wayne State University Press, 1969), 14–74.

2. See e. g., David Hounshell, *From the American System to Mass Production, 1800–1932: The Development of Manufacturing Technology in the United States* (Baltimore: Johns Hopkins University Press, 1984), especially 303–330; and Thomas P. Hughes, *American Genesis: A Century of Invention and Technological Enthusiasm 1870–1970* (New York: Viking, 1989), 203–220. Ovshinsky used the Chaplin sequence as an example of the dehumanizing effects of repetitive work in his review of Wiener's *The Human Use of Human Beings* (see chapter 3).

3. "History of General Motors," https://en.wikipedia.org/wiki/History_of_General_Motors.

4. Dominic Capeci, *Detroit and the Good War* (Lexington: University of Kentucky Press, 1996), 1; Steven Klepper, "Disagreements, Spinoffs, and the Evolution of Detroit as the Capital of the Automobile Industry," *Management Science* 53, no. 4 (April 2007): 618.

5. John Hartigan Jr., "Remembering White Detroit: Whiteness in the Mix of History and Memory," *City and Society* XII, no. 2 (2000): 20–21; "Detroit Race Riot (1943)," *BlackPas.org: Remembered and Reclaimed*, http://www.blackpast.org/?q=aah/detroit-race-riot-1943; Capeci, 21; Cameron McWhirter, "One Street Mirrors City's Fall: Racial Fears Trigger White Flight in '50s," *Detroit News*, June 17, 2001, http://www.detnews.com/specialreports/2001/elmhurst. Ovshinsky helped organize the Detroit chapter of the Congress of Racial Equality (CORE) and was active in other civil rights work.

6. For an account of the recent history of unions and automation, with particular attention to the automation of machine tools, see David F. Noble, *Forces of Production: A Social History of Industrial Automation* (New York: Oxford University Press, 1984). Ovshinsky's involvement with the UAW would in time yield different benefits for him. Almost three decades later, Jack Conway, formerly the number two man in the UAW (under Walter Reuther), would send Atlantic Richfield's Thornton Bradshaw to him, which led in 1979 to the first large research contract for Energy Conversion Devices (see chapter 6). That, in turn, made possible the development of some of Ovshinsky's most important energy technologies (see chapters 8 and 9).

7. The previous year, they had lived in a duplex at 17350 Greenfield.

8. Ralph Jr., in turn, was "like an uncle to me," recalled Ben Ovshinsky, who added, "we called him 'Uncle Ralph.'" The familial as well as work relationship between Ovshinsky and Ralph Geddes Jr. is partly documented in a collection of letters written to Ovshinsky during the 1950s by Ralph Jr., e.g.: September 13, 1954; December 14, 1954; September 24, 1955; November 13, 1955; January 24, 1956; February 2, 1956, Ovshinsky papers.

9. U.S. Patent 2,807,964 (1957), S. R. Ovshinsky, "Automatic Transmission."

10. The rejection of the window safety device also typified the attitude of all American automakers toward safety improvements. Not until 1968, after the outcry raised by Ralph Nader's *Unsafe at Any Speed* (1965), and the National Traffic and Motor Vehicle Safety Act (1966), were car companies required to install such basic safety devices as seat belts. See Ruth Swartz Cowan, *A Social History of American Technology* (New York: Oxford University Press, 1997), 240.

11. Herb explained that in New Britain they had "designed the mechanical part, and we built a mechanical device that connected to the steering wheel of a small tractor, but we never finished the controls for it. And that's what Stan did at Hupp."

12. Geddes, who would routinely fire his most capable executives, was not the first in American automobile history to suffer from paranoia. Henry Ford would also fire staff members who showed initiative and imagination. See Hughes, *American Genesis*, 210–212.

13. Electric power steering has recently become more common, mainly because it saves power and so improves fuel economy. The new systems are, like many other current automotive functions, computer controlled.

14. Ovshinsky also learned that Hupp had stolen his invention. His original patent application for his electric power steering device (his "steering assist mechanism") was given the number

383,041; a copy of the continuation application 433,647 is in the Ovshinsky papers. After withdrawing Ovshinsky's applications, Hupp claimed the flexible coupling as its own. The patent that was actually filed on September 14, 1954, and granted on March 5, 1957, US Patent 2,783,627, "Resilient Coupling for Servomechanism Control," is in the name of Peter F. Rossmann, assignor to Hupp Corporation. It admits, "This application is related to copending applications Nos. 383,041 and 433,647, filed September 29, 1953, and June 1, 1954, respectively by Stanford R. Ovshinsky," but it falsely claims the rights were "assigned to the assignee of the present application."

15. The automatic tractor also entertained Herb's children, who remember driving it in the early 1960s down the back alley on 6 Mile Road. Herb rebuilt the tractor in the 1970s for one of Stan's birthdays.

16. Herb recalled that Stan chose the name "General Automation" in 1950, before they went to New Britain.

17. Watkins would work with Stan for the next two decades, helping to promote several of his inventions. Two examples of his striking graphic work are the ads for the Ovitron (figure 4.7) and the threshold switch (figure 5.8).

18. Ovshinsky later offered jobs to both men; Swigert accepted and worked for him for a time.

19. The programmable automatic lathe would have indeed been "a marvel" for its time, advancing the existing principles of tracer control to a fully electronic form. With the later use of digital computers, lathes and other machine tools would be automated by numerical control. See Noble, *Forces of Production*, esp. 79–105.

20. In that period Ovshinsky became a lifetime member of the Society of Automotive Engineers, an organization that, as he would later joke, was "the most primitive group in society."

21. Ovshinsky addressed his letter of resignation from Hupp to D. H. Gearheart, August 15, 1955, Ovshinsky papers.

22. John D. Cooney, "Multiple-Ball Relays," *Control Engineering*, February 1958, 2.

23. Ovshinsky began studying the brain as a control device even before he moved to New Britain and, as we later explain, pursued the topic much further in the mid-1950s.

24. The suit and countersuit concerned a false claim by the Tann Corporation that Ovshinsky's patent on his later invention, the Ovitron switch (discussed later in the chapter), infringed the sales agreement the Ovshinskys had made. Documents include the "Settlement Agreement" of October 1963, in which both parties agree to withdraw their suits, each paying the other a dollar in token damages, Ovshinsky papers.

25. Emma Goldman, *Anarchism and Other Essays* [1910] (New York: Dover, 1969), 62. See also Paul Avrich, *Anarchist Voices: An Oral History of Anarchism in America* (Princeton, NJ: Princeton University Press, 1995).

26. Education was important to the anarchists. Ferrer believed in early education as the key to social change, and the Ferrer Association founded the first "Modern School" in New York in 1911. It included among its principals and teachers several important writers, including Jack London and Upton Sinclair, and the philosopher Will Durant. See Paul Avrich, *The Modern School Movement: Anarchism and Education in the United States* (Princeton, NJ: Princeton University Press, 1980), in which chapter 9 is devoted to the Mohegan Colony. Anita and André Herrault Miroy, as well as Henri Dupré, appear in the index. Thanks to Robin Dibner for the reference.

27. He had been in the United States illegally from the beginning and needed to become a citizen to get the visa.

28. Iris's daughter Robin Dibner was told that Henri apprenticed with Escoffier in Paris during the time Ho Chi Minh was supposed to be doing the same in London.

29. Andy's uncle, Bern Dibner, the first in the family to attend college, became a successful engineer, inventor, author of books about the history of science, and philanthropist. He became well known in the history of science for his generous support of the field, establishing the Dibner Fund, the Dibner Institute for the History of Science, and the Burndy Library. The libraries at the Polytechnic Institute of New York University and the engineering school of the Technion in Haifa also bear his name. Bern's son David, later the director of the Dibner Fund, attended the Putnam Valley Central School with Iris. Andy Dibner's twin brother, a year older than Bern's son, was also named David.

30. They cut a $2 bill, the cost then of a marriage license, in two; each kept half, promising to tape it together when they were ready to buy their marriage license. Many years later, Robin said, she found Iris's half of the $2 bill in her jewelry box, "and she let me have it, that half of a $2 bill."

31. Similarly, neuropsychologists have learned much about how the brain makes memories by studying cases where traumatic injury has deprived someone of this ability.

32. Epilepsy had been postulated as a brain wave phenomenon in the 1920s, when Hans Berger's invention of the electroencephalogram showed different wave patterns associated with different kinds of seizures. Letters Ovshinsky exchanged in this period with Humphrey Osmond and Abram Hoffer reveal that schizophrenia was more than a purely intellectual concern for him. In the spring of 1956 he had suffered an episode of severe psychological disturbance, including disordered thoughts, depression, and panic. Iris believed there must have been some physical cause for such uncharacteristic distress, and in searching the psychiatric literature she found an article by Hoffer and Osmond that reported producing "model psychoses" in subjects by giving them metabolic breakdown products of epinephrine. That was the clue. Ovshinsky sometimes inhaled adrenalin for his asthma, and a few months earlier he had used some that was discolored, probably containing chemicals like those Hoffer and Osmond had used in their experiments and producing similar psychological effects. Iris went on to investigate these substances for her dissertation research, and Ovshinsky gave more prominence to brain chemistry in his account of epilepsy and schizophrenia. See letters to Osmond and Hoffer, August 30, 1956; February 13, 1957; February 14, 1958; March 5, 1958; March 6, 1958, Ovshinsky papers.

33. Ovshinsky's most general formulation of this view appears in "A Concept of Schizophrenia," cited in the following note. From the perspective of twenty-first century neuroscience, Ovshinsky's view of both epilepsy and schizophrenia as problems of control is not actually wrong but too general to be very meaningful. Both disorders are now understood to include several varieties; epileptic seizures, for instance, can take different forms, and not all involve motor control. Thanks to Carol Baym for this clarification.

34. Among Ovshinsky's many published or unpublished papers on these themes in the 1950s are: "The Use of Electromechanical Motion to Replace the Loss of Human Movement" (1950); "Nerve Impulse" (1955); "The Cerebellum and Its Error-Detecting Function in the Establishment of Meaningful Nerve Impulses" (1956); "A Concept Regarding Acting Transport of Epinephrine Through the Blood-Brain Barrier" (c. 1956); "Electrical Activity of the Brain" (1955); "Nerve Activity in Models, and the Physiological Processes which Produce Thoughts, Abstractions, and Activities, Both of the Normal and Abnormal Variety" (c. 1955); "Combined Cortical and Cerebellar Stimulation" (with F. Morin and G. Lamarche), Department of Anatomy, Wayne State University, College of Medicine, *Anatomy Records* 127 (1957): 436; "A Concept of Schizophrenia," *Journal of Nervous and Mental Disease* 125, no. 4 (1957): 578; "Cortical and Cerebellar Stimulation in Walking Cats" (with H. Portnoy and F. Morin), presented before the Detroit Physiological Society (December 19, 1957); "Functional Aspects of Cerebellar Afferent Systems and of Cortico-Cerebellar Relationships" (with F. Morin and G. Lamarche), *Laval Médical* 26 (1958): 633–643; "Suggested Biochemical Factors in Schizophrenia," *Journal of Nervous and Mental Disease* 127, no. 2 (August 1958): 180–185; "The Physical Base of Intelligence—Model Studies," presented at the Detroit Physiological Society (December 17, 1959); "The Reticulo-Endothelial System and its Possible Significance in Schizophrenia," *Journal of Neuropsychiatry* 3 (1961): 38–48.

35. The Worcester Foundation is best known for the development of the contraceptive pill in the 1950s.

36. At one point when Iris was helping Stan track down the references for a paper on the "Conditioned Reflex," she could not find an article by two Russian scientists, Livanov and Poliakov, and suggested that Stan write to the Soviet embassy in Washington. His request instead prompted a visit from an FBI agent, who, as Iris recalled, "came to talk to Stan about animation." (The agent meant automation.) From then on FBI agents again came regularly to see Stan, adding to their already lengthy file on him (see chapter 2). By December 13, 1955, Ovshinsky had a copy of this and other papers by Livanov, supplied to him by Dr. W. Grey Walter of the Burden Neurological Institute in Bristol. Ovshinsky had earlier been studying Walter's work in cybernetics (see chapter 3). Ovshinsky, letter to W. Grey Walter, December 13, 1955, Ovshinsky papers.

37. Ernest Gardner, letter to S. R. Ovshinsky, June 17, 1955, Ovshinsky papers.

38. Ovshinsky's recollections here are clearly colored by his later disillusioning experience with the hostile reception of his chalcogenide switches (see chapter 6).

39. Key Sweeny, secretary of R. W. Gerard, letter to Ovshinsky at General Automation, Inc., October 18, 1956, Ovshinsky papers.

40. Indeed, the ability of filmed tantalum to hold a charge made it work in an electrolytic condenser; Bell Labs had recently introduced just such a tantalum electrolytic capacitor.

41. See, for instance, "Nerve-Cell Studies Lead to New Static Component," *Electronic News*, July 23, 1958, 10: "Like the neuron (a single nerve cell), the Ovitron unit changes from nonconducting to conducting states when triggered by a low-energy stimulus. … According to modern theory, the neuron is surrounded by a semipermeable membrane charged positively on the outside, negatively on the inside. Permeability of the membrane surface increases and it becomes conducting when stimulated by a nerve impulse. In the Ovitron, two load-carrying electrodes immersed in an electrolyte are coated with oxide films that form semipermeable surfaces. Application of a small DC potential to a grid control element polarizes the load electrodes, switching surface films from nonconducting to conducting states. When the control signal is removed from the grid—or when a signal of opposite polarity is applied—conducting surfaces of the load-carrying electrodes revert to their original nonconducting states." Similar accounts appear in "The Electrochemical Relay: A Remarkable New Switching Form," *Control Engineering*, July 1959, 121–124; and "How Liquid-State Switch Controls A-C," *Electronics*, August 14, 1959, 76–80.

42. The Ovitron did involve a kind of threshold switching, though not in the usual sense of merely increasing the voltage to overcome resistance. Guy Wicker, who years later repeated Ovshinsky's Ovitron experiment, explained that electrically biasing the electrodes negatively or positively makes the oxide become thicker or thinner and that electrons can quantum-mechanically tunnel through an oxide if it is thin enough. It is this de-plating, thinning effect that allows the system to act as a switch analogous to a nerve cell.

43. The history of science offers several examples of important discoveries made on the basis of faulty assumptions. Ampère postulated circular molecular currents, which did not actually exist, but the notion enabled him to produce the first unified theory of electromagnetism. His student Carnot thought of heat as a kind of invisible substance, but that mistaken belief enabled him to conceive his theory of the ideal heat engine, which was a step toward the formulation of the second law of thermodynamics.

44. The distinction and relationship between science and technology are recurrent and unresolved issues in the history of technology. For an analysis of different views, see John M. Staudenmaier, SJ, *Technology's Storytellers: Reweaving the Human Fabric* (Cambridge, MA: MIT Press, 1985), 86–101. For an argument that technology constitutes a separate, autonomous way of knowing, see Walter G. Vincenti, *What Engineers Know and How They Know It: Analytical Studies from Aeronautical History* (Baltimore: Johns Hopkins University Press, 1990).

45. The letters in the Ovshinsky papers documenting the young Ralph Geddes's role in connecting Stan with Allen are dated March 4, 1956, and April 11, 1956.

46. Charles Allen had worked himself up to being a multimillionaire, starting out as a runner on the New York Stock Exchange after dropping out of school at age fifteen. In 1922, Charles and his brother Herbert founded Allen and Company, which became prominent in the entertainment field and is still among the leading investment companies in media and entertainment.

47. The Preliminary Prospectus of an Ovitron Stock offering in December 1959 says the company "was organized under the laws of the State of Delaware on November 7, 1958."

48. See note 24 above.

49. In addition to those already cited, articles appeared in the *Journal of Commerce, American Machinists, Chemical Week, Product Engineering, Machine Design, Business Week,* and *Mechanical Engineering*.

Chapter 5: New Beginnings in the Storefront (1960–1964)

1. Robin remembered these cards accumulating "all over the house, hundreds of them." Many remain in the Ovshinsky papers; one reads, "Happy 10th Anniversary, let's make it ten to the tenth!" Stan also made cards for other occasions, like birthdays or when Robin would come home from college, and sometimes he added one of his drawings (see the interlude).

2. They also didn't avoid showing their love in public. Harvey recalled a time when they were taking the boys to a public beach and a policeman broke up their passionate displays of affection. "Dad and Iris thought it was funny. I thought it was embarrassing and gross."

3. However Stan and Iris conceived of the new company, the earliest promotional materials make no mention of its social mission—understandably, since such declarations would be unlikely to attract investors.

4. Distinctive in many ways, the storefront fits a general profile that historians of technology have identified for a successful "place of invention," including an inspiring mission, a flexible work space with ample resources, strong leadership, a sense of freedom in the work itself, and good communication. See Arthur P. Molella, ed., "Places of Invention," report on the first Lemelson Institute, held in Incline Village, Nevada, August 16–18, 2007, and Arthur P. Molella and Anna Karvallas, eds., *Places of Invention* (Washington, DC: Smithsonian Institution Scholarly Press, 2015).

5. Harley Shaiken, then fifteen and one of the earliest part-time employees, recalls the powerful impression the storefront space made on him, especially Ovshinsky's "extraordinary science books," as well as his "books about politics, about the arts. The range would still take my breath away now." Shaiken is now a professor of social and cultural studies and the chair of the Center for Latin American Studies at the University of California, Berkeley.

6. "An Introduction to Energy Conversion Laboratories" (1961), 3, Ovshinsky papers.

7. In later years Ovshinsky spoke of their concern with the more general problem of fossil fuels, but that broad category seems to have become common only later. From the earliest company documents, however, it is clear that the focus on alternative energy sources was foundational.

8. The company was incorporated in August 1961.

9. The close connection Ovshinsky sensed between energy and information would be described in physics and information theory as the relation between information and entropy.

10. There is, however, no indication that Ovshinsky anticipated the more general problem of global warming, although some scientists were already then discussing the possibility. See Spencer R. Weart, *The Discovery of Global Warming* (Cambridge MA: Harvard University Press, 2003).

11. The concept gained wider popular attention with the publication of E. F. Schumacher, *Small is Beautiful: A Study of Economics as if People Mattered* (London: Blond & Briggs, 1973).

12. Because thin-film panels are light in weight, unlike heavy glass ones, such transportation was possible. One anonymous reviewer has criticized Ovshinsky's general concern with developing countries and his use of this image in particular as paternalistic and even racist. That is not how they have been perceived by those directly engaged with the issues. Cuauhtémoc Cárdenas, the former mayor of Mexico City, and Michelle Bachelet, the former and current president of Chile, found both Ovshinsky's message and the photo inspiring. The anthropologist Beatriz Manz, who has studied Mayans in Guatemala for decades, has showed the image to the Nobel laureate Rigoberta Menchú and also to Mayan villagers in the Lacandon rain forest, who were equally enthusiastic about the possibilities of the decentralized use of solar energy it represents. Thanks to Harley Shaiken for this information. See also the account of Ovshinsky's visit to Chile at the end of chapter 12.

13. Stanford Ovshinsky letter to Mr. Fenner Brockway, M. P., House of Commons, London, May 27, 1960, Ovshinsky papers. In later years, Ovshinsky would not have included nuclear energy in the list of alternatives.

14. "Physical Analogs of Physiological Communication and Control Systems," Proposal, c. November 1960, from Energy Conversion Laboratories, Detroit, Michigan, Ovshinsky papers. The proposal requests a six-month contract "to design a simple computing device composed of a number of our electrochemical switches wherein thresholds will be set by the internal environment of the system interacting with outside stimuli."

15. Not until Ovshinsky's first trip to Japan in 1964 did he find a patron who could provide adequate support. During the trip, he met the heads of Sony, including Akio Morita, who became a lifelong friend. Fuji Film and Nippon Steel were among the other Japanese companies that then funded ECL. These were among the Japanese firms that, unlike their American counterparts, were quick to recognize the potential of Ovshinsky's discoveries of the threshold switch and phase-change memory. See the 1987 NOVA documentary, *Japan's American Genius*.

16. Stanford R. Ovshinsky, "Thermoelectric Device," US Patent 3,508,968 (filed May 28, 1962; granted April 28, 1970). The patent describes "a pair of spaced apart metallic electrodes and a lithium compound interposed between and in contact with those metallic electrodes." A "substantial" DC potential occurs when there is a temperature differential between the electrodes.

17. He did, however, cover it with a Canadian patent: Stanford R. Ovshinsky, "Thermoelectric Generator Using Free Radicals as a Heat Source," Canadian Patent 686,092 (filed August 15, 1960; granted May 12, 1964). Much later, prompted by the furor over the claims in 1989 by Pons and Fleishman to having achieved "cold fusion," Ovshinsky asked Guy Wicker to repeat this early hydrogen experiment. Wicker determined that the bright light Ovshinsky had observed was caused by a faulty vacuum in his apparatus.

18. In later years, especially after the energy crisis of the early 1970s, there were many such schemes for creating a "hydrogen economy," some entirely hypothetical and some pursued with significant research funding. For a survey focusing on the hydrogen fuel cell, see Matthew N. Eisler, *Overpotential: Fuel Cells, Futurism, and the Making of a Power Panacea* (New Brunswick, NJ: Rutgers University Press, 2012), esp. 98–124. Ovshinsky's early hydrogen work came well before this trend.

19. Ben remembers Max telling him that when he was young he'd also seen a black man lynched in Selma.

20. Stan described Max as "a poet" with "natural leadership ability" and later made him the company's Vice President for African Affairs, a title that meant little to others but made perfect sense to Max and the people who knew and loved him.

21. Eventually, Stan and Norma reconciled, especially around their shared desire for Dale to become independent and earn a living. After remarrying, Norma moved to California and later to Miami. She died in 1985, at the age of sixty-two.

22. One might recall that Stan showed no such antimilitary views during World War II, when he tried to enlist and then worked on making bombers, or during the Korean War, when he was proud of how his lathe helped supply badly needed artillery shells. Perhaps his and Iris's later concerns stemmed from the Cold War era.

23. Stan became chairman of the Detroit chapter of SANE in 1960. See his "Statement Concerning My Public and Private Activities during the Period December 3, 1953 to Date," written March 15, 1970, Ovshinsky papers. Another instance of Stan's political activities in this period was his role in organizing a testimonial dinner to raise funds to cover medical and hospital expenses for Walter Berman after he was brutally beaten during a "Freedom Ride" in the South. The principal speaker at the dinner was Norman Thomas. See "Liberal Group in Dinner-Fund for 'Freedom Rider,'" *Redford Record*, March 15, 1962, clipping in the Ovshinsky papers. Stan and Iris were involved in the Freedom Ride of 1961 (see, e.g., mimeographed document, "Freedom Ride 1961") and in demonstrations at Woolworth/Kresge stores (e. g., undated copy, "Instructions to Kresge Demonstrators," Ovshinsky papers).

24. It is not clear now why concerned parents in that period thought that by using powdered milk they could avoid the traces of strontium 90, but it was a very common practice. For an early 1960s take on the problem see B. L. Larson, "Significance of Strontium 90 in Milk," *Journal of Dairy Science* 43, no. 1 (1960): 1–21, http://www.journalofdairyscience.org/article/S0022-0302(60)90106-5/pdf. Thanks to Bo Jacobs for this reference.

25. Harvey believes that the man Stan helped to acquire a home was Reverend Milton Henry, a noted civil rights lawyer and black separatist.

26. Stan had become acquainted with both Hoffer and Osmond in the 1950s, when he corresponded with them about research on brain chemistry (see chapter 4). In the 1960s, their work had acquired cultural cachet, and today Osmond seems to be best known for coining the term

"psychedelic." Pauling and Teller were among the group of distinguished scientists who became ECD consultants (see chapter 7).

27. Robin remembered that some of Stan and Iris's trips were international ones for a week or more, including Sweden, England, and later Japan. Sometimes Anita and Henri stayed with the children, or sometimes their devoted (live-out) housekeeper Mrs. King.

28. The disparaging phrase originated with Wolfgang Pauli, who dismissed the statistical approximations used in the study of solids. See Lillian Hoddeson, Ernest Braun, Jürgen Teichmann, and Spencer Weart, eds., *Out of the Crystal Maze: Chapters from the History of Solid-State Physics* (New York: Oxford University Press, 1992), 181, n. 458. After solid-state physics based largely on the study of crystals became well established, the same "dirty" dismissal was applied to amorphous and disordered materials.

29. Frederick Seitz, *The Modern Theory of Solids* (New York: McGraw Hill, 1940). See also Charles Kittel, *Introduction to Solid State Physics* (New York: John Wiley & Sons, 1953). As late as 1983, in the first systematic study of amorphous solids, Richard Zallen listed three of their attributes as "Structure, Solidity, and Respectability," where the third quality is "only recently attributed to glasses in conventional attitudes about what constitutes the discipline of solid-state physics." Richard Zallen, *The Physics of Amorphous Solids* (New York: John Wiley & Sons, 1983), 11.

30. The regular, long-range periodic order of crystals allowed precise calculations and explanations of their behavior, unlike the short-range order of amorphous solids. In time, the emerging body of research on non-crystalline materials joined crystallography and quantum theory as the third pillar of the solid-state physics edifice. See Spencer Weart, "The Solid Community," in *Out of the Crystal Maze*, 617–669, esp. 622–629.

31. Thomas James Gray, *The Defect Solid State* (New York: Interscience Publishers, 1957).

32. Michael Riordan and Lillian Hoddeson, *Crystal Fire: The Birth of the Information Age* (New York: W. W. Norton and Co., 1997), 125–141. The original point-contact transistor, however, did not depend on the oxide layer.

33. Even before stopping his work on the Ovitron, he had begun to minimize the use of the electrolyte and was trying to get a switching effect from the oxides alone. "When I set up ECL," Ovshinsky recalled, "I said I was going to do only solid state. Instead of having intervening things like electrolytes, I was going to reproduce the brain cells with their synapses in solid state matter." (Replacing the liquid electrolyte in Bardeen and Brattain's device by an oxide film was also a crucial step in the invention of the transistor. See *Crystal Fire*, 134).

34. Although oxygen is a chalcogen, the term "chalcogenide" is typically not used in speaking about oxides and is reserved for referring to sulfides, selenides, and tellurides. Ovshinsky recalled that it was Gray's book that gave him the hint leading to his choice of the chalcogenides. The most likely part seems to be Gray's discussion of zone refining, *Defect Solid State*, 159.

35. As Hellmut Fritzsche explained, with Ovshinsky's thin films the breakdown electric field of half a million volts per centimeter could easily be reached by voltages between 50 and 200 volts.

With bulk samples of 0.1 centimeter or more, however, it would require enormous voltages of several hundred thousand volts. See N. A. Goriunova and B. T. Kolomiets, "New Glassy Semiconductors," *Bulletin of the Academy of Sciences of the U. S. S. R.*, Physical Series 20, no. 12 (1957): 1372-1376. [A translation of "Novye Stekloobraznye Poluprovodniki," *Izvestiya Akademiia Nauk*, Seriia Fizicheskaia, 20, no. 12 (1956): 1496-1500.]. In studying chalcogenides, Kolomiets's laboratory was essentially alone in the Soviet Union, and his work was considered as only academic, with no industrial significance. As Kolomiets himself said, all this changed after Ovshinsky's visit to Leningrad in 1967 (see chapter 6), when the growing attention to his discoveries led to a dramatic increase in chalcogenide research. For further references, see B. T. Kolomiets, "Vitreous Semiconductors," *Physica Status Solidi* 7 (1964): I, 359–372; II, 713–730. Writing in 1970 about the work of Kolomiets in the 1950s, Ovshinsky noted, "At about the same time I independently began to investigate the electrical properties of disordered materials with the avowed purpose of finding switching effects." Stanford R. Ovshinsky, "Amorphous Semiconductors," *Science Journal 5A, no.2 (1969), 73-78.*

36. Unlike selenium, which was commonly used in making glass, tellurium, found in the sludge from copper refining, was not then a well-known element. Iris remembered visiting a copper smelting company on the river in downtown Detroit and "getting the sort of scrap that they had."

37. To stabilize and strengthen rubber for tires, cross-links are created through the addition of sulfur in the process of vulcanization discovered by Charles Goodyear and others in the nineteenth century.

38. See David Adler, "Amorphous Semiconductor Devices," *Scientific American* 236 (May 1977): 36–48, 45. As Adler explains, the cross-links "provide a structural stability that retards crystallization of the glass."

39. The "cross," repeatedly traced over billions of AC cycles, appears on the screen of an oscilloscope that is in a circuit with the switch and an AC source. A bright spot on the screen indicates the value of the voltage and the current. As the voltage alternates sixty times a second, the spot moves, but the voltage changes so rapidly that the eye sees the spot's path as a linear trace. Amorphous material normally has such a high resistance that the current remains very small and the trace looks horizontal. But, as the voltage increases and reaches either a positive or negative threshold value, the switch suddenly becomes so highly conducting that the spot jumps almost vertically up or down.

40. We are indebted to one of the anonymous reviewers for calling our attention to the significance of this working-class genealogy.

41. Although the new switch continued to be called the Quantrol in Great Britain, it soon became known in the United States as the Ovonic Threshold Switch.

42. "Electronic Machine's Licensing Pact," *The Times* (London), January 1, 1963. See also "Controlling Alternating Current," *Automation*, April 1963. A transcript of Ovshinsky's talk at the London press conference is in the Ovshinsky papers.

43. Part of the communication problem was that Ovshinsky was unwilling to reveal the switch's composition when it was not yet patented. There were, however, other gains. Ron Neale, a young engineer at the Electronic Machine Company, became an early ECL staff member, joining after Ovshinsky's London press conference. Another acquisition at this time was a 1954 Bentley, a gift from one of the investors in the British company.

44. He would win his second Nobel Prize in 1972, for the theory of superconductivity. See Lillian Hoddeson and Vicki Daitch, *True Genius: The Life and Science of John Bardeen* (Washington, DC: National Academy Press, 2002).

45. In later years, Ovshinsky relied on others to place his phone calls. Even this early, it may have been Iris who dialed.

46. Hellmut Fritzsche, "Interpretation of the Double Reversal of the Hall Effect in Tellurium," *Science* 115, no. 2995 (May 23, 1952): 571–572.

47. Years later, Ovshinsky felt relieved that he had chosen Fritzsche when Holonyak testified against him in an unsuccessful suit that the United Nuclear Company brought against ECD (see chapter 6).

48. Hellmut Fritzsche, "A Life with Stan," in *Reminiscences and Appreciations: Presented to Stanford R. Ovshinsky on the Occasion of his 80th Birthday*, electronic manuscript, Ovshinsky papers.

49. It was actually Lionel Robbins. Powell did not become the regular driver until spring 1964, after which he often picked up Fritzsche on his frequent visits. The point is trivial, but it illustrates how memories can become conflated and rewritten, an important consideration in using oral history—as this book must.

50. As the physicist Steve Hudgens later observed, "People had seen a situation where if you apply a voltage something breaks down and burns out, but Stan's switch switched and then went back to the insulating state, and switched and went back and forth and back and forth and back a zillion times."

51. "I was the first customer of this guy who later made a business out of [scientific] papers," Ovshinsky explained. "He only had [a case full of index] cards and he was doing it himself obviously. You could ask for something on science subjects, and he would get you a paper." Every week the man calculated the most cited papers and gave them ratings. This service was immensely useful to scientists before the era of photocopying and citation indexes.

52. Fritzsche, "A Life with Stan," in *Reminiscences and Appreciations*.

53. In later years, Ovshinsky's US Patent, 3,271,591 (filed September 30, 1963; granted September 6, 1966), would be referred to as ECD's principal patent. See, e.g., ECD's Form 10-K report, June 1970.

54. "That's it: a layer of magic and two contacts," Steve Hudgens later observed. Of course, the "magic" was based on the complex new physics of amorphous semiconductors, for which scientific explanations did not appear until several years later, and which is still not completely explained (see chapter 6).

55. Not until later would the difference between the two devices be scientifically explained. While the threshold switch returns to the insulating state when the current drops below a small critical value, in the memory switch a phase change from the amorphous to the crystalline state occurs due to heating when a medium current is applied, and the material remains conductive until a higher current pulse heats it more, melting it and restoring the amorphous state.

56. Its potential is already suggested in the 1964 article (Mason P. Southworth, "The Threshold Switch: New Component for AC Control," *Control Engineering*, no. 4 (April 1964): 69–72, on page 72). "Note that the memory cannot be destroyed by loss of power. Also, readout is nondestructive: when in the conducting state, changes in input voltage cannot switch the device to its nonconducting state. ... Thus they can be used in logic and switching circuits ... and are suitable for computer memory and crosspoint switching applications."

57. Harvey recalls Stan was nearly electrocuted when a visitor ignorantly connected two wires.

58. Robin recalled, "His EKG was never normal after that. Every new doctor seeing it would ask when he had had a heart attack, because the shock must have left some degree of scarring."

59. The name was changed on September 3, 1964, as shown in the "Amended Certificate of Incorporation," December 11, 2007, once included on the former ECD website.

60. Wayne State University opened a special room for storing Morin's papers, including Ovshinsky's notes on their joint research. The Wayne archivist told LH on December 1, 2009, that these papers had been destroyed.

Chapter 6: The Birth of ECD: An Invention Factory (1965–1979)

1. The date of the move is confirmed by the address on Hellmut Fritzsche's letters to Ovshinsky in the period, as well as an anonymous handwritten and undated document about the early history of ECD in the Ovshinsky papers.

2. Ovshinsky's handwritten notes about ECD's organization, Ovshinsky papers. The actual corporate documentation disappeared with the destruction of ECD's records at the time of its 2012 bankruptcy.

3. Edison was a pioneer in the move from individual to collaborative invention. Paul Israel locates this change in Edison's work on electric lighting, which was "not a single invention emanating from an inspired genius. Instead it was a complex network of inventions produced by one of the first institutions of organized corporate research." Paul Israel, *Edison: A Life of Invention* (New York: John Wiley & Sons, 1998), 167.

4. Suits in themselves were hardly a new addition to Ovshinsky's wardrobe. Earlier, his father Ben and Stan himself had them made by a relative who was a tailor and, when he worked as a manager at New Britain and Hupp, Ovshinsky wore suits. Now, however, it marked his new role as executive and fundraiser.

5. Stanford R. Ovshinsky, "Reversible Electrical Switching Phenomena in Disordered Structures," *Physical Review Letters* 21, no. 20 (November 11, 1968): 1450–1453.

6. For instance, *Electronics* magazine (November 30, 1964) and *Control Engineering* (April 1964) included descriptions of Ovshinsky's chalcogenide switches.

7. The paper was initially rejected as the work of an amateur, but because highly reputable physicists like Fritzsche and Sir Nevill Mott backed it, the paper was finally accepted. See the referee reports and related correspondence with the editor Samuel Goudsmit, Ovshinsky papers.

8. Mimeographed program for a benefit concert featuring a performance by Isaac Stern, Sunday, November 10, 1968, Ovshinsky papers. See also, in the interlude, Ovshinsky's drawing of Stern on the back of the program.

9. William K. Stevens, "Glassy Electronic Device May Surpass Transistor," *New York Times*, November 11, 1968, 1, 42, http://query.nytimes.com/mem/archive-free/pdf?res=9B06E3D61530E 034BC4952DFB7678383679EDE.

10. Ibid., 42.

11. The reasons for the controversy have been perceptively analyzed from a science studies point of view by Michael Gibbons and Philip King, "The Development of Ovonic Switches: A Case Study of a Scientific Controversy," *Science Studies* 2, no. 4 (October 1972): 295–309. Gibbons and King describe the controversy as a striking example of the hostile response likely to meet an outsider who challenges scientific norms: "Ovshinsky had mixed the exchange system of the entrepreneurial-industrial community with that of science, and in so doing violated the norm of disinterestedness." Yet, as they note, researchers in large industrial laboratories, who are not seen as outsiders, mix those systems with impunity. See also Philip M. Boffey, "Ovshinsky: Promoter or Persecuted Genius," *Science* 165 (August 15, 1969): 673–677.

12. The *Times* was sufficiently influenced by Bell Labs, General Electric, and Texas Instruments, however, so as not to mention Ovshinsky's work again until his death. He learned from a friend who was an editor for the *Times* that they had received instructions never to publish anything about him again. Even after his death in 2012, his *Times* obituary was grudging and inaccurate (see chapter 13).

13. Hellmut Fritzsche, letter to Nevill Mott, February 1969, Ovshinsky papers.

14. The name, a portmanteau coinage combining OVshinsky and electrONIC, had actually been chosen by Robin and Steven, but Ovshinsky obviously liked it, for from then on the threshold switch was called the Ovonic switch, and the memory switch was the Ovonic Memory Device. Ovshinsky's insistence on this self-promotional labeling irritated even many who acknowledged his inventive genius.

15. ECD stock had closed at 57 bid (60 asked) on Friday, November 8, reached a high of 145 bid (165 asked) in the trading frenzy on November 12, closing that day at 85 bid (100 asked), and then ending the week back down at 70 bid (80 asked). Gene Smith, "Electronics Stock Issue Has

Its Day," *New York Times*, November 18, 1968, p. F-18. See also Gene Smith, "Glassy Electronics Device Stirs Stock Market," *New York Times*, November 13, 1968, 61.

16. See the editorial in *Electronics Review* 41, no. 24 (November 25, 1968). This dismissal was based on earlier work at Bell Labs that had been reported in A. David Pearson, W. R. Northhover, Jacob F. Dewald, and W. F. Peck, Jr., "Chemical, Physical, and Electrical Properties of Some Unusual Inorganic Glasses," in *Advances in Glass Technology, Technical Papers of the VI International Congress on Glass*, Washington, DC, July 8–14, 1962 (New York: Plenum Press, 1962), 357–365. Working with a compound of tellurium, arsenic, and iodine, Pearson's group had observed the current-voltage characteristic that later marked Ovshinsky's memory switch, but they did not report observing threshold switching and did not seem to recognize the important implications of their findings. Moreover, the material they used was unstable and its behavior irreproducible; Bell Labs therefore discontinued the work. Ovshinsky learned of Pearson's work only after he published his own discoveries. Fritzsche investigated, and in a letter concluded that Pearson's "memory device might be considered a close miss. He might have discovered the right one if he had worked on it for a few years more." Fritzsche, letter to Ovshinsky, February 1969. Ovshinsky papers). For Ovshinsky's account, see S. R. Ovshinsky, "An Introduction to Ovonic Research," *Journal of Non-Crystalline Solids*, 2 (1970): 99–106, 100–101. For Pearson's account, see A. David Pearson, "Memory and Switching in Semiconducting Glasses: A Review," *Journal of Non-Crystalline Solids*, 2 (1970): 1–15.

17. According to Richard Zallen, the Bell Labs attitude toward Ovshinsky eventually changed.

18. Ovshinsky grew increasingly philosophical as the years went by about the hostile responses to his work. "That's an old story," he said. "Everybody who makes a revolutionary advance has to expect it. And a lot of them crumble under the attack." Ovshinsky didn't crumble, but he suffered considerably. When Mott invited the leading solid-state theorist Philip Anderson, then based at Bell Labs, to have lunch with the Ovshinskys and himself, Anderson firmly refused, adding, "I don't want to know him." Retelling the story years later, Ovshinsky still clearly felt the pain of this rejection, and this was but one example of the hostility that he and his work on amorphous and disordered materials met with during and after the 1960s.

19. The graph does not seem to have been published, but an early review article by Adler makes the same point in its opening lines. "There has been a recent surge of interest in the subject of amorphous semiconductors, a field which only a few years ago was generally considered to be about as scientific as witchcraft or alchemy. The major reason for the change in attitude is not difficult to pinpoint: the turning point was the publication by Ovshinsky detailing the various types of switching phenomena that characterize a large class of amorphous solids, and the subsequent publicity describing many potential applications of these phenomena." David Adler, "Amorphous Semiconductors," *Critical Reviews in Solid State and Materials Sciences* 2, no. 3 (1971): 317–465.

20. Summarized in the review article by B. T. Kolomiets, "Vitreous Semiconductors," *Physica Status Solidi* 7 (1964): part I, 359–372, part II, 713–733.

21. Despite this recognition, Ovshinsky apparently had to remind Kolomiets two years later of his priority in discovering chalcogenide switches. Ovshinsky, letter to B. T. Kolomiets, December 17, 1969, Ovshinsky papers. The point was repeated in a letter Ovshinsky sent to R. Grigorovich in December 1988, excerpted in the Ovshinsky papers.

22. Morrel H. Cohen, H. Fritzsche, and S. R. Ovshinsky, "Simple Band Model for Amorphous Semiconductor Alloys," *Physical Review Letters* 22 (May 19, 1969): 1065–1068. Later in 1969, there was a memorable symposium in New York City on semiconductor effects in amorphous solids. John de Neufville said that Ovshinsky and his colleagues created the meeting to "assert the scientific legitimacy of what they were doing." Afterward, he gave a fancy dinner for his team at the Plaza Hotel. The tradition of an Ovshinsky dinner continued well into the 1980s at the annual American Physical Society meetings held in March, always devoted to solid-state physics. Richard Zallen later deposited in the Virginia Tech archives records documenting the March meetings he attended. The Ovshinsky dinners, which were arranged mainly by David Adler, consisted of 15 to 20 physicists (always including Adler, Fritzsche, and Cohen), and took place "at the best restaurant in the city in which the March meeting was being held," Zallen said.

23. The physics underlying this mobility edge had already been discussed a decade earlier by Philip Anderson. See P. W. Anderson, "On the Absence of Diffusion in Certain Random Lattices," *Physical Review* 109, no. 5 (1958): 1492–1505. Anderson's paper was a crucial theoretical contribution to the field of amorphous semiconductors, but it was abstruse and thus hardly understood or even noticed until Mott portrayed the phenomenon in a more accessible way, "and then," Richard Zallen observed, "everyone understood it." See Richard Zallen, *The Physics of Amorphous Solids* (New York: John Wiley & Sons, 1983), 233–239.

24. American industrial research laboratories divide between those established by individual inventors to support their own inventive work and create new industries and the typically much larger laboratories established and supported by corporations to advance their goals. ECD fits the former model, for which Edison's Menlo Park (1876) was the prototype, aimed not at supporting an existing industry but at creating new industries. See Paul Israel, *Edison: A Life of Invention* (New York: John Wiley & Sons, 1998). Labs that fit the other model include those established by General Electric (1900), DuPont (1902), Eastman Kodak (1913), and Bell Labs (1925). Those interested in the rich existing literature on early industrial research in American can begin with Leonard Reich, *The Making of American Industrial Research: Science and Business at GE and Bell, 1876–1926* (Cambridge: Cambridge University Press, 2002); Thomas P. Hughes, *American Genesis: A Century of Invention and Technological, 1870–1970* (New York: Viking, 1989); Jon Gertner, *The Idea Factory: Bell Labs and the Great Age of American Innovation* (New York: Penguin, 2012); David A. Hounshell and John Kenly Smith, *Science and Corporate Strategy: Du Pont R and D, 1902–1980* (Cambridge: Cambridge University Press, 1988); and Lillian Hoddeson, "The Emergence of Basic Research in the Bell Telephone System, 1875–1915, *Technology and Culture* 22, no. 3 (July 1981): 512–544.

25. As Harvey recalled, Ovshinsky "hated being referred to as a 'maverick' inventor." Iris agreed: "It's dismissive and judgmental. And it's wrong!" But some of his admirers disagreed. Robert R. Wilson, a physicist and a sculptor, insisted, "Stan certainly is a maverick. Being a maverick in

science is a little like being a pioneering artist. Before Van Gogh or Picasso painted the kinds of pictures that they made, somebody doing that would not be considered a true artist." Ovshinsky took pride in being a pioneer, but his dislike of being called a maverick seems to reflect his need for acceptance.

26. The membership of the group changed over time. In the early years (circa 1970), it included Ed Fagen, Julian Feinleib, Jesus Gonzalez-Hernandez, Sato Iwasa, Simon Moss, John de Neufville, Howard Rockstadt, Charles Sie, Robert Shaw, and John Thompson.

27. Feinleib overlapped at Harvard with David Adler, who later became one of the most important ECD consultants.

28. See Ovshinsky's September 1969 talk at the Museum of Science and Industry in Chicago discussing his idea for an optical mass memory, "Applying Emerging Technologies," Ovshinsky papers. See also Lawrence Lessing, "The Printed Word Goes Electronic," *Fortune*, September 1969, 116–190, esp. 189–190; and "Great Hopes from Ovshinsky's Little Switches Grow," *Fortune* (April 1970), 110–114, 122–124.

29. Ovshinsky's early broad patent filed in January 1969 in his own name covered the then-evolving ideas resulting in ECD's rewritable optical storage disks. US Patent 3,530,441, "Method and Apparatus for Storing and Retrieving of Information." The patent was granted in September 1970 before the papers by Feinleib et al. (referenced in note 30) appeared. Because of this broad patent, Ovshinsky would benefit from licensing Japanese companies when they applied his optical memory inventions (see chapter 10). ECD's patent attorney Larry Norris once told Chet Kamin that with a slight change in wording this patent could also have covered all subsequent DVD technology, including Blu-Ray.

30. J. Feinleib, J. de Neufville, S. C. Moss, and S. R. Ovshinsky, "Rapid Reversible Light-Induced Crystallization of Amorphous Semiconductors," *Applied Physics Letters* 18, no. 6 (March 15, 1971): 254–257; and also J. Feinleib, S. Iwasa, S. C. Moss, J. P. de Neufville, and S. R. Ovshinsky, "Reversible Optical Effects in Amorphous Semiconductors," *Journal of Non-Crystalline Solids* 8–10 (1972): 909–916.

31. When Ovshinsky asked Turnbull, by then an ECD consultant, what he really thought of de Neufville, Turnbull replied, "I think it'd be fine if you don't mind that he wants to tell you how to run your company." "That's certainly not a problem," Ovshinsky replied, and offered de Neufville a job.

32. While running the research-sputtering machine that laid down layers of the material, de Neufville also worked in what was then called the bomb room, because with the earlier method of making the layers using Ovshinsky's heated powders, "every now and then one of these would blow up."

33. De Neufville's work in the 1970s was not the end of his relationship with ECD. In 1980 he became vice president and then, from 1983 to 1985, CEO of ECD's Ovonic Battery Company. He also served on ECD's board.

34. See Michael Riordan and Lillian Hoddeson, *Crystal Fire: The Birth of the Information Age* (New York: Norton, 1987), 225–253.

35. R. G. Neale, D. L. Nelson, and G. E. Moore, "Amorphous Semiconductors 1. Nonvolatile and Reprogramable, Read-Mostly Memory Is Here," *Electronics* 43, (September 28, 1970): 56–70.

36. After graduating from high school Shaiken had studied for about fifteen months at the University of Chicago before dropping out to learn a trade and spending a four-year stint as an apprentice at Cadillac. Later, he went back to school to get his degree at MIT. He later became Professor of Social and Cultural Studies and Chair of the Center for Latin American Studies at the University of California, Berkeley.

37. The experimental and numerical work was done by Shaw's graduate student Jim Kotz. See J. Kotz and M. P. Shaw, "A Thermophonic Investigation of Threshold and Memory Switching Phenomena in Thick Amorphous Chalcogenide Films," *Journal of Applied Physics* 55 (1984): 427–439. See also M. P. Shaw, "Electrical Switching and Memory Effects in Thin Amorphous Chalcogenide Films," in David Adler, ed., *Physics of Disordered Materials*, Institute for Amorphous Studies Series (New York: Springer, 1985), 793–809.

38. Thus selenium, for example, can switch from a valence of 2 to a valence of 4 by sharing its two lone pairs. As Guy Wicker put it, "chalcogenides have two personalities, they are schizophrenic. They are a valence 2 and they are valence 4, and they don't know which one they are." Both want to capture atoms, but the valence 4 needs more energy, and Ovshinsky exploited this "personality" in designing his materials, in which the lone pairs also bond. See also Hellmut Fritzsche, "Why Are Chalcogenide Glasses the Materials of Choice for Ovonic Switching Devices?" *Journal of Physics and Chemistry of Solids* 68 (2007): 878–882.

39. Einstein also stressed his reliance on visualization. "My power, my particular ability, lies in visualizing the effects, consequences and possibilities. ... I grasp things in a broad way easily. I cannot do mathematical calculations easily. I do them not willingly and not readily." See Gerald Holton, *The Scientific Imagination: Case Studies* (Cambridge: Cambridge University Press, 1978), 279.

40. M. Kastner, D. Adler, and H. Fritzsche, "Valence-Alternation Model for Localized Gap States in Lone-Pair Semiconductors," *Physical Review Letters* 37, (1976): 1504–1507. The basic picture of the switching was that the increased electrical field allows the lone pairs to participate in conduction, until they recombine and return to their original orbitals, switching off again.

41. David Adler, Heinz K. Henisch, and Sir Nevill Mott, "The Mechanism of Threshold Switching in Amorphous Alloys," *Review of Modern Physics* 50 (1978): 209–221. See also David Adler, "Amorphous Semiconductor Devices," *Scientific American* 236 (May 1977): 36–48.

42. Stephen Hudgens, "Progress in Understanding the Ovshinsky Effect: Threshold Switching in Chalcogenide Amorphous Semiconductors," *Physica Status Solidi B*, vol. 10 (2012): 1951–1955. The whole issue of the journal was dedicated to Ovshinsky for his ninetieth birthday and contains memoir papers by Hellmut Fritzsche and Genie Mytilineou, as well as papers on recent work in phase-change memory.

43. The word "orbital" simply acknowledges the quantum mechanical nature of the energy levels. The s, p, d, and f orbitals are conceptualized as clouds of probability of finding an electron or hole in a particular place.

44. S. R. Ovshinsky and K. Sapru, "Three Dimensional Models of Structure and Electronic Properties of Chalcogenide Glasses," *Proceedings of the Fifth International Amorphous and Liquid Semiconductors Conference*, Garmisch-Partenkirchen, Germany (1974): 447–452.

45. See "Energy Conversion Devices Names Cunningham President," *Wall Street Journal*, November 10, 1969.

46. OIS was based at 7250 Clairemont Mesa Boulevard, San Diego, and OMI in Los Angeles at 5261 West Imperial Highway.

47. See "Energy Conversion Files 'High Risk' Offering," *Detroit News*, June 9, 1971. Articles in the *Detroit Free Press* ("ECD Completes Interim Financing," October 15, 1971) and the *Detroit News* ("Electronic Firm Stock Sale," October 13, 1971) report that ECD had to withdraw a stock offering and arrange for private financing.

48. There was no connection between this Ovonic Imaging Systems and the company renamed Ovonic Imaging Systems in 1985, which had been originally called Ovonic Display Systems when Ovshinsky set it up in 1984. In 1986 the company was again renamed and became Optical Imaging Systems (see chapter 10).

49. Klose was recruited by Hellmut Fritzsche, who had known him as an outstanding technician in Germany and helped him come to Purdue, where he earned an MS in solid-state physics. Klose started in the storefront in 1964, where he worked on making numerous switching alloys and was soon traveling widely to represent the company. His technical skills were extremely important in the company's early years, when he worked to turn Ovshinsky's ideas into functional devices, e.g., the first threshold switch that could actually be used as an electronic component. Serving as project manager for the microfiche camera, he went with the project to California. See Peter Klose, "High- and Low-Lights from My Time at Energy Conversion Devices (1964–1983)" in *Reminiscences and Appreciations: Presented to Stanford R. Ovshinsky on the Occasion of his 80th Birthday*, electronic manuscript, Ovshinsky papers.

50. See "Energy Conversion Devices Reaches into Micrographic Markets," *Micrographic Weekly*, June 21, 1971, 2.

51. 3M terminated its agreement with ECD in September 1979. As Strand recalled, the rising price of silver gave the Ovonic film a price advantage, but users continued to buy silver-based film rather than switch.

52. "Disc Drive to Have Glass and Lasers," *Datamation*, October 1, 1971. Also, Memorandum, "Addendum to Supplemental Data—28 March 1972: Present State of Ovonic Film Development," Ovshinsky papers.

53. ECD's Form 10-K report for 1975 blamed "the unavailability of outside financing required to market the system," 13.

54. Cunningham's need for control may have also weakened the management of OMI. Feinleib and Iwasa wanted to be included in running the business, but they were rebuffed. As John de Neufville later put it, Cunningham felt that the scientists "just figured out the science," and after that his attitude was, "we don't really need you." Feinleib and Iwasa quit and soon found good jobs.

55. The difficulty in finding people to help with managing ECD was not so much a question of their getting along with Ovshinsky as of their understanding the complex business he was trying to build, with its many concurrent and interrelated programs. As Chet Kamin observed, "Conventional managers did not do well at ECD."

56. "He let us all go," Ed Fagen recalled, but added that Ovshinsky was always generous with severance pay.

57. As Johnson observed, "Time after time" Ovshinsky was able to convince people to put money into his projects, "but it was always different people."

58. The contract, for $304,000, was completed in December 1973. See ECD's 1974 Form 10-K report, 3.

59. The only competing memory at the time was magnetic core memory, used in the early IBM computers. This was before Intel brought out its electrically erasable EEPROM, the precursor of flash memory, which became dominant and remains so today. The advent of flash memory just as Ovshinsky was hoping to commercialize electrical phase-change memory was a major obstacle to the success of his invention, so he focused his attention in the 1980s on the optical application of phase-change memory (see chapter 10).

60. N-type material has a higher concentration of negative charge carriers (electrons), while P-type has a higher concentration of positive charge carriers (holes). A hole is the absence of an electron in a particular place in an atom. Although it is not a physical particle in the same sense as an electron, a hole can move from atom to atom in a semiconductor material, producing the effect of a positive current.

61. W. Spear and P. LeComber, *Solid State Communications* 17 (1975): 1193–1196. (Ironically, as Marc Kastner noted, Spear himself "refused to accept the idea that his films had hydrogen in them.") Although Spear and LeComber are generally given the credit for this discovery of the high photoconductivity of hydrogenated amorphous silicon, the first to see it were R. C. Chittick and his colleagues at the Standard Telecommunication Laboratory. See R. C. Chittick, J. H. Alexander, and H. F. Sterling, "The Preparation and Properties of Amorphous Silicon," *Journal of the Electrochemical Society* 116, no. 1 (1969): 77–81. But Chittick's company then chose to focus on crystalline silicon, so he and his team discontinued their amorphous silicon research and transferred their equipment and know-how to Spear. See also the retrospect article of Chittick and Sterling, "Glow Discharge Deposition of Amorphous Semiconductors: The Early Years," in David Adler and Hellmut Fritzsche (eds.), *Tetrahedrally-Bonded Amorphous Semiconductors* (New York: Plenum Press, 1985), 1–10.

62. Hereafter, when we refer simply to "amorphous silicon" in discussing photovoltaics, it should be understood as "hydrogenated amorphous silicon."

63. "Chemical Modification of Amorphous Chalcogenides," in *Disordered Materials: Science and Technology. Selected Papers by Stanford R. Ovshinsky*, ed. David Adler, Brian B. Schwartz, and Marvin Silver (New York: Plenum Press, 1991), 48–50.

64. Chemical modification of chalcogenides did, however, prove helpful in developing thermoelectric energy. And more recently, chalcogenides have been used successfully in thin-film (but not amorphous) solar cells made of cadmium telluride or copper indium selenide.

65. D. E. Carlson and C. R. Wronski, "Amorphous Silicon Solar Cell," *Applied Physics Letters* 28 (1976): 671–673. After persuading Ovshinsky to begin working with amorphous silicon, Fritzsche brought plasma deposition equipment from Chicago and trained the physicist Larry Christian to operate it. Christian, Fritzsche recalled, was severely handicapped by muscular dystrophy, but "his mind and brain functioned very well." ECD's hiring him was one of many examples of the way Ovshinsky gave opportunities to talented people who might not otherwise have had them. For other examples, see chapter 7.

66. In addition to patent considerations, Ovshinsky seems to have believed that using fluorine would also solve the problem of the Staebler-Wronski effect (discussed further in chapters 8 and 12). David Staebler and Christopher Wronski, who worked with Carlson at RCA, had discovered in 1977 that the efficiency of amorphous solar cells decreased over time by 10–30% through exposure to sunlight.

67. Later, however, fluorine became necessary to make microcrystalline rather than amorphous material when that was adopted for the P-layer (see chapter 8).

68. For a broader look at the history of solar power in America, see Jay Inslee and Bracken Hendricks, *Apollo's Fire* (Washington DC: Island Press, 2008), especially chapter 3, "Waking Up to the New Solar Dawn," 66–87.

69. According to Nancy Bacon, it was Ovshinsky's use of publicity that caught ARCO's attention. In March 1979, he, Iris, and Dave Adler had held a press conference in London announcing ECD's new photovoltaic research results. Ovshinsky said he needed $10 million to develop the technology. The story was picked up by newspapers including the *Chicago Tribune*, whose science editor, Ronald Kotulak, wrote an article that appeared on March 26, 1979: "Advances Seen in Solar Power." The article paired RCA, "one of the giants in solar energy conversion," with ECD, "one of the smallest companies in solar energy research, but one that has played a pioneering role."

70. Conway, aside from having been Walter Reuther's right-hand man in the United Auto Workers, had also served in several posts in the Kennedy and Johnson administrations, where he helped launch housing, Head Start, and Job Corps programs.

71. Kamin was later lead counsel in the lengthy antitrust litigation by MCI Communications Corporation against AT&T. The landmark 1980 trial resulted in a $1.8 billion antitrust judgment for MCI and eventually in the breakup of AT&T in 1982.

72. ECD's 1983 Form 10-K report, 7.

73. The trial of about three weeks took place in the courtroom of the astute Chicago judge Joseph Wosik. The possibility that Iris might be called as a witness was remote because her testimony was expected to match Ovshinsky's, but hoping to upset him, Cunningham had his lawyers move to exclude her from the courtroom. The maneuver backfired. Grasping its intent, Kamin recalled, the judge said, "Mrs. Ovshinsky, I think that rather than sit out in the hall you would be much more comfortable sitting in my chambers." As Kamin observed, "If you're on the other side, that's not a good sign."

74. In spite of the favorable judgment, there was still the prospect of a retrial, which would have been expensive and time-consuming, so ECD opted to negotiate a settlement and paid UNC a large sum even though they were in the right. Notably, one of the expert witnesses that UNC brought in was Nick Holonyak, who in 1963 had been the other physicist besides Hellmut Fritzsche that Bardeen had recommended to Ovshinsky to examine his threshold switch (see chapter 5). Ovshinsky now felt very glad that he had chosen Fritzsche.

75. That was indeed the approach taken by ECD's directors once Ovshinsky was no longer in control, with ruinous results (see chapter 11 and the epilogue).

76. Also, because the large research programs had to be considered as expenses rather than additions to capital, ECD's revenues almost never yielded a profit. For an instance of how that sometimes caused serious problems, see the account of Ovshinsky's legal conflicts with William Manning in chapter 11.

Chapter 7: The ECD Community: A Social Invention (1965–2007)

1. Mike Fetcenko, "Opportunity, Generosity and Perseverance," in *Reminiscences and Appreciations: Presented to Stanford R. Ovshinsky on the Occasion of his 80th Birthday,* electronic manuscript, Ovshinsky papers.

2. Not all were happy about those demands, however. See Dick Flasck's stories below.

3. To orient Ben Ovshinsky when he joined the company, Stan once drew the organization on a whiteboard, "a spider web with him in the middle." The lines radiated from Stan to all the offices and divisions. "Even though people had formal titles, there was no formal chain of command."

4. Iris compared skepticism about ECD with what was said about her marriage to Stan. "They'd see us being very loving and they'd say, 'How long have you been married?' And then they'd say, 'Just wait another year or two. It will be ruined.'"

5. Ovshinsky found it heartbreaking to lay people off, even more so when "they came to kiss and console me." Ghazaleh Koefod said, "You could tell it really hurt Stan and Iris more than it actually hurt the employee, the fact that he had to let them go."

6. Rabi received the 1944 Nobel Prize in Physics "for his resonance method for recording the magnetic properties of atomic nuclei."

7. John S. Rigden, *Rabi: Scientist and Citizen* (Cambridge, MA: Harvard University Press, 2000), 181. Ovshinsky frequently quoted these words.

8. By 1967, the ECD shares held by Ovshinsky (and later also by Iris), designated "Class A," had been weighted three votes to one against ordinary shares. Later, in 1979, the ratio increased to 10:1 and eventually, in 1982, to 25:1. It was not unusual to have a separate class of shares for the founders of a company, "founder's stock," that was treated differently so the founder could maintain control while raising money. The arrangement had to be periodically reauthorized by a vote of all the shareholders.

9. Thanks to Harvey Ovshinsky for a copy of the outtake. While he was the director of production at Detroit Public Television, Harvey first introduced the idea of a documentary about his father to NOVA.

10. Ovshinsky's friendship with Rabi led to ECD's early $250,000 license to IBM of the memory switch, for Emanuel Piori, IBM's director of research, was a close friend of Rabi's. (Piori had helped advance American industrial research and also helped to establish the system of peer review in government-supported science.) On hearing from Rabi about the threshold switch and phase-change memory, Piori invited the Ovshinskys to lunch at the University Club in midtown Manhattan. Ovshinsky always remembered Piori's comment about his switch: "I don't know what IBM would do with it, but I know IBM should be part of it." As the license was being formalized, Piori asked where to send the check. Iris simply opened her purse. Ovshinsky's discussions with IBM also brought Jim Birkenstock, IBM's vice president and director of commercial development, to ECD's consulting staff and board of directors. Jim Birkenstock, "Stan Ovshinsky: Inventor, Negotiator, Inspiring Leader," in *Reminiscences and Appreciations*, 2002, electronic manuscript, Ovshinsky papers.

11. For Ovshinsky's account of his relationship with Mott, see "Mott's Room," in *Nevill Mott: Reminiscences and Appreciations*, ed. E. A. Davis (Abingdon: Taylor and Francis, 1998), 282–285.

12. Mott once told the physicist Arthur Bienenstock, "A lot of my best ideas came from Stan. He just gave them to me." See "Bienenstock on Ovshinsky," *Berkeley Review of Latin American Studies*, Spring 2008, 25.

13. Pauling won a Nobel Peace Prize for his antiwar activism, opposing both nuclear weapons and the Vietnam War, views Ovshinsky shared.

14. Morrel H. Cohen, H. Fritzsche, and S. R. Ovshinsky, "Simple Band Model for Amorphous Semiconductor Alloys," *Physical Review Letters* 22 (May 19, 1969): 1065–1068.

15. As seen in chapter 6, Adler also made major contributions to explaining the Ovshinsky effect. See David Adler, Heinz K. Henisch, and Sir Nevill Mott, "The Mechanism of Threshold Switching in Amorphous Alloys," *Review of Modern Physics* 50 (1978): 209–221. As an instance of his ability to explain complex ideas to a general audience, see David Adler, "Amorphous-Semiconductor Devices," *Scientific American* 236 (May 1977): 36–48.

16. Much detail about Adler can be found in Nevill Mott, Stanford Ovshinsky, and Brian B. Schwartz, "David Adler," Obituary, *Physics Today* 41, no. 9 (February 1988): 104–108.

17. Much later, Ross recalled, when he was writing an article for *Science* with Ovshinsky and Mike Fetcenko, "Stan wanted to write this in the way that he writes. You know, some Romanian said wonderful things about him. And I said, look, none of that junk goes in here." Shortly after the conversation, Iris phoned Ross and said, "You write the article just the way you think it should be done." See S. R. Ovshinsky, M. A. Fetcenko, and J. Ross, "A Nickel Metal Hydride Battery for Electric Vehicles," *Science* 260 (April 9, 1993), 176–181.

18. A selection of the talks: David Adler, "Physics of Amorphous Semiconductors" (September 29, 1982); Barry Commoner, "Transition to a Solar Economy" (April 10, 1984); Leon Cooper, "Understanding the Brain: Faith and Science Today" (October 25, 1984); Robert R. Wilson, "Art, Intuition, and Science" (September 20, 1985); James W. Cronin, "Do the Macroscopic and Microscopic Asymmetries in Nature Have the Same Origin?" (January 17, 1986); Linus Pauling, "Covalent Bond Theory" (April 21, 1986); Sir Nevill Mott, "The Mobility Edge Since 1969" (April 28, 1986); Harold E. "Doc" Edgerton, "Strobe Lights and Their Uses" (June 9, 1986).

19. As some of Momoko's colleagues have noted, her upper-class background contributed to her success. She came from a provincial samurai family that lost its property in the American occupation but retained its status. Besides the respect this gained her, her knowledge of correct manners and speech was also important, but her fearlessness and resourcefulness in dealing confidently with the powerful came from her own strong character.

20. Some of this story appears in the PBS NOVA documentary *Japan's American Genius*. See also chapter 5.

21. ECD's Form 10-K report, June 1983, 28.

22. Reischauer, who had recently published *The Japanese*, thought he knew all about the culture. "But when he met Momoko," John de Neufville recalled, "he realized that she didn't fit into his account and was immediately attracted to her." Reischauer later became a member of ECD's board of directors, and after his death in 1990 Haru replaced him.

23. Joi is now a venture capitalist and the director of the MIT Media Lab. Mimi is a cultural anthropologist at the University of California at Irvine.

24. Steven believes this ability made Ovshinsky unable to grasp the toll that stress took on others. "When he listened to music or read pure science, he let go of the really pressing issues related to business and raising money. My mother was not able to." In time, Iris broke down under the cumulative strain (see chapter 11).

25. "The most important thing that I remember," said Steven, "was how much we were intentionally exposed to everything," for example, to "different types of cuisine," or to the ballet or a concert directed by Leonard Bernstein. "A lot of these times we'd either be in our pajamas under our clothes and we were sort of falling asleep or we'd leave halfway through," but he and Robin appreciated the experiences.

26. In one period, when Steven was studying the violin under Misha Mishakoff of the Detroit symphony, who had been Toscanini's concert master, the Old World teacher would "yell, scream and insult" the boy, and even hit his hands when he made mistakes. He would accuse Steven of

not practicing even when Steven had practiced a great deal. One day Iris reached her limit, and the lesson ended abruptly when "a knitting needle flew across the room. 'Don't you ever talk to him that way!'" Mishakoff later called to apologize, and there was one more lesson. But "it was disgusting," said Steven, for the master said nothing but "very good, very nice." For a time Steven lost his love for the violin.

27. Music also formed a different kind of bond between Steven and Stan, who felt an affinity between his son's dedication to art and his own scientific work (see the interlude).

28. As mentioned earlier, Iris had wanted to become a doctor herself but was turned down at the University of Michigan (see chapter 4).

Chapter 8: Solar Energy: Working at the Edge of Feasibility (1979–2007)

1. Thornton Bradshaw retired at ARCO; Sohio was taken over by BP; the Canon venture ended because Tanaka lost control. In each case, the new management did not continue their predecessors' support of ECD. Similar interruptions befell the battery program when ANR was taken over by Coastal Industries and the hydrogen program when Texaco was taken over by Chevron (see chapter 9).

2. Roll-coating was a familiar idea to those who had worked on the Micro-Ovonic Fiche film (MOF) for OIS (see chapter 6). But producing solar cells that way was a far more radical idea because of the danger of contamination between the deposited layers. Vin Canella remembered that around 1978, "Stan showed a reporter the drum coater for the MOF and said, 'This is how we're going to make solar cells.' I choked."

3. 1 megawatt = 1,000 kilowatts. Each solar panel is rated by its DC output power in watts (measured under standard test conditions); the number of megawatts describing a solar panel production machine designates the output of all the panels it produces in a year. The output of a 5-megawatt machine would be enough panels to power about 1,000 homes. Since, unlike power plants, the output of a solar panel machine is added each year, in ten years a 5-megawatt machine would produce enough panels to power 10,000 homes. (And the output of each of the 30-megawatt machines ECD later built would in ten years power 60,000 homes.)

4. PVD is a method for depositing thin films used in other applications like semiconductors as well as solar cells. In the ECD process, the silane gas was ionized by radio (later microwave) frequency radiation, creating a plasma.

5. Doehler, a German-born physicist, had grown up in France and later worked at Bell Labs before joining ECD.

6. Ovshinsky was seldom discouraged when others saw insurmountable problems. In fact, he welcomed hearing about them. "Don't just tell me the good news," he would tell his researchers, "I want to hear the bad news. Give me the bad news so we can figure out how to solve these problems."

7. As explained in chapter 6, Ovshinsky had divided the solar group into two concurrent research programs. Arun Madan's group, which included Wally Czubatyj, Jeff Yang, Steve Hudgens, and Mel Shaw, worked systematically to understand the physics of thin-film solar cells; Izu's group, which included Herb Ovshinsky, Vin Canella, and Joe Doehler, aimed at developing methods of commercial production.

8. Norris worked for ECD as patent counsel from 1980 to 1987 and then worked at a law firm in Washington, DC, until he retired.

9. Among the most important patents for the roll-to-roll concept are US Patents 4,519,339, 4,410,558, and 4,609,771.

10. Siskind told about the time when Ghazaleh Koefod, who as Ovshinsky's secretary could take shorthand, was angry enough with him to write up exactly what he said. "He called her in and he screamed, 'No one talks like this! This doesn't make any sense!'"

11. According to page 4 of ECD's Form 10-K report dated June 30, 1982, the initial $3.3 million negotiated in May 1979 concluded in April 1981, but the additional $25 million to further the development of ECD's energy technology was to be concluded in June 1983.

12. The terms of the agreement allowed ECD to keep not only its photovoltaic inventions but also the nickel metal hydride battery, also begun under the ARCO grant (see chapter 9).

13. Sohio, Standard Oil of Ohio, had been the original piece of John D. Rockefeller's Standard Oil Company and remained as a separate corporation after it was broken up; others were Exxon, Chevron, and Amoco.

14. Sharp got its name in the 1930s, when they had held the license for selling Eversharp pencils.

15. The degradation occurs naturally when the absorption of sunlight creates electron-hole pairs in the intrinsic layer. Some pairs recombine, preventing the electrons from reaching the N layer to create a current, and the energy released can break silicon-hydrogen bonds, leaving behind unsaturated bonds (dangling bonds) that stimulate more such re-combinations, further shortening the lifetime of the electron-hole pairs. Such degradation has been minimized by the use of hydrogen dilution of silane in the source gas for depositing the material, but even the best cells degrade. See Subhendu Guha, Jeffrey Yang, and Bao Jie Yan, "High Efficiency Multi-junction Thin Film Silicon Cells Incorporating Nanocrystalline Silicon," *Solar Energy Materials and Solar Cells* 119 (December 2013): 1–11.

16. Alloys with higher concentrations of germanium absorb longer wavelengths, at the red end of the spectrum.

17. See, e.g., S. Guha, K. L. Narasimhan, and S. M. Pietruszko, "On Light-induced Effect in Amorphous Hydrogenated Silicon," *Journal for Applied Physics*. 52, no. 2 (February 1981): 859–860.

18. Stanford R. Ovshinsky and Arun Madan, "A New Amorphous Silicon-based Alloy for Electronic Applications," *Nature* 276 (November 30, 1978): 482–484.

19. The group also included Yang and the Oxford theorist Michael Hack, who Guha considered "a very, very creative scientist." This was the group formerly headed by Arun Madan. By this time, the tensions between Ovshinsky and Madan had reached the breaking point, and Madan had resigned to pursue solar energy research elsewhere. Madan died on January 2, 2013, and so could not be interviewed for this book.

20. Guha recalled that after he reported this result to Ovshinsky, who had called from Japan to ask if there was anything new, "my wife and I were surprised the next day to find a bottle of champagne and a bouquet of flowers from Stan and Iris delivered to our home with a congratulatory note."

21. This design increased conductivity by accelerating the movement of the electrons and holes. The electronic structure of solids is conceptualized as a set of bands, ranges of energy where conductivity can occur. These are separated by gaps, where conductivity cannot occur. Varying the proportions of silicon and germanium, which have different size band gaps, so as to make it more likely for electrons to cross the gaps helped to increase conductivity.

22. There were also setbacks in the battery and display programs at this time, which aggravated the crisis. See chapters 9 and 10. Even during the cutback period, ECD continued its solar research, developing relationships with government agencies and the military, especially through its work on portable solar products.

23. Unlike Sharp, Canon was interested in making solar panels for buildings, though later it also used small ones on some of its cameras for recharging batteries.

24. Hellmut Fritzsche recalled that in negotiating the venture, Ovshinsky suggested "49.5% for us, 49.5% for Canon and 1% for Reischauer. The Japanese could not say no." For the story of how Momoko Ito had recruited former US ambassador Edwin Reischauer and his wife Haru, see chapter 7.

25. Eventually, Ovshinsky lost control of ECD itself (see chapter 11).

26. The discipline of manufacturing demanded by Canon also extended to the ECD crew, who had to wear the obligatory blue coats with the United Solar logo, to arrive early in the morning, and to punch the clock. To say ECD didn't have a clue about manufacturing, however, seems overstated and overlooks the company's successful battery production (see chapter 9). Ovshinsky always wanted to move beyond R&D to manufacturing, but it was only with the Canon venture that a sustained, large-scale manufacturing program became possible.

27. President Clinton speaking at the inauguration of PATH (Partnership for Advancing Technology in Housing), May 4, 1998.

28. Stempel, who had been CEO of General Motors, became chairman of the board and executive director of ECD through his involvement with the EV1 electric car (see chapter 9).

29. Bekaert had started out in the nineteenth century as a company that coated barbed wire and later moved into sputtering on glass. They now manufacture numerous wire products, including reinforcing wire for tires.

30. Ovshinsky had planned it as a 30-megawatt machine, and it did reach that capacity.

31. By now, ECD was making solar panels by much more than "the mile." The web from each roll was 14 inches wide and 8,500 feet long.

32. Nancy Bacon, "Testimony to U.S. House Committee on Science and Technology Subcommittee on Energy and Environment," July 14, 2009, https://science.house.gov/sites/republicans.science.house.gov/files/documents/071409_Bacon_0.pdf.

33. Until ECD's bankruptcy, United Solar held all the world records for thin-film silicon solar cell efficiency.

34. *The Michigan Daily*, "Bush to Stop at Auburn Plant," February 20, 2006.

35. Unlike crystalline panels, whose photovoltaic material was deposited on glass, Ovshinsky's thin-film panels could function even when damaged. He liked to show how one still worked after a bullet had been fired through it, which recalls his showing the astonished Hellmut Fritzsche how the threshold switch still worked when contaminated (see chapter 5).

Chapter 9: Hydrogen and Batteries: The Genie and the Bottle (1980–2007)

1. For an enthusiastic celebration of Ovshinsky's hydrogen vision, see George S. Howard, *Stan Ovshinsky and the Hydrogen Economy: Creating a Better World* (Notre Dame, IN: Academic Publications, 2006). For a more general, equally celebratory account see Jeremy Rifkin, *The Hydrogen Economy: The Creation of the Worldwide Energy Web and the Redistribution of Power on Earth* (New York: Jeremy P. Tarcher / Penguin, 2002). Whereas in 1960 Ovshinsky had been well in advance of the 1970s vogue for such schemes, by 1980 visions of a hydrogen economy were much more common. See Matthew N. Eisler, *Overpotential: Fuel Cells, Futurism, and the Making of a Power Panacea* (New Brunswick, NJ: Rutgers University Press, 2012), esp. 98–124.

2. The hydrides, as the electrochemist Dennis Corrigan pointed out, were the third class of amorphous and/or disordered materials that Ovshinsky examined at ECD, after the chalcogenides and amorphous silicon.

3. To Ben Chao, who joined ECD in October 1980 as a fresh PhD from Syracuse University, the two buildings, with equipment scattered everywhere and numerous experiments underway, seemed like the environment of a university lab. Chao would perform material characterizations for all three sections of the hydrogen group.

4. Fetcenko's story appears in chapter 7 as an example of how the ECD community nurtured individuals' professional development.

5. Unlike disposable batteries, NiMH batteries are rechargeable with current from an external DC power source, which triggers a series of electrochemical reactions. In charging, molecules from the positive nickel hydroxide electrode change to nickel oxyhydroxide, releasing a hydrogen ion (i.e., a proton) and an electron. The hydrogen ions, being positive, move through the electrolyte to the negative electrode, made of Ovshinsky's disordered metal alloy, causing it to absorb

hydrogen and change into metal hydride. (A separator prevents short-circuiting.) In discharging, electrons move through the external circuit and can do work. The metal hydride gives up hydrogen ions, which move back through the separator to the positive electrode, and the nickel oxyhydroxide reforms back into nickel hydroxide.

6. These results were reported in the group's quarterly meeting with ARCO on January 27, 1982. The NiMH batteries that ECD later developed had twice the energy density of nickel-cadmium batteries and could be safely recycled because they did not contain toxic metals (lead, cadmium, or mercury).

7. Marvin Siskind pointed to a previous version invented in the 1960s or 1970s during the ComSat (communications satellites) program. At General Electric, the inventor Fritz G. Will had patented a rechargeable battery of this kind in April 1975 based on a lanthanum-nickel compound. Early in 1977, Philips Corporation patented a similar battery. See the following links (or links within the links): http://www.google.com/patents/US4004943 and http://www.google.com/patents/US3874928.

8. Corrigan, an electrochemist who had been working on lead acid batteries for GM, joined ECD in 1992 to work on developing the NiMH battery for the EV1.

9. The elements changed as time went on, as did their specific proportions.

10. As Fetcenko explained, these disordered materials offered several advantages, like multiplying surfaces: "Think about a garbage can filled with baseballs. There's a lot of empty space, which we can fill with marbles or BBs. That gives us the surfaces we need to make a balance of oxidation and corrosion. If this thing corrodes, it's no good. If it oxidizes and forms a hardened steel, nothing can get through it. We need to find a balance to that. And disorder is the way to do that. So, some of the elements that are going in there are designed to corrode. Some of them are designed to passivate [prevent corrosion]. Some of them are designed to hold the hydrogen tightly, while others are designed to hold it more loosely, so we can tune in exactly where we want on the strength of the metal-to-hydrogen bond. Simultaneously, we need to tune in that oxide, so that it's the right degree of porosity, the right degree of stability, the right degree of catalytic activity, all those things are what disorder can do. It can control the level of defects, the alignment of the atoms. Nickel metal hydride batteries could not exist without this deliberately complex alloy system that Stan really was the pioneer of."

11. US Patent 4,623,597 (1986), Krishna Sapru, Benjamin Reichman, Arie Reger, Stanford R. Ovshinsky, "Rechargeable Battery and Electrode Used Therein."

12. Reger died of leukemia on October 28, 1983 (obituary in Ovshinsky papers), but Reichman continued to work with the batteries.

13. Dennis Corrigan, who did not join ECD until ten years later, commented that had he seen the demonstration and heard Ovshinsky's prediction about using the battery in an electric car, "I wouldn't have found that to be at all credible."

14. John de Neufville was the CEO; Dhar was the COO. Sapru and Hong stayed on in hydrogen storage.

15. According to Dhar, Ovshinsky would have preferred to call it the Nickel-Ovonic hydride battery.

16. The battery partnership with ANR began in October 1982 and concluded in November 1985. See ECD's 1985 Form 10-K report, 10.

17. ECD's 1990 Form 10-K report, 16.

18. Michael Shnayerson, *The Car That Could: The Inside Story of GM's Revolutionary Electric Vehicle* (New York: Random House, 1996); Chris Paine, director, *Who Killed the Electric Car?*, film (Culver City, CA: Sony Pictures, 2006). For a social history of the electric car from its beginnings in 1897, see David A. Kirsch, *The Electric Vehicle and the Burden of History* (New Brunswick, NJ: Rutgers University Press, 2000).

19. Shnayerson, *The Car That Could*, 174. The shifting political climate for alternative energy was symbolized on the White House roof, where Jimmy Carter had solar panels installed. After Ronald Reagan took office, they were removed. Under Barack Obama they were installed again.

20. The USABC included the three major auto manufacturers as well as the California Electric Power Research Institute, local utilities, and the Department of Energy. ECD's 1992 Form 10-K report, 2, 3, 8.

21. US Patent 5,536,591 (1996), Michael A. Fetcenko, Stanford R. Ovshinsky, Benjamin S. Chao, and Benjamin Reichman, "Electrochemical Hydrogen Storage Alloys for Nickel Metal Hydride Batteries." The patent describes the electrochemical storage alloys that enabled using the NiMH battery for electric and hybrid vehicles, discussed later in this chapter. See S. R. Ovshinsky, M. A. Fetcenko, and J. Ross, "A Nickel Metal Hydride Battery for Electric Vehicles," *Science* 260 (April 9, 1993): 176–181; also S. R. Ovshinsky and M. A. Fetcenko, "Development of High Catalytic Activity Disordered Hydrogen-Storage Alloys for Electrochemical Application in Nickel-Metal Hydride Batteries," *Applied Physics A* 72 (2001): 239–244.

22. As vice president for R&D, Venkatesan directed the step-by-step development that led from small C-size cylindrical cells to the much larger battery packs that powered several electric vehicle prototypes made by GM, Honda, Toyota, and Hyundai.

23. See Shnayerson, *The Car That Could*, 170–180.

24. Stempel explained that "at the time it was thought that CO_2 was fairly harmless" and not until about 1980 did GM start to recognize the seriousness of the problem of CO_2.

25. S. F. Brown, "Chasing Sunraycer across Australia, *Popular Science* 232, no. 2 (February 1988): 64–114.

26. Ovshinsky also received by accident a recording of a winter 1997 voice memo from one of the GM engineers reporting on the battery's "very encouraging" performance in cold weather,

which made it superior to the lead acid alternative in another way. Transcript in Ovshinsky papers.

27. The partnership between Ovshinsky and Stempel took a while to develop because Stempel had a noncompetition agreement after leaving GM and could at first only serve as an advisor. But in December 1995 he became ECD's chairman of the board and executive director and served for nearly twelve years. Ovshinsky kept the titles of president, CEO, and director. See ECD's 1996 Form 10-K report.

28. See draft of "Key Elements of GM-ECD/OBC Joint Business Venture," February 24, 1994, Ovshinsky papers.

29. Both GM statements are from the Advanced Technology Vehicles presentation to the press at the North American International Auto Show, January 4, 1998 (transcript from Harvey Ovshinsky). Former Chrysler chairman Lee Iacocca was also impressed, stating that "at some point we've got to realize that you're going to have to face up to a new millennium which for the young people is going to be an electric world."

30. Rick Wagoner, the CEO of GM who ordered the destruction, later said that "axing the EV1 electric car program and not putting the right resources into hybrids" was the worst decision of his tenure at GM. See "Interview With Rick Wagoner," *Motor Trend,* June 2006, 94.

31. "GM R&D chief Larry Burns ... now wishes GM hadn't killed the plug-in hybrid EV1 prototype his engineers had on the road a decade ago: 'If we could turn back the hands of time,' says Burns, 'we could have had the Chevy Volt 10 years earlier.'" *Newsweek,* March 13, 2007.

32. Another successful suit was against the French battery company SAFT, with which ECD had a joint development agreement, after which SAFT tried to use some of the technology in violation of a confidentiality agreement. As Chet Kamin noted, it was another case of dealing with "these giant companies. They'd look at ECD and their weak financials and how small they were and say, 'Well, we can just do whatever we want.'" But in litigation during the late 1990s, SAFT, like Matsushita and others, learned that was not so.

33. Baotou has more recently become notorious as a site of horrendous pollution caused by toxic byproducts of refining the rare earth minerals for which it is the leading source. See http://www.bbc.com/future/story/20150402-the-worst-place-on-earth and http://www.news.com.au/travel/world-travel/asia/baotou-is-the-worlds-biggest-supplier-of-rare-earth-minerals-and-its-hell-on-earth/news-story/371376b9893492cfc77d23744ca12bc5.

34. Vijan joined ECD in 1976 as a part-time lab technician and with help from the company's education support got an MS in chemical engineering. After working for several years in the OIS display program and moving with it to Guardian (see chapter 10), she returned and joined the battery program in 1996.

35. See Eisler, *Overpotential,* esp. 112–115.

36. Indeed, Ovshinsky later enjoyed showing how safe it was by carrying a sample around in his shirt pocket. There had been some work in the 1980s by Daimler-Benz that also included plans to use metal hydride storage. See Eisler, *Overpotential,* 114–115.

37. Born in China, Young grew up in Taiwan, came to the United States in 1966, and earned a 1971 PhD from Rensselaer Polytechnic in experimental solid-state physics studying radiation damage to solar cells on space vehicles. After teaching at the University of Kentucky for four years, she joined the solid-state physics division at Oak Ridge National Laboratory, where she started a photovoltaic program.

38. A few years earlier, the Ovonic Battery Company had received its first US ATP award to develop a nickel-metal hydride battery for electric vehicles.

39. Young had recently returned from Taiwan, where she was promoting electric vehicles, an effort that yielded commitments from two major companies and a $10 million battery licensing agreement.

40. Oil companies were already producing hydrogen from natural gas to use in the refining process. If hydrogen were to replace gasoline, they could provide it.

41. On the promotion of fuel cells over batteries, see Eisler, *Overpotential*, esp. 146–149, 171–172; also Eisler, "Cold War Computers, California Supercars, and the Pursuit of Lithium-Ion Power," *Physics Today* (September 2016): 30–36.

42. Both are electrochemical devices with two electrodes in an electrolyte, but while a battery stores energy, a fuel cell generates energy by converting fuel (usually hydrogen).

43. A heat exchanger was necessary because the processes of absorbing and releasing hydrogen are accompanied by releasing and absorbing heat.

44. More than one well-informed observer has reported that Dhar, who Ovshinsky had hoped would one day be his successor in heading ECD, did not leave voluntarily but was forced out by rivals. Ovshinsky's inability to prevent this was another sign of his waning power in ECD.

45. While the stories of the crushing of Ovshinsky's blue hydrogen car and the crushing of GM's many leased EV1s are similar in that both involved large corporations destroying alternative energy vehicles, they differed in that the hydrogen car was never sold or leased.

46. Flyer announcing the event and ad in the *Akron Beacon Journal*, Ovshinsky papers. The evening before he had been warmly welcomed at the Portage Country Club: "Reception Honoring Akron Native and Inventor Stanford Ovshinsky."

47. Wilhite had encountered Ovshinsky earlier while serving in a management position at the National Inventors Hall of Fame, then based in Akron.

48. City of Akron News Release, May 22, 2002.

49. Goodyear Tire and Rubber still had headquarters there, but was only manufacturing racing tires.

Chapter 10: Information: Displays and Memory Devices (1981–2007)

1. Ovshinsky appears to have seen this future himself. Chet Kamin recalled that when Ovshinsky was once challenged by a management consultant to name his single most important

technology, he at first resisted choosing, because for him all of ECD's programs were interrelated, interdependent, and of comparable importance. But when pressed, he named his information technologies.

2. Elsewhere, there had been earlier developments of flat-panel displays, such as those made by RCA for hand-held calculators, but they were too small for most purposes. Thin-film semiconductors, however, can cover large areas; their capacity for expansion allowed making displays that have continually grown larger, as well as the thin-film solar panels that ECD manufactured. For a general history, see Joseph A. Castellano, *Liquid Gold: The Story of Liquid Crystal Displays and the Creation of an Industry* (Singapore: World Scientific Publishing, 2005). Earlier examples using amorphous silicon diodes in various structures had been demonstrated in Japan by Togashi and in Europe by Szydlo; these were improved by the OIS team. See William den Boer, F. D. Luo, and Zvi Yaniv, "Microelectronics in Active-Matrix LCDs and Image Sensors," in Mohammad A. Karim, ed., *Electro-Optical Displays* (New York: Marcel Dekker, Inc. 1992), 69–119.

3. Castellano, *Liquid Gold,* 183. Active matrix displays, in which pixels are individually controlled and frequently refreshed, were invented in the 1970s at RCA.

4. Flasck had earlier advocated using amorphous silicon for displays when Ovshinsky was still hoping to use chalcogenides. He left ECD soon after Ovshinsky placed Johnson above him in the display program, for he felt it unlikely that he could then rise beyond his present position. Moving to the Bay Area, he and Scott Holmberg founded Alphasil, one of the first thin-film transistor (TFT) amorphous silicon LCD fabrication lines. Alphasil did reasonably well until 1989, when insufficient military demand forced it to close. See Castellano, *Liquid Gold,* 185–186.

5. Yaniv knew the name Ovshinsky from having written his 1972 master's thesis on amorphous semiconductors, but he had no idea that he was now the president of ECD. Ovshinsky was clearly intent on hiring Yaniv, for when Yaniv could not afford the $30,000 down payment on the house he had found for his small family, Ovshinsky simply wrote him a check for the amount and told Yaniv to pay him back when he asked for the money, which he never did. Zvi Yaniv (as recounted to Debra L. Winegarten), *My Life on the "Mysterious Island" of Nanotechnology: An Adventure through Time and Very Tiny Spaces* (New York: Page Publishing, 2017). Also, Yaniv, email to Hoddeson, August 20, 2015.

6. Vijan was put in charge of photolithography, and McGill took charge of deposition and the semiconductor end. At OIS Vijan also set up a clean room and equipped it with equipment for making silicon wafers.

7. Although the name was the same as the earlier company created by Keith Cunningham in 1971 (see chapter 6), there was no connection between the two. In 1986, the name was changed again to Optical Imaging Systems.

8. See OIS President's Letter, 1987, Ovshinsky papers.

9. By this time, Yaniv had become president of OIS, after Johnson left to become chairman of the Computer Science Department at the University of Utah in Salt Lake City. See Rick Reif, "Breaking Away," *Forbes*, December 25, 1989, 132; Yaniv, *Mysterious Island*, 100–101.

10. Under the ARCO program a way to use multilayer materials to create both an x-ray and a neutron mirror had been accidentally discovered. All medical x-ray systems now use the technology.

11. As noted in chapter 7, from the time when ECD went public in 1969, Ovshinsky's shares had been designated as Class A and weighted to have three votes per share, automatically converting to common stock, which had one vote per share, in September 1979. At first Ovshinsky held all of the Class A stock; he later transferred some of it to Iris. The load was increased to ten votes per share by vote of the shareholders in September 1979 and the conversion date to common stock deferred to 1988. The load was again increased to twenty-five votes per share in January 1982 and the conversion date extended to 1993 (see ECD's 1979 and 1983 Form 10-Ks). Later extensions of the conversion date brought it to 1999 and then 2005. The Manning agreement would have required Ovshinsky to surrender twenty-four of his twenty-five votes per share by irrevocable proxy to a committee of five independent directors whom Manning would appoint. Details about some of the suits may be found at http://openjurist.org/833/f2d/1096/manning-v-energy-conversion-devices-inc-r and http://openjurist.org/13/f3d/606/manning-v-energy-conversion-devices-inc-r. See also ECD's 1987 Form 10-K report, 49–51.

12. Jonathan Fahey, "Repeat Pretender," *Forbes,* November 24, 2003, 86.

13. The family arranged to have Steven's live bassoon playing piped into Stan's room. As the surgery was delayed for a long time, Steven recalled playing "for five hours or something, right to the minute they rolled him away."

14. The license extended a technical assistance agreement with Samsung negotiated in January 1987 "to participate in the joint development of certain liquid crystal television products for the hand-held television market." The initial one-year agreement with Samsung for $250,000 was soon extended to two years with an additional $273,000 (ECD's 1988 Form 10-K, 19). Other similar agreements for technologies developed by OIS were negotiated with Nippon Steel, Canon, and the US Air Force.

15. Yaniv, *Mysterious Island,* 105.

16. OIS President's Letter, 1989, 6, Ovshinsky papers.

17. Sharp was another large manufacturer of LCDs that got its start with ECD's technology. Their engineers and scientists had learned about making thin-film amorphous silicon while working on the roll-to-roll solar cell machine that ECD built for Sharp in 1983 (see chapter 8). Their ECD license for the deposition system helped them go on to manufacturing flat panel displays. Meera Vijan noted that there were other licensee companies OIS trained as well, including Unipac in Taiwan.

18. Yaniv, *Mysterious Island,* 125–126.

19. Vijan and Canella went with the company to Guardian. Yaniv was replaced with one of Guardian's executives.

20. As explained earlier, there were two forms of phase-change memory. The original, and more important, was electrical, in which a switch remained set until a second pulse returned it to its initial state. The second, though first to be commercially developed, was the optical, in which the phase change from amorphous to crystalline is triggered by a laser; again, a second pulse switched it back, making the CDs and DVDs rewritable.

21. "Erasure Means," US 4,667,309 A, inventor Michael Hennessey, filed July 8, 1985; granted May 1987.

22. See N. Akahira, N. Yamada, K. Kimura, and M. Takao, "Recent Advances in Erasable Phase-Change Optical Disks," *SPIE* 899 (1988): 185–195. This $Ge_2Sb_2Te_5$ alloy was later used to produce the PD drive that led to the CD-RW rewritable optical disks that are still in use. It also, as discussed below, led to a great improvement in ECD's electrical phase-change memory.

23. The storage density rose to several hundred megabytes for CDs and gigabytes for DVDs and Blu-Ray.

24. To manage the phase-change memory program, Ovshinsky hired new staff. They included Roger Pryor, a former student of the ECD consultant Heinz Hennish; George Cheroff, a processing facility expert from IBM brought in to construct a clean room; and Dan O'Donnell, a computer architect from IBM hired to develop a computing paradigm based on chalcogenide switches and memories. O'Donnell in turn recruited Guy Wicker, then a memory system designer at IBM, who had known about Ovshinsky's technology since he was a boy and eagerly seized the opportunity to work at ECD.

25. Flash memory had been invented by Fujio Masuoka at Toshiba around 1980 and was later commercialized by Intel.

26. Solving the problem of current reduction was less dramatic, achieved by several modifications to the insulation and connectors. See Wally Czubatyj, Tyler Lowrey, Sergey Kostylev, and Isamu Asano, "Current Reduction in Ovonic Memory Devices," *Proceedings of the European Symposium on Phase Change and Ovonic Science* (Berlin: John Wiley & Sons, 2006), 143–152.

27. Cross-point switching (interconnecting components in three-dimensional rather than two-dimensional arrays), not only made the switching between components extremely fast but the extra dimension greatly increased the scale of the memory capacity. Chalcogenide memories work much better than silicon-based ones for building three-dimensional structures (see comments on the recent Intel/Micron work in the epilogue).

28. The group also included: Dave Jablonski, Pat Klersy, Dave Beglau, Guy Wicker, and later Boil Pashmakov and Sergey Kostylev.

29. Micron had been funded in its early days by rich local businessmen who had made their money from Idaho potatoes.

30. Guy Wicker wrote his PhD dissertation on the physics of electrical phase-change memory, showing why the smaller these devices became, "the lower the capacitance becomes, the more manageable the temperature becomes, and the more predictable the device behavior becomes."

In silicon-based memories, when the depletion region between the P- and N-layers becomes too small it starts to leak, eventually to an intolerable level. Vertical pillar transistors have been designed in an effort to prevent this leaking, but their usefulness seems to be limited to about 5 nanometers.

31. With such small structures, the polymer breakdown layer ECD inadvertently made was not necessary to reduce the current.

32. The actual shares were: ECD 38.5%, Lowrey 38.5%, Intel 10%, Ward Parkinson 4%. The remaining 9% was held by Bob Jecman, Boise Investors, and employee options. OUM was also called Phase Change RAM, or PCRAM, or PRAM.

33. A solid-state physicist, born and educated in Bulgaria, Boil Pashmakov took his PhD in the Bulgarian Academy of Sciences and came to Chicago to work with Hellmut Fritzsche, who later recommended him to Ovshinsky, and he joined ECD in November 1994. Although Pashmakov worked with Ovonyx, he remained an ECD employee.

34. As the leading manufacturer of memory for space and defense, Lockheed Martin was especially interested in the fact that the Ovonyx memories could tolerate exposure to cosmic rays and other radiation.

35. Those who joined Tyler there included Wicker, Hudgens, and several others from Micron. Another Ovonyx team including Pashmakov, Klersy, Kostylev, and Czubatyj, worked at the new facility in Rochester Hills, where ECD had recently moved from Troy. Pashmakov explained that those who went to Santa Clara mostly worked with Intel, while those at ECD tested devices made by all the partners. To get around the problem that the various partners were competitors, Tyler assigned different team members to work with different partners to prevent intellectual property contamination.

36. As Wicker explained, Intel restricted the work with Ovonic memory switches: "New materials are forbidden in a modern clean room, and without them the phase-change memory couldn't be developed."

37. "Most computer people just thought he was crazy," Ito added. Not only did he speak his own language, which "became more and more divergent from the main stream of computer culture," but also by "not being in Silicon Valley, it was very hard to get computer people to come out."

38. Ovshinsky also noted, "DuPont used my phase-change memories for their blood chemical analyzers. They loved it so much that when I ran out of money and I couldn't make the product anymore, they threatened to sue me."

39. A number of scientists, including Bob Johnson and Steve Hudgens, agree that when Ovshinsky says that information is encoded energy he is really talking about entropy, a probability measure. But for Ovshinsky, the microphysics of the cognitive device offered a concrete physical example: the accumulating microcrystallites, each caused by a discrete energy pulse, each encoded a piece of information. For a discussion of percolation in amorphous solids, see Richard Zallen, *The Physics of Amorphous Solids* (New York: John Wiley & Sons, 1983), 135–204.

40. Stanford R. Ovshinsky, "Ovonic Chalcogenide Non-Binary Electrical and Optical Devices," *Proceedings of SPIE, Seventh International Symposium on Optical Storage*, China, 5966 (2005), 1–6. Reprinted in Hellmut Fritzsche and Brian Schwartz, *Stanford R. Ovshinsky: The Science and Technology of an American Genius* (Singapore: World Scientific Publishing, 2008), 104–109.

41. As Pashmakov observed, such a device had been predicted by Leon Chua, "Memristor—The Missing Circuit Element," *IEEE Transactions Circuit Theory* 18, no. 5 (1971), 507–519. It would provide the missing link between magnetic flux and electric charge. The other three circuit elements are resistor (the link between current and voltage), capacitance (the link between voltage and charge), and inductance (the link between current and magnetic flux). The memristor was conceived as a device whose resistance depends on the total amount of charge passed through it; it thus stores information as a nonvolatile memory. That, Pashmakov noted, is exactly what the ECD team showed its cognitive device does, as they reported with Ovshinsky at the European Phase-Change and Ovonic Science (E\PCOS) conference in Milan, September 2010. Pashmakov also referred to seemingly independent research by IBM, Xerox, and University of California groups on developing neural networks based on memristors using other materials than chalcogenides (e.g., oxides). This came out after 2007, when Ovshinsky had left ECD (see chapter 11). Ovshinsky was quite upset that the publications often did not reference the ECD work. Pashmakov suggested this was because ECD mainly had patents and had published very little. For other, more recent work, see the epilogue.

42. Several others helped with the cognitive computer project. Pat Klersy returned from California to set up and run the fabrication lab. Dave Beglau also worked on fabricating and testing devices. Alistair Livesey joined the group a bit later to work on algorithms and architecture. Sasha Shevchenko, who in the Ukraine had been a professor of applied mathematics and computer science, developed algorithms and computer simulations. Morrel Cohen was also an important adviser on algorithms. Others who worked with the group included Hassan Mia, whose field was finance, and Takeo (Ted) Ohta from Kyoto, visiting from Matsushita.

Interlude: Science, Art, and Creativity

1. Autobiography, "Journeyman of the Imagination," July 2, 1995, p. 38, Ovshinsky papers. This is one of many fragments of Stan's dictated memoirs, in which versions of this quote appear several times. He considered "Journeyman of the Imagination" as a working title for his autobiography, which he never completed.

2. Autobiography, "Creativity and Innovation," September 3, 1998, 6, Ovshinsky papers.

3. Ovshinsky also used the poetry of others to express himself. LH remembers how during interviews he would often pull a volume of poetry off his shelf and read a few lines to her.

4. Reprinted in *Stanford R. Ovshinsky: The Science and Technology of an American Genius* (Singapore: World Scientific, 2008), 359–360. The translation of Yukawa's book appeared in 1973. That Ovshinsky chose to write about it eighteen years later suggests that he was by then reflecting more about the sources of his own creativity.

5. Many of these Einstein quotes are from Gerald Holton, "On Trying to Understand Scientific Genius," *The American Scholar* 41 (1971–72): 95–110.

6. The passage from Einstein is frequently quoted. See, e.g., Harold A. Popp, *Discovering the Creative Impulse* (Bloomington, IN: West Bow Press, 2014), 8.

7. Their letters were gathered in a large compilation that Fritzsche presented at Ovshinsky's ninetieth birthday celebration, "Six Years of Correspondence / 1994-2000 / Stan and Hellmut / Thoughts on Cosmology: Gravity / Dark Matter / Vacuum Energy / Vacuum Fluctuations / Expansion of the Universe," to which Fritzsche added, "These letters complement and summarize our discussions in Stan's home which served as his relaxation from the stresses of his day," Ovshinsky papers.

8. Another area where Ovshinsky tried to make an important scientific contribution was superconductivity. With Fritzsche, he used his typical method of visualization to develop a novel theory of the Cooper pairing phenomenon basic to the BCS (Bardeen, Cooper, Schrieffer) theory of superconductivity. Unlike his cosmological speculations, his superconductivity work also had a practical potential, and he worked on the problem experimentally. Around 1987, when there was great excitement and competition among researchers over achieving superconductivity at higher temperatures, Ovshinsky and his team reached a very high transition temperature of 154 K by adding a small amount of fluorine to copper oxide.

9. Report included in P. R. Holland, editor, email to Hellmut Fritzsche, re Manuscript: Ho 3312, "The Origin of Dark Matter in the Universe," March 16, 2000. In "Six Years of Correspondence," Ovshinsky papers.

Chapter 11: Losing Iris, Losing ECD

1. Ovshinsky was one of the "Heroes for the Planet" named in a *Time* feature focusing on electric car design. See Margot Hornblower, "Listen, Detroit: You'll Get a Charge Out of This," *Time* 153, no. 7 (February 22, 1999), 80.

2. Robin recalled the increased burden on Iris after Stan broke his hip in January 2006. A series of infections and other problems after that required long hospital stays. "My mom would stay the whole time with him in the hospital, never getting an uninterrupted night's sleep."

3. Max Powell later commented to me that the water in the lake was "really was too cold" for swimming, even in August, but that "they were used to it."

4. Max Powell later noted, "She did *everything* for him. He's the genius, but she did the thinking."

5. Iris had not wanted to go on living without Stan. "She said she'd kill herself if I died," Stan recalled. "I always tried to talk her out of it, but there was suicide in her family and I was afraid." Iris's mother Anita had committed suicide, as had Anita's father after her mother's death. Anita's death in late 1966 had been traumatic for the whole family. She had become depressed after Henri insisted they live half of the time in Florida, where, as Robin explained, "she didn't like

being around a bunch of old people." When her depression grew worse after her doctor put her on Valium, Stan and Iris arranged for her to see a prominent New York psychiatrist. But after the three had checked into the St. Moritz hotel and Stan and Iris had left for a business meeting before taking her to lunch and her afternoon appointment, Anita jumped to her death from the twenty-eighth floor. Knowing that Robin, Steven, and all three of the Ovshinsky boys had loved Anita, Iris tried to shelter them by not letting them attend her funeral, but that only made things worse for the children because they had no closure. The only good result was that after Anita's death Iris would no longer frighten the children by saying that if Stan died before her she would commit suicide.

6. Robin suspects that Stan revised his memory of scheduling the stress test. "He seemed shocked when we found out about the appointment. Even her primary care doctor didn't know about it. I suspect she had secretly made the appointment herself with the cardiologist."

7. Robin reported that after Iris died "Steven found 3x5 cards in various purses and also on her desk, with comments like, 'I'm having chest pain.' 'The chest pain is getting worse.' 'I don't want to worry Stan.' And she never did tell anybody about it." Steven was also haunted by the memory of Iris telling him that she no longer enjoyed swimming, even with Stan, because she didn't feel strong enough to lift her head the way she used to.

8. Ovshinsky's loaded vote had been repeatedly increased and authorized for fixed periods of time (see note 11 in chapter 10), of which the last ended on September 30, 2005, when his Class A shares automatically converted into common stock (ECD's 2007 Form 10-K report, 72). By that time, there were nearly 28 million common shares outstanding, and it would have been impossible to get another extension approved by shareholders.

9. It seems clear that the new directors were not actually planning to develop the company that the Ovshinskys had created but rather to realize quick gains. Cutting most research and laying off so many of the staff made no sense as a long-term strategy, but it would have boosted profits and made it possible to sell the company. As one knowledgeable observer surmised, "The play was keep solar, cut out everything else, show profits for a few quarters, and sell the company." Whatever their intentions, in the end, they only drove ECD into bankruptcy (see the epilogue).

10. A few months later Ovshinsky received a reply from his colleagues in the model shop, written on the stationery of the International Association of Machinists and Aerospace Workers. "The Model Shop colleagues would like to take this time to express our gratitude and our appreciation for the opportunity to work with you over the past years. Your guidance and support has catalyzed our professional development and for that we will be forever grateful. We would like also to convey our immense sorrow at the passing of Iris who was truly a pioneer and will hold a special place in all our hearts forever. Let it be known that if you ever have further need of us we will be there at a moment's notice. It has truly been a pleasure to work with you." The letter included a quote from Shirley Chisholm: "When morality comes up against profits, it is seldom that profit loses." It was signed, "Sincerely, your colleagues in the Model Shop," with the signatures following. Model shop colleagues, letter to Ovshinsky, July 16, 2007, Ovshinsky papers.

11. The ECD news release of March 5, 2007, says that the change was made "at his request," one of several false or misleading statements in this and the subsequent August 23 news release; both releases are among the Ovshinsky papers.

12. On September 1, 2007, the day after Ovshinsky's forced retirement, Mark Morelli became president. A former helicopter attack pilot who had been a division president at Carrier, he had little understanding of ECD's technologies, and as Hellmut Fritzsche remarked, he "brought from his previous job a number of people to serve as vice presidents, people who had no experience and knowledge of solar panels and their unconventional production process."

13. Jay B. Knoll, General Counsel, letter of agreement to Ovshinsky, August 23, 2007, Ovshinsky papers.

Chapter 12: New Love, New Company

1. She had lived separated and then divorced from her husband for some twenty-five years.

2. Because the apartment was on a hallway, walking down a hotel hallway in later years would often trigger fond memories for Stan, and whenever they drove past the apartment building he would wave.

3. Harvey and Robin had both been aware of the relationship earlier. Harvey recalled, "In the middle of winter, I saw Rosa and Dad get out of the car. And he looked good. His scarf was blowing in the wind, his coat was blowing in the wind; he looked healthy, alive, and so did she." Robin suspected in November when she visited for his birthday and Thanksgiving. "I had noticed how he was mentioning Rosa a lot and saw the way he looked at her at a little birthday gathering the hydrogen group had for him." Later, during the Hawaii New Year's trip, "when we were sitting by the pool he told me he was interested in her. It was hard to take, but I remember thinking it was his life, and I was going to be an adult about it, and it had been awful seeing him so miserable, crying so much."

4. Dale was especially grateful for Rosa's assistance in managing his finances and helping, as Iris had, to ease his relationship with Stan.

5. Ever since his heart surgery in 1987, Stan had been on blood thinners, which made such bleeding a serious danger.

6. Ting is Rosa's middle name.

7. Ovshinsky had previously increased annual production of his roll-to-roll machines to thirty megawatts (see chapter 10). Jumping to a gigawatt would increase it over thirty times more. One such machine would produce enough solar panels each year to provide electricity for 250,000 homes.

8. Strand recalled, "Stan confided to me several times that Rosa didn't believe the project would succeed." But, he explained, "She knew Stan loved the work and that it made him happier and

healthier." Similarly, Ben, like Harvey, saw Ovshinsky Innovation and Ovshinsky Solar as essentially a way for Stan "to stay alive and engaged."

9. Strand hired all but Pashmakov, whom Ovshinsky personally recruited.

10. In the competition between the deposition and fluorine's etching effect, Fritzsche observed, the deposition had to win.

11. Measuring the density of states can reveal the presence of regions in the material (predominantly caused by defects) where electrons could be captured, thus lowering the current.

12. To publicize his achievement and help get more support, Ovshinsky gave an interview with the *Bulletin of the Atomic Scientists* that presents his case for a more enlightened clean energy policy. "Stanford Ovshinsky: Pursuing Solar Energy at a Cost Equal to or Lower Than That of Coal Electricity," *Bulletin of the Atomic Scientists* 67, no. 3 (2011): 1–7.

13. By this time Ovshinsky had resigned as president, but his separation agreement entitled him to bring visitors to United Solar and OBC.

14. For the whole clip, see https://www.youtube.com/watch?v=EXkFhiabmYY.

Chapter 13: Last Days

1. The fact that Rosa started planning the party in May argues against the notion some have expressed that she planned it as a memorial for Stan to enjoy while he was still alive. No one in the family, least of all Rosa, suspected that Stan's death was imminent.

2. Sasha and Stan had been talking since Sasha's arrival a few days earlier. Sasha recalled an intense scientific discussion at the Institute about the role of lone pair electrons in chalcogenides. "He was so excited. His eyes were shining so bright." Sasha was impressed by how sharp Stan's mind still was, how he grasped Sasha's points before he could even finish a sentence. Sasha asked Stan to be co-author of the paper he was working on; Stan agreed, and it became his last, posthumous publication. Alexander V. Kolobov, Paul Fons, Junji Tominaga, and Stanford R. Ovshinsky, "Vacancy-Mediated Three-Center Four-Electron Bonds in GeTe-Sb_2Te_3 Phase-Change Memory Alloys," *Physical Review B* 87, no. 16 (2013): 165206-1-165206-9.

3. Harvey had stepped in several days before to help "executive produce" the event, after receiving a call from Stan: "We're desperate, we need your help. Rosa can't do this alone."

4. Some of the speeches can be found on Forever Missed, Stan's memorial website, http://www.forevermissed.com/stanford-r-ovshinsky/#about.

5. The talks by Senator Levin, Hellmut Fritzsche, and Harley Shaiken were entered into the *Congressional Record* 158, no. 121 (September 11, 2012), https://www.congress.gov/crec/2012/09/11/CREC-2012-09-11-pt1-PgS6103.pdf.

6. Irina noted that Stan's back pain had started before the Canada trip, and his physician had suggested doing an MRI in May. But when they returned his finger was the primary concern, and they never did the MRI.

7. Prostate cancer usually grows so slowly that men with it most often die of other causes, and Stan, like many in his situation, didn't want to risk losing his sexual potency.

8. Robin said she had "found reams of notes of my mother's about various people they had consulted about Dale. And Stan supported him his whole life, fully and comfortably."

9. It was clear to Rosa that Stan was still troubled by not having told her about his cancer and was asking her to reassure him that she had forgiven him. Robin added, "That was part of the confirmation to me, that they really, really did have a deep love, that those would be his last words." Irina noted that Stan had started to trust only Rosa, and reacted to her presence until the end: "If she's coming into the room, when he's already in this coma, if he just saw Rosa, he'd start to smile, just for her."

10. The pin indicated that he was a fellow of the American Association for the Advancement of Science.

11. For the viewing, they had "paid a fortune to have ivy brought in and festooned on all the pews to cover up the stars of David," Robin recalled.

12. These errors in the *New York Times* included saying that Steven and Robin were Norma's children. "I was so angry," Robin said. "And they had the wrong city for the company. And how about calling him Stanley on the front page?" The *Times* corrected some of these errors a few days later. Links to this and other obituaries can be found on the Forever Missed website (see note 4 above).

13. The rare books from Stan's collection went to the Joseph A. Labadie Collection and the Harlan Hatcher Graduate Library of the University of Michigan. His papers went to the Bentley Historical Library at the University of Michigan.

Epilogue: Deaths, Survivals, and Revivals

1. Jay Inslee and Bracken Hendricks, *Apollo's Fire: Igniting America's Clean-Energy Economy* (Washington, DC: Island Press, 2008), 68.

2. Nancy Bacon and Marv Siskind believe that even with the downturn in the US photovoltaic industry, ECD could have pulled through the crisis had ECD's new managers not chosen such a risky financial strategy. Bacon remarked that if she had still been CFO, ECD would not have issued the convertible bonds. Convertible bonds are good when the stock rises, but if it declines there may not be sufficient funds to make the required cash repurchase of the bonds.

3. The callous disposal of ECD's property extended to the corporate files, which went into dumpsters, making historical studies like this one much more difficult.

4. Due to continuing financial problems, Ovonic Hydrogen Storage was sold to Vodik, a small Texas company, and the assets of Ovonic Fuel Cell were also sold, both for very small amounts.

5. Fetcenko said that in today's battery industry, BASF Ovonic is "maybe the only successful example of such a licensing business model." But ECD was the pioneer in making nickel metal hydride batteries, and BASF now has the patents. "Whether it's a Duracell or an Energizer consumer battery, or whether it's a Prius with a Panasonic battery, you have to take our license, but we have the know-how and the skill to teach you how to make profitable batteries."

6. Toyota has now (2016) begun offering the option of lithium ion batteries.

7. "Intel and Micron Produce Breakthrough Memory Technology," *Intel News Release*, July 28, 2015, https://newsroom.intel.com/news-releases.

8. See the lengthy "Daily Tech" analysis the next day: "If Intel and Micron's 'Xpoint' is 3D Phase Change Memory, Boy Did They Patent It," http://www.dailytech.com/Exclusive+If+Intel+and+Microns+Xpoint+is+3D+Phase+Change+Memory+Boy+Did+They+Patent+It/article37451.htm.

9. "3D Xpoint Steps into the Light," *EE Times*, January 14, 2016, http://www.eetimes.com/document.asp?doc_id=1328682.

10. See Guy Wicker, "A Review of Recent Phase Change Memory Developments," presented at E\PCOS, September 2016. In 2017, Intel began marketing its new memory chip under the name Intel Optane. A recent analysis, based on taking apart one of the memories, showed that its Xpoint memory is a phase-change memory with a "doped chalcogenide" access switch, i.e. "a type of Ovonic Threshold Switch" (http://techinsights.com/abouttechinsights/overview/blog/intel-3D-xpoint-memory-die-removed-from-intel-optane-pcm). This shows conclusively that, as Steve Hudgens observes, "Stan's original idea for a high-performance, multi-layer thin-film memory based on amorphous materials is now officially a mainstream product!" (Steve Hudgens, email to LH, June 15, 2017).

11. See Aviva Rutkin, "Crystal Mimics Brain Cell to Sift through Giant Piles of Data," *New Scientist*, August 3, 2016, https://www.newscientist.com/article/2099913-crystal-mimics-brain-cell-to-sift-through-giant-piles-of-data.

12. Stanford R. Ovshinsky, "Ovonic Chalcogenide Non-Binary Electrical and Optical Devices," *Proceedings of SPIE, Seventh International Symposium on Optical Storage* 5966, China (2005): 1–6. Reprinted in Hellmut Fritzsche and Brian Schwartz, *Stanford R. Ovshinsky: The Science and Technology of an American Genius* (Singapore: World Scientific Publishing, 2008), 104–109.

13. The *New Scientist* account ("Crystal Mimics," see note 11 above) gets this backward: firing occurs with the phase change from amorphous to crystalline.

14. For a broad review of the history and recent developments in the effort to make electronic devices that emulate the brain, see Navnidhi K. Upadhyay, Samuel Joshi, and J. Joshua Yang, "Synaptic Electronics and Neuromorphic Computing," *Science China: Information Sciences* 59 (May 11, 2016): 061404, doi: 10.1007/s11432-016-5565-1. The article notes that among emerging nonvolatile memory technologies that are candidates for creating neuromorphic components,

phase-change memory is "the most mature," and it cites Ovshinsky's original 1968 "Reversible Electrical Switching" paper and the one on cognitive computing he wrote with Boil Pashmakov in 2003. S. R. Ovshinsky and B. Pashmakov, "Innovation Providing New Multiple Functions in Phase-Change Materials to Achieve Cognitive Computing," *Proceedings of the Materials Research Society*, 803 (2003), https://doi.org/10.1557/PROC-803-HH1.1.

15. Wicker also built a working Ovitron switch, so that his barn housed both the earliest and the latest instances of Ovshinsky's amorphous devices.

16. After Ovshinsky's death, OI and its assets, including equipment and patents, entered the Ovshinsky Foundation, headed by Steven Dibner. Rather than pay to protect the patents, the family sold them to Wicker, and the foundation was liquidated.

17. Tom Henderson, "Ovshinsky's Dream Shines on in Solar Firm," *Crain's Detroit Business*, September 14, 2014, http://www.crainsdetroit.com/article/20140914/NEWS/309149980/ovshinskys-dream-shines-on-in-solar-firm.

Conclusion

1. National Inventors Hall of Fame, "Stanford R. Ovshinsky," http://www.invent.org/honor/inductees/inductee-detail/?IID=514.

2. Ovshinsky's discoveries have, however, fed the development of normal science. There are now annual conferences held by E\PCOS (European Phase Change and Ovonics Symposium), where researchers from many countries report work in the new area he pioneered. "Normal science" is a term coined by Thomas S. Kuhn in *The Structure of Scientific Revolutions* (Chicago: University of Chicago Press, 1970), 35–42. It describes the kind of science that slowly accumulates knowledge within the framework of established theory. In Kuhn's terms, Ovshinsky was a "revolutionary" scientist who triggered a paradigm shift in condensed matter physics. (But as we have seen, he also depended on normal scientists to translate and carry out his ideas.)

3. See S. R. Ovshinsky, "Amorphous and Disordered Materials—The Basis of New Industries," *Materials Research Society Symposium Proceedings* 54 (1999): 339–412.

4. Harley Shaiken, family press release announcing Stan's death, October 18, 2012.

5. On Fordism and the related concept of Taylorism, see Thomas P. Hughes, *American Genesis: A Century of Invention and Technological Enthusiasm 1870–1970* (New York: Viking, 1989), 203–220, and David A. Hounshell, *From the American System to Mass Production 1800–1932: The Development of Manufacturing Technology in the United States* (Baltimore: Johns Hopkins University Press, 1984), 249–253.

Index

Note: ECD = Energy Conversion Devices; ECL = Energy Conversion Laboratories; SO = Stanley Ovshinsky

Active-matrix liquid crystal displays (LCDs), 210, 344n3
Adcock, Willis, 89
Adler, David (Dave)
  communication skills, 155
  consulting work for ECD, 155–156
  friendship with SO, 155–156
  model for the Ovshinsky effect, 134–135
  on research on amorphous and disordered materials, 127
Adominis, Al, 138
Adrenalin, psychoses-like effects, 315n32
Advanced Research Group, ECD, 178
Advanced Research Projects Agency (ARPA), 130
Advanced Technology Program (ATP), National Institute of Standards and Technology (NIST), 199
Agnew Machine Company, Milford, Michigan, General Automation work, 71
  antisemitism in hiring, 32–33
  early years in, 3, 22–24
  efforts to move ECD hydrogen program to, 206–208
  father's settling in, 17–18
  Goodyear Airdock, 45
  immigrant population, 15
  impact of the Depression in, 26
  Iris's burial in, 247
  Munitz family move to, 19
  SO's burial in, 279
  SO's family return to, 42–43
Akron Central Industrial Union Council, 30
Akron Standard Mold
  first job at, 29–32, 226
  ongoing work for, 47, 49
Alcoholics Anonymous, funding for ECL from members of, 99
Alda, Alan, 203
Alkaline fuel cells project, 200
Allen, Charles, 87, 317n46
Allen, Robert (Bob)
  Ovitron lawsuit, 91, 99
  Ovitron partnership, 86–86, 89–90
Alphasil, 344n4
Alternative energy, affordable technology, 10, 97–98, 193, 294–295, 318n7, 319n12. *See also* Thin-film solar panels
American Natural Resources Company (ANR), 192, 211, 336n1, 341n16
Amorphous and disordered materials. *See also* the Ovitron; Ovonic thin-film amorphous threshold switch
  attributes, 321n29
  capacity for hydrogen storage, 188
  chalcogenide alloys, 110, 141, 304n4

Amorphous and disordered materials (cont.)
  comparison with crystalline materials, 108–109
  and development of flat panel displays, 4–5
  as "dirt materials," 108
  and ECD's NiMH batteries, 190–191
  electronic properties, explorations of, 3–4
  growing body of research on, 127
  hydrides as, 339n2
  mobility edge, 327n23
  recognition of potential of, 292–293
  Sapru's models of, 135–137
  SO's pathway to understanding, 2, 107–108
Amorphous chalcogenide semiconductors
  and cross-point switching, 346n27
  growing body of research on, 326n19, 327n22
  mechanisms of action, 108, 133–137, 330n43
  New York Times article about, 124–126
  scientific explanations, 124, 323–324nn54–55
  SO's interest in neural network applications for, 221
Amorphous silicon solar panels. *See also* Hydrogenated amorphous silicon
  ability to work despite damage, 338n35
  contracts with SOHIO and Sharp to produce, 175
  improvements to efficiency, 177–178
  PVD method, 173, 336n4
  roll-to-roll production approach, 171–173
  Staebler-Wronski effect, 176
Analogical thinking
  in cognitive science, 305n13
  linking of disordered materials and cosmology, 236–237
  linking of tellurium and DNA, 110
  neurophysiology studies, 56–57, 311n22
  Pauling's use of, 305n11
  and the relationship of science and music, 234
  role in innovation, 5–6, 294, 305n13

and SO's view of threshold switch as electronic, 133–134
Anarchism, 7–8, 75–77, 315n26
Anderson, Philip, 327n23
Annual Physical Society meeting dinners, 327n22
ANR. *See* American Natural Resources Company
Antimilitary views, 103, 320n22
Antisemitism, 32, 94
Antiunion violence, 8, 34–35
Applied Physics Letters, paper on optical memory, 129
Armstrong, Edwin, 306n16
Artillery shell production, 71
Assembly-line, psychological consequences, 65, 295
Atlantic Richfield (ARCO)
  allocations for hydrogen research, 187
  ARCO Solar, 143–144, 171–172, 174–175, 345n10
  final settlement with, 337n12
  support for alternative energy development, 143–144, 332n69, 337n11
Atomic bomb/hydrogen bomb, 47, 154
Automatic tractor, 61–63, 71, 314n15
Automatic transmissions, 68
Automation
  and closed loop systems, 312n33
  early focus on, 3, 63
  and the "industrial computer," 63
  SO's advocacy of, 66, 295
  union opposition to, 66

Bachelet, Michelle, 265–266, 319n12
Bacon, Nancy
  as CFO at ECD, 144, 211
  on financial errors by new ECD board, 353n2
  negotiation/management of grants and contracts, 174–175, 178–179
  replacement of, on ECD board, 249
Baker, Ken, 196
Baker Brothers, Toledo, 55

Band-gap profiling, 178–179, 338n21
Baotou Rare Earth Manufacturing, 197–198, 346n27
Baranoff, Barney, 46–47
Baranoff, Francis Wolinsky, 46
Bardeen, John, 113–114, 155, 323n44
Bar mitzvah, 27, 105, 152, 166
Barnard, Tim, 262–263
BASF Ovonic, 286, 354n5
Batteries, 99, 188, 343n42. *See also* Nickel metal hydride battery (NiMH)
Bedaux, Charles, 308n2
Beglau, Dave, 217, 346n28, 348n42
Bekaert, Belgium, 182–184, 338n29
Benjamin Center Drive Lathe
  artillery shell production using, 58
  impacts, 51, 310n10
  innovative design, 50–51
  invention and development, 48–52
  Kronenberg's evaluation, 53–54
  limit switches, 107
  patents, 56, 309n9
  production changes, 63–64
  production for Norris Thermador, 60–61
Berger, Hans, 315n32
Berman, Walter, 320n23
B.F. Goodrich, Akron, Ohio
  founding, 15
  Miller Plant 1, 35–36
  Miller Plant 2, 19, 33–35
  union organizing at, 8, 34–35
Bienenstock, Arthur, 6–7, 155, 156, 334n12
Birkenstock, Jim, 334n10
Birmingham, Michigan, Ovshinsky home in, 106–107
Blieden, Richard (Dick)
  on benefits to ECD from dissolution of relationship with Sohio-BP, 178
  response to plan to mass produce amorphous silicon solar panels, 172
  solar energy work, 143–144
  work on the SOHIO contract, 175
"Bloody Sunday," 16

Bloomfield Hills, Michigan, Ovshinsky home in, 118–119
Bodman, Samuel, 183
Bradshaw, Thornton, 144, 313n6, 336n1
Brain studies. *See* Neurophysiology
Brockway, Fenner, 97–98
Buchholzer, Frances Seiberling (Fran), 206
Bund culture, 26
Burroughs Corporation, Detroit, 139–140
Bush, George W., 185–186

Calgary, Canada, trip to, 270–272
Calgory (Kalvarija), Lithuania, 15
California Air Resource Board, Zero Emission Vehicle mandate, 193, 196–197
Canella, Vincent (Vin)
  on ECD corporate culture, 149–150
  on SO's complexity, 163
  on SO's enabling others to flourish, 160
  on working with Samsung, 214
  work on amorphous silicon solar cells, 141–142, 210–211, 336n2, 337n7
Canon, Japan. *See also* United Solar Systems
  agreement to develop amorphous silicon copier drums, 179
  interest in developing solar technology, 179–180, 338n23
  learning manufacturing from, 180–181
  manufacturing discipline imposed by, 338n26
Carlson, David, 141
The Car That Could (Shnayerson), 192–193
Case Institute of Technology, Cleveland, 50
Catalytic converters, 194
Chalcogenides/chalcogenide alloys
  chemical modification, 141, 332n64
  cross-link structures, 110–111
  and electrical memories, 5
  Sapru's visualizations of, 136–137
  "schizophrenic" nature of, 329n38
  studies of, 3–4, 109
  as term of reference, 321n34

Chalcogenide switches, in Intel's 3D Xpoint memory chip, 287–288, 354n10
Chao, Ben
  background and work at ECD, 339n3
  on ECD layoffs, 178–179
  on impacts of new management at ECD, 249–250
  on relationship with Chevron, 203–204
Cheroff, George, 346n24
Chevron, joint venture with, 201–204
Chile, trip to, 265–266
Christian, Larry, 156, 332n65
Chrysler, 65–66
Civil rights, political/union activism. *See also* Workmen's Circle
  blacklisting because of, 32
  and SO's children's education, 102–104
  civil rights activism, 320n23
  early activism, 8
  father's influence on, 24–25
  responses to racism, 41, 100
  and socialist background/sewer socialism, 7, 10, 26, 308n3, 308n16
  union activities, 27, 34–37, 308n3, 313n6
  and using science for social good, 2–3, 7–8, 63, 96, 100–101, 106, 147–149, 294–295, 310n18
Clinton, Bill, 181–182
Closed loop systems, 61, 69–70, 312n33
Cobasys (Chevron Ovonic Battery Systems), 203
Cognitive computer
  focus on during early 2000s, 221–223
  and IBM's "artificial neuron," 288
  team members, 347n42
Cohen, Morrel
  as consultant at ECD, 155, 327n22
  role explaining science underlying threshold switching, 124, 126–127
  role in cognitive computer project, 348n42
  work on cosmology project, 247–248
Cold fusion experiments, 319n17

Communism/Stalinism, 26, 68
"A Concept of Schizophrenia" (Ovshinsky), 316n33
Condensed matter physics, 303n1
Cone Automatic Machine Company, Windsor, Vermont, 72–73
Congress of Industrial Organizations (CIO), 30
Continuous production methods, 138, 142–143
Control Engineering, article about the Ovonic threshold switch, 116–117
Conway, Jack, 144, 313n6, 332n70
Corrigan, Dennis, 190–191, 200, 340n8, 340n13
Cosmology studies, 234–237, 247–248
Covalent bonding, 136
Creative process, creativity. *See also* Analogical thinking; Intuition
  and ability to handle multiple lines of thought simultaneously, 5
  and analogic thinking, 5–6
  appreciation for in all fields of endeavor, 233–234
  drawings and paintings, 27–28, 225–230
  integrative, cross-fertilizing approach, 292–295
  and the invention of the Ovitron, 86
Creativity and Intuition: A Physicist Looks at East and West (Yukawa), 234, 348n4
Cross-point switching, 346n27
Crystals
  as basis for solid-state physics and transistors, 108
  crystalline semiconductors, 126
  precise measurements associated with, 321n30
Cummings, Richard, 118
Cunningham, Keith, 137–139, 145, 331n54, 333nn73–74
Cybernetics; or Control and Communication in the Animal and the Machine (Wiener), 56, 311n24

Index

Cybernetics studies, 3, 56–57, 61
Czubatyj, Wally
  electronics group leadership, 217
  physics group member, 156
  work with the Ovonyx team, 219–220, 346n26, 347n35
  work on solar energy, 337n7

Dark matter, skepticism about, 235
Dauschotz (Dokshytsy) *stetl*, Daitch family from, 19
Davidson, William Morse (Bill), 214–215
The Defect Solid State (Gray), 108
Democratic socialism, 26
De Neufville, John
  on Cunningham, 331n54
  at ECD structures lab, 130, 139, 328n26, 328nn31–32
  on Momoko's negotiating skills, 160–161, 335n22
  other roles at ECD, 328n33, 341n14
  on the Ovshinsky meetings on amorphous materials, 327n22
Density of states measurements, 263, 352n11
The Depression, impacts in Akron, 26
Desktop computers, 304n2
Detroit, Michigan. *See also* Benjamin Center Drive Lathe; Energy Conversion Devices (ECD)
  antisemitism in housing, 94
  auto industry in, 3
  civil rights activism in, 100
  Iris's return to, 93-95
  labor movement in, 66–67
  Ovshinsky homes in, 65–66, 71, 98
  renting storefront on McNichols Road, 72
Detroit Physiological Society, 89
Dhar, Subhash
  and Baotou joint manufacturing venture, 197–198
  and Ovonic Battery Company (OBC), 191–192, 341n15
  departure from ECD, 203, 343n44
  with Texaco Ovonic Battery Systems, 200, 341n14
Dibner, Andrew (Andy)
  divorce from Iris, 93
  family move to Worcester, 81
  marriage to Iris, 78
  SO's debates/arguments with, 79, 102
Dibner, Bern, 315n29
Dibner, David, 315n29
Dibner, Iris Miroy. *See* Ovshinsky, Iris Miroy Dibner
Dibner, Richard (Dick)
  cancellation of patent application for electric power steering, 70
  introduction of Iris to Stan, 79
Dibner, Robin
  childhood memories, 104, 106, 166–167
  family trip to Hawaii, 254–255
  at Mashie's 80th birthday celebration, 168
  and pressure to achieve, 167
  relationship with "Stan-Dad" and Iris, 102, 166–167
  response to move to Detroit, 93
  response to SO's relationship with Rosa, 256–258, 260, 351n3
  trip with SO and Rosa to Calgary, 270–272
Dibner, Steven (Steve)
  childhood memories, 104, 106–107, 166, 335n25
  leadership of the Ovshinsky Foundation, 355n16
  at Mashie's 80th birthday celebration, 169
  on SO's appreciation for music, 233, 345n13
  on SO's use of anger, 164
  and pressure to achieve, 166–167
  relationship with "Stan-Dad," 102
  response to move to Detroit, 93
  response to SO's relationship with Rosa, 256–257
  support for musical talents, 166–167, 335–336n26, 335n24
  visit to in Santa Fe by SO and Iris, 242

Diesel engines, high school exam on, 31
Dirt, role in discovering the Ovshinsky effect, 113
Dissent quarterly, 66–67
Doehler, Joe
  background, 336n5
  role in producing Ovonic solar panels, 171, 173, 337n7
Dover, Ohio, move to, 51–52
Duluth, Minnesota, father's work in, 17
DuPont, 347n38
Dupré, Henri, 75–77
Dykstra, John, 54
Dynamic random-access memory (DRAM), 218

Ecological agriculture, 312n34
Edison, Thomas
  and collaborative invention, 123, 324n3
  comparisons of SO with, 2, 105, 154, 191
  as inventor as opposed to scientist, 304–305n8
  Menlo Park laboratory, 327n24
Einstein, Albert
  admiration and appreciation for, 234
  amended field equations, 349n10
  as a visual thinker, 305n14
Electrical automatic transmission, 3, 68
Electrical phase-change memory
  development of, 1, 5, 131, 217
  magnetic core memory vs., 331n59
  optical phase-change memory vs., 346n20
  revival of at ECD, 216–217
  three-dimensional, modelling and development of, 217–218
Electric cars. *See* EV1 car; Nickel metal hydride (NiMH) batteries
Electric power steering, 69–71, 313nn13–14
Electroencephalogram, 315n32
Electrolytes, elimination of, in SO's solid-state approach, 321n33
"Electro-Mechanical Motion" (Ovshinsky), 311n26

Electromechanical switches, 107–108
Electronic Machine Control, Ltd., 113
ELPIDA, 220
Encryption, threshold switching for, 222
Energy, complementarity with information, 96, 318n9, 347n39
Energy Conversion Devices (ECD). *See also specific inventions, subsidiary companies, and research units*
  business model, 212
  closing of the machine shop, 250, 350n10
  cognitive computer project, 347n42
  commercial successes, 4
  commitment to exploring SO's ideas, 151
  consultants, 152–157
  Cunningham lawsuit, 145
  design for alkaline fuel cells, 200
  dismantling of, 353n3
  education program, 130
  electrical phase-change memory division, 131
  expansion of following ARCO grants, 144
  expansion of solar thin-film production under Bekaert collaboration, 183
  and failure to reach agreement with Tatung for thin-film transistors, 215
  financial difficulties and layoffs, 139, 149–150, 178, 211, 249, 331n56, 333n5, 338n22
  financing, 127, 143–144, 178–183, 192–193, 195, 197, 218, 331n57, 347n32
  hiring and promoting of women at, 149
  hydrogen research programs, 187–192, 199–202, 207–208
  Institute for Amorphous Studies, 157–159
  integrative, cross-fertilizing approach to research, 146
  international flavor, 149
  as an invention factory, 123
  joint venture with Canon, 179–180
  and the mass production of amorphous silicon solar panels, 171–172

Momoko Ito's roles at, 160–163
move to Rochester Hills facility, 220
naming of Morelli as president, 351n12
new directors and bankruptcy, 249, 285, 350n9, 353n2
NGEN (Next GENeration of computers) program, 216–217
optical phase-change memory technology, 215
organizational plan, 148, 333n3
patent department, 174
phase-change memory groups, 217–218, 346n28
photovoltaic program, 139–142
physics department, 127–128, 130, 328n26
progressive corporate goals and culture of cross-fertilization, 96, 123, 146–152, 292–295
purchase of USSC, 183–185
relationship with IBM, 334n10
renaming and expansion, 4, 118–119
research environment, 127–128, 146, 152–154
silicon germanium alloy production, 177
SO's changing role from ECL, 123
SO's forced retirement from, 248–249
and SO's need for control, 162–163
stock ownership, 334n8, 345n11, 350n8
Troy, Michigan facility, 118–120, 123, 324n1, 324n59
tuition reimbursement program, 147
Energy Conversion Laboratories (ECL)
early research, 99–100
family environment, 100
funding, 98–99
research environment, 318n4
underlying concept and goals, 2, 7–8, 95–96, 100–101, 318n8
Environmental health concerns, 96–97
E\PCOS (European Phase Change and Ovonics Symposium), 349–350n5, 355n2
Epilepsy and schizophrenia studies, 82–83, 315n32, 316nn33–34

Escoffier, Auguste, 75–76, 315n28
European Inventor Award, 272–273
EV1 car, 194–196. *See also* Nickel metal hydride (NiMH) batteries
Exchange Auto Parts, 46–47

Fagen, Ed, 130, 328n26, 331n56
Falls, Bruce, 201–202, 204–205
Federal Bureau of Investigation (FBI) surveillance, 68, 316n36
Feineison, Joseph, 34
Feinlieb, Julius, 128–129, 155
Fetcenko, Mike
  acceptance of European Inventor Award for SO, 273
  on benefits of disordered materials, 340n10
  on the ECD culture, 147–148
  on Iris's role at ECD, 164–165
  on relationship with Chevon, 203
  on Siskind's role at ECD, 174
  work for BASF Ovonic, 286, 354n5
  work with the hydrogen storage program, 188, 190–192
Fitzpatrick (Akron Standard Mold shop superintendent), 30, 33
Flasck, Richard (Dick)
  on annual "Christmas crisis" at ECD, 163–164
  departure from ECD, 210, 344n4
  responsibilities at ECD, 130, 139
  on SO's visual thinking process, 6
Flash memory
  competition with electrical phase-change memory, 217, 331n59
  corporate support for improving, as challenge for Ovonyx, 221
  invention, 346n25
  limitations, 218
  precursors, 331n59
Flat-panel displays
  change from diodes to transistors, 213
  development of using diodes, 4–5, 210–211
  ECD's failure to benefit from, 209

Flat-panel displays (cont.)
  inadequacy of chalcogenide materials for, 210
  and thin-film transistors, 175
Fluorine, failed efforts to use in devices, 142, 264, 332nn66–67
Fontana, Georgina, 253, 267
Ford, Henry, 65, 294–295, 313n12
Fordism and Taylorism, 308n2, 355n5
Forge work, love of, 31
Francisco Ferrar Association, "Modern School," 315n26
Frenchy (Imperial Electric foreman), 36
Frisbee, Bob, 60
Fritzsche, Hellmut
  on the firing of Pashmakov, 250
  first meeting with SO, 114–115
  friendship with SO, 237
  and lone pair semiconductors, 134–135
  management of ECD Physics Department, 128
  nondisclosure agreements, 116
  paper on properties of amorphous semiconductors, 127
  on photoconductivity measurements at Ovshinsky Innovation, 263–264
  replacement of, on ECD board, 249
  response to demonstration of threshold switch, 115
  response to negative publicity about SO, 126
  at SO's ninetieth birthday party, 268, 269
  on SO's use of public media, 124–126
  on Teller's visits to ECD, 154
  on understanding film conduction, 107
  on voltages required by thin-film switches, 321–322n35
  work as consultant to ECL, 116
  work at Ovshinsky Innovation, 262–263
  work on tellurium and disordered systems, 114
  work with SO on cosmology, 234–237, 247–248, 338n24
  work with SO on superconductivity, 349n87
Fritzsche, Sybille, 114
Fuel cells. *See* Hydrogen fuel cells

Gardner, Ernest, 83–84
Garlovsky daughters, 16
Gasiorowski, Paul, 262–263
Geddes, Ralph
  brokering solution to Stanford Roberts's financial problems, 55–56
  business ventures, 48–49
  erratic support from, 54, 70–71
  familial relationship with, 313n8
  managerial talents, 68
  role during SO's Detroit years, 63, 67–68
  startup funding for General Automation, 71
Geiss, Dick, 201
General Automation Corporation, Detroit, 71–74
General Motors (GM)
  EV1 car, 192–194, 342nn30–31
  fight against the California air standards, 196
  founding and growth, 65–66
  open loop hydraulic steering, 69
Gerard, Ralph W., 84
Germanium telluride, 6, 337n16
Gernsback, Hugo, 26
Gide, André, book report on, 28
Gigawatt machine
  abandonment of after SO's death, 286
  estimated productivity, 351n7
  as goal at Ovshinsky Innovation, technical approach, 260–264
  initial concept, 185
Glasses, chalcogenide. *See also* Amorphous and disordered solids; Chalcogenide switches; Flat panel displays
  as an amorphous solid, 108–109
  characteristics, 3–4
  studies of, at ECD, 127, 130
  use of tellurium, 322n36
"Glassy Electronic Device May Surpass Transistor" (*New York Times*), 1, 124–126
GM-Ovonic, 195–196, 200

Goldman, Emma, 75
Goodrich, Benjamin Franklin, 15
Goodyear Aircraft bomber plant, Litchfield Park, Arizona, 40–41
Goodyear Airdock, 45
Goodyear Zeppelin Corporation, 45
Goudsmit, Samuel, 126
Gray, Thomas James (*The Defect Solid State*), 108
Guardian Industries, and Guardian OIS, 215
Guha, Subhendu
  during Bush visit to United Solar, 186
  negotiations with Canon, 180
  roles at United Solar, 180–181, 182–183
  work on Ovonic solar panels, 177–178, 338n20

Hack, Michael, 338n19
Hawaii, family trip to, 254–255
Haywood, Bill (Big Bill), 17
Heckeroth, Steve, 177, 181
Heidrich Tool and Dye Corporation, Detroit, 55
Heinnig, Ruth, 49
Hennessy, Mike, 262–263
Henry, Milton, 320n25
Herrault, André (André Miroy), 75–76
Hitachi Maxell, 192, 211, 220
"Hold the Fort" (union song), 233
Holmberg, Scott, 344n4
Holonyak, Nick, 114, 323n47, 333n74
Hong, Kuochih, 188–189, 191, 341n14
Howe, Irving, 66–67
Hower Trade School, Akron, 31
Hudgens, Stephen (Steve)
  on amorphous semiconductors, 323n54
  on corporate development of flash memory, 221
  on explanations of the Ovshinsky effect, 133–135, 329n29
  management of Ovonyx in California, 220
  perfection of "spectrum splitting" process, 177
  on plan to mass-produce silicon solar panels, 172
  on revolutionary nature of SO's switch, 323n50
  on Tyler Lowrey, 218
  on versatility of the OUM device, 219
Human intelligence, cognitive computer as analog to, 223
The Human Use of Human Beings (Wiener), 56
Hupp Motorcar Corporation, Detroit
  artillery shell and rocket tube production, 71
  Benjamin Lathes for, 72
  cybernetic components for, 68
  Geddes's purchase of, 54
  Geddes's sale of, 73
  hiring of SO by, 63
  and SO's electric power steering invention, 69–71, 313–314n10
Hydrogen
  commercial sources, 343n40
  early work with, at ECL, 99–100
  generation, storage, and utilization of, 187–188
Hydrogenated amorphous silicon. *See also* Amorphous silicon solar panels
  development for use in LCDs, 210
  early research, 331n60
  photoconductivity, 141
  use in the ECD photovoltaic program, 141–142
Hydrogen cars
  ATP funding for, 199
  collaborators, 199
  development process, 198–202
  second and third version, 205
  SO's approach to, 198–199
  SO's promotion of, 203
Hydrogen fuel cells
  batteries vs., 343n42
  development of by ECD researchers, 188–189
  joint venture with Texaco to develop, 199–200

Hydrogen loop concept
  early work with, at ECL, 100
  energy storage issues, 187
  and hydrogen cars, 198–199
  ongoing research, 320n18
  as three-part system, 2, 205–206
Hydrogen research program, ECD. *See also* Nickel metal hydride (NiMH) batteries
  development of storage technologies, 187–188, 201
  move to OBC, 192
  researchers associated with, 187–188, 190–192, 201–202

Iacocca, Lee, 342n29
IBM
  "artificial neurons," 288, 348n41
  licensing of threshold memory switch to, 334n10
  work with ECD on optical memory, 138, 215, 331n59
Imperial Electric, Akron, 36–37
Industrial computer, 61, 311n21
Industrial research laboratories, types, 327n24
Industrial Workers of the World (IWW, "the Wobblies"), 17
Information, complementarity with energy, 96, 318n9, 347n39
Information technology
  as focus at ECD, 209
  as focus of ECL, 96
  SO's contributions to, 11
  SO's recognition of importance of, 343–344n1
  as way of addressing social problems, 2
Instant-imaging technologies, 137–138
Institute for Amorphous Studies, ECD
  Board of Directors, 158
  colloquia and lectures at, 158–159
  move of SO's office to, 250
  sale of after SO's death, 280
  vision for, 157

Intel
  collaboration with ECD, 131
  EEPROM memory, 331n59
  flash memory, 346n25
  and Ovonyx, 220, 287–288, 347n32, 347nn35–36
  3D Xpoint memory chip, 346n27, 354n10
Intellectual freedom, SO's lifelong commitment to, 293–294
Intuition
  and analogical thinking, 57, 110, 293–294
  and capacity for visualization, 135
  and communication challenges, 128–129
  and creativity, 233–234, 236
  reinforcement by reading and experience, 108–109
  SO's application to cosmology, 236
  trust in, 4, 7, 51, 86, 192
Inventions. *See also* Patents *and specific inventions*
  early interest in, 26
  and goal of benefitting society, 3, 10, 294–295
  incorporation into larger systems, 304n7
  and patent litigation, 306n16
  range of, 1–5, 123, 291
Ionic bonding, 136
Ito, Joichi (Joi)
  career, 335n23
  on challenges of working with SO, 221
  on parental role filled by SO and Iris, 166
  on SO's divergence from mainstream computer culture, 347n37
  on SO's idiosyncratic scientific explanations, 137
  at SO's ninetieth birthday party, 268, 269
  on SO's not moving to Silicon Valley, 294
Ito, Mizuko (Mimi), 161–162, 163, 165, 335n23
Ito, Momoko
  background, upbringing, 335n19
  capacity to hold alcohol, 162

Index 367

hiring of, 160
marriage to Masat Izu, 138
move to Japan with children, 163
negotiating skills, 160–161, 180, 215–216
promotion to vice president, 161
"I want more" story, 9
Iwasa, Alice, 129
Iwasa, Sato, 128–129, 131, 138, 328n26, 328n30, 331n54
Izu, Masatsugu (Masat), 138, 142, 159, 337n7

Japan. See also Ito, Momoko and specific Japanese companies
Japan-ECD, 162–163
and negotiations with the Japanese, 8
sources of funding for ECL in, 319n15
Jewish Labor Bund, 16
Jobs, Steve, 304–305n8
John R. Buchtel high school, Akron, 28, 31
Johnson, Robert, 139, 157, 210

Kamin, Chester (Chet)
on cross-fertilization of activities at ECD, 146
defense of ECD against UNC lawsuit, 145, 333nn73–74
on hiring of Cunningham, 137
legal career, 332n71
on battery litigation with Matsushita, 197
on SO's lack of fear, 8
on SO's mental capacity, 5
on ECD's patent litigation, 342n32
on SO's relationship with Stempel, 195
Kastner, Marc, 134–135, 157, 158
Klersy, Patrick (Pat), 217, 219–220, 262–263, 346n28, 347n35, 348n42
Klose, Peter, 138, 330n49
Kluver, Heinrich, 84
Knudsen, William Signius, 52–53
Koefod, Ghazaleh, 150, 158, 159, 333n5, 337n10
Kolobov, Alex (Sasha), 267, 352n2
Kolomiets, Boris T., 110, 127, 321–322n35, 327n21

Kotz, Jim, 329n37
Kronenberg, M., 53–54, 310n13
Kuhn, Thomas S., 355n2
Kumar, Arun, 181–183, 285
Kvant, joint venture in Soviet Union, 181–183

Lagos, Ricardo, 265
Lark-Horovitz, Karl, 114
Laser-related research, 128–129
Lathes. See also Benjamin Center Drive Lathe
basic principles, 50
belt-driven, 34
center-driven, evolving ideas about, 42
high-speed automatic lathe, 3
programmable automatic lathe, 72–73, 314n19
SO's adaptations, 30, 47
Lawsuits. See also Patent litigation
with Bill Manning, 213–214, 245n11
related to the Ovitron, 91
with Tann Corporation, 74, 87
LeComber, Peter, 141, 331n61
Leff, Harvey, 18, 175
Lipscomb, William, 155
Litchfield Park, Arizona, Goodyear Aircraft bomber plant, 40
Lithium Corporation of America, 99
Lockheed Martin, 347n34
Logic studies, 63
Lone pair semiconductors, 134–136, 236, 329n38, 329n40, 352n2
Lou (uncle), 17
Lowrey, Tyler, 218–221, 347n35

Machines, machinery
early fascination with, 25–26
SO's intuitive understanding of, 30
SO's lifetime love for, 7, 28, 31
Machinist/toolmaker work
and early tool purchases, 31
early work experience, 3
and the importance of the Benjamin Lathe, 51

Madan, Arun, 141–142, 157, 337n7, 338n19
Magnetic core memory, 331n59
Manning, William (Bill), 211–214, 345n11
Massachusetts Institute of Technology (MIT), 61
Masuoka, Fujio, 346n25
Mathematics
  and SO's visual capacity, 18n14
  trouble understanding in school, 25
Matsushita patent infringement lawsuit, 215–216
McCarthy, Walter, 194
McCarthy era investigations, 68
McGill, John, 211, 344n6
Memories, as historical sources, challenges of using, 303n6, 323n49
Memory switches, phase-change. *See also* Ovshinsky effect; Threshold switches
  comparison with threshold switches, 324n55
  neural network applications, 220–221
  and read-mostly memory, 131
  understanding, as early focus of ECD Physics Department, 128
Metal hydride batteries. *See* Nickel metal hydride (NiMH) batteries
Metallurgy, education in, 31
"Method and Apparatus for Storing and Retrieving of Information" (patent), 328n29
Metzger, Jim, 249
Micron Semiconductors. *See also* Intel
  joint venture with ECD, 218–219
  purchase of Ovonyx, 287
Micro-Ovonic Fiche (MOF), 137–138
Miroy, André (André Herrault), 75–76
Mishakoff, Misha, 335–336n26
Mobility edge, in threshold switches, 127, 236, 327n23
Mohegan Colony, Westchester, NY, 75–76
Mongolia, OPBC-Baotou battery manufacturing plant, 197–98
Moore, Gordon, 131
Morelli, Mark, 351n12

Morin, Ferdinand A., 83–84, 120
Morita, Akio, 319n15
MOSFETs (metal-oxide semiconductor field-effect transistors), 221
Moss, Howard, 130–131, 328n26
Mott, Nevill Francis, 124–126, 135, 154, 157, 325n7, 326n18, 327n23, 334nn11–12
Muni, Paul (Frederick Weisenfreund), 24, 307n12
Multiple-ball switch, 73–74
Munitz, Bertha. *See* Ovshinsky, Bertha (Teibel) Munitz
Munitz, Rebecca Daitch, 19
Music
  father's love for, 24
  SO's love for, 233
  science as analogous to, 233
  Steve Dibner's talents, 165, 233, 335n24, 335–336n26
Musk, Elon, 304–5n8
Myasnikov, Vitaly, 201
Mytilineou, Eugenia (Genie), 216, 227, 273–274, 280
National Inventors Hall of Fame, posthumous induction of Ovshinsky into, 291–292
Navy, rejection by, 37–38
Neale, Ron, 131, 323n43
Nemanich, Gene, 201, 203
Nerve cell analogy, neural networks. *See also* Cognitive computers
  as basis for the Ovitron, 6, 85, 293–294, 317n41
  as contribution to inventing the threshold switch, 108
  and cognitive computing, 220–223, 354–355n14
  IBM's "artificial neuron," 288
  mechanical analogs to nerve cell behavior, 3, 6
  modeling, 3
  and signal transmission between neurons, 3, 84–85
"Nerve Impulse" (Ovshinsky), 57, 311n26

Neurophysiology
  epilepsy and schizophrenia studies, 82–83
  papers on nerve impulses, 83
  studies related to, 3, 56–57, 83–84, 119–120
New Britain Machine Company Group, Connecticut, 55–59, 61, 312n31
New York Times
  SO's obituary in, 325n12
  Steven's article about chalcogenide switches, 124–126
  Sunday edition, habit of reading, 25
NGEN (Next GENeration of computers) program, 216–217
Nickel metal hydride (NiMH) batteries
  development of, 1, 4, 8–9, 188–191
  and the electric car, 9, 193–194
  manufacturing of, OBC-Baotou approach, 198
  marketing using licensing model, 192
  patent for, 190–191
  patent infringement lawsuits, 197
  posthumous honoring of Ovshinsky for, 291–292
  recharging process, 339–340n5
  safety of, 342n36
  unique features, 190–191
  use of catalytic oxides to enhance, 193
  as a use of disordered material, 339n2
"Normal" vs. "revolutionary" science, 335n2
Norris, Larry, 8, 174, 179, 328n29, 337n8
Norris Thermador, 60–61
North American International Auto Show, Advance Technology Vehicles presentation, 196, 342n29
Norway, cruise to, 273–274
Noyce, Robert, 131

OBC. *See* Ovonic Battery Company (OBC)
O'Donnell, Dan, 346n24
ODS. *See* Ovonic Display Systems (ODS)
OI/OS. *See* Ovshinsky Innovation/Ovshinsky Solar
OIS. *See* Ovonic Imaging Systems, Inc. (OIS)

Open loop hydraulic steering, 69
Optane (Intel), 354n10
Optical phase-change memory
  comparison with electrical phase-change memory, 346n20
  development of at ECD, 128–130, 215
  devices dependent on, 5
  increasing speed and life cycles, 216
  225 alloy, 216
OSMC. *See* Ovonic Synthetic Materials Company (OSMC)
OUM. *See* Ovonyx (Ovonic Unified Memory, OUM)
The Ovitron (amorphous switch)
  commercial development, 86–89
  electrochemical operating principles, 85–86
  as example of SO's creative process, 86
  lawsuits related to, 87, 107, 314n24
  mechanism of action, 317nn41–42
  and the nerve cell analogy, 3, 6, 85
Ovitron Corporation, 86–91
  public announcement of, 87–88
Ovonic Battery Company (OBC)
  early history, 191–192
  joint manufacturing venture with Baotou, 197–198
  licensing business model, 192
  purchase by BASF and ongoing operations, 286
  US ATP grant award, 343n38
Ovonic Display Systems (ODS), 210–211
Ovonic Hydrogen Storage, 354n4
Ovonic Imaging Systems (OIS), formerly Ovonic Display Systems, 211–215
Ovonic Imaging Systems, Inc. (OIS), 137–139, 330n48
The Ovonic Link, IRIS feature, 149
Ovonic Memories, Inc. (OMI), 137–139
Ovonic Memory Device, naming, 325n14
Ovonic Quantum Control Device, 222–223
Ovonic solar panel
  "band gap profiling," 178
  cell structure inversion, 178

Ovonic solar panel (cont.)
  production of through United Solar, 180–182
  roll-to-roll production approach, 173
  spectrum splitting, 177
  underlying technology, 173
Ovonic Synthetic Materials Company (OSMC), 212
Ovonic thin-film amorphous threshold switch.
  discovery of, 111–113
  licensing of, 113
  mechanisms of action, 133–135, 325n11, 329n40, 330n43
  naming, 325n14
  patents and journal articles about, 116–118
  promotional efforts, 113–117
Ovonyx (Ovonic Unified Memory, OUM), 219–220, 287, 347n35
Ovshinsky, "Alter," 16
Ovshinsky, Benjamin (Ben, father). *See also* Workmen's Circle
  background and immigration to Ohio, 15–16
  breadth of reading and interests, 24
  childhood in Russia, 15–16
  death and funeral, 48, 309nn4–6
  drawing of, 18, 277
  generosity, 21
  health problems, 42–43, 46
  influence on son, 7, 24
  love for theater and music, 24, 233–234
  marriage to Bertha Munitz, 19–20
  move to Akron, 17–18
  railroad work and travels, 17
  relationship with son, 24, 46
  religious practice and cultural identity, 308n18
  scrap metal business, 18
  social values, activism, 7, 24
  work with horses, 15, 17–18
Ovshinsky, Benjamin (Ben, son)
  birth, 51
  childhood memories, 69, 71, 310n15
  civil rights activism, 103–104
  education, 106

  on growing hostility to SO among ECD board members, 241
  recollections of the Hupp plant in Detroit, 72
  relationship with father, 167
  on relationship with Geddes, 67, 313n8
  response to SO's relationship with Iris, 102
  response to SO's relationship with Rosa, 257
  SO's letter to about studying, 311n28
Ovshinsky, Bertha (Teibel) Munitz (mother)
  childhood in White Russia (Belarus), 19
  conflicts with SO over religion and politics, 27
  conflicts with husband over politics, 21
  friction with Norma, 45
  immigration to the US, 19
  kindness, hospitality, 21
  marriage to Ben Ovshinsky, 19–20
  SO's financial support for, 48
  religious observance, 21
  separation of SO from his father, 46, 48
  visit to Phoenix, 41–42
  wedding gift to son, 39
Ovshinsky, Cathie Kurek (Mrs. Harvey)
  on Iris's declining physical and mental health, 242
  marriage, 167
  response to SO's relationship with Rosa, 256, 258
Ovshinsky, Dale (son)
  birth, 51
  developmental challenges, 81, 104
  with family, 167
  gratitude to Rosa, 351n4
  later life and religious convictions, 104
  support from father and Iris, 167, 353n8
Ovshinsky, Harvey (son)
  arranging obituaries for SO, 280
  birth, 51
  career, 168–169
  childhood memories, 71, 104–105, 309n1, 309n6

Index

with family, 168
on Iris's declining health, 242
response to SO's relationship with Iris, 103
response to SO's relationship with Rosa, 95, 257–258
at SO's ninetieth birthday party, 268, 352n3
on test drive of electric vehicle, 193–194
Ovshinsky, Herbert (Herb, brother)
on benefits to ECD from dissolution of relationship with Sohio-BP, 178
birth, 21
collaborations with brother, 91, 100, 261
and the development of the Benjamin Lathe, 50
and the development of instant imaging technologies, 138
marriage and move to New Britain, 58
move to Detroit, 71
polio, 54
relationship with nephew, Ben, 51, 54
response to SO's relationship with Iris, 82
on SO's marital difficulties, 54–55
visit to Phoenix, 41–42
work at Exchange Auto Parts, 47
Ovshinsky, Iris Miroy Dibner (Mrs. Stanford)
biochemistry degree, 83
childhood and upbringing, 7–8, 75–78
community-building at ECD, 147–150, 164–165
death, 244–246
divorce from Andy Dibner, 93
with family, 168
family history of suicide, 349n5
family life, role in sustaining, 95, 102–106, 164–167, 242, 353n18
funeral and memorial, 247
health issues, 241–242, 350nn6–7
idealism/dedication to societal justice, 7–8, 95, 97, 100–101, 105–106, 123–124, 147–148, 193, 232–233
impact of SO's health problems on, 349n2
living arrangements in Detroit, 93–95
marriage to Andy Dibner, 78–79

marriage to SO, 95, 102–103
move to Michigan, studies and degrees, 78
partnership role at ECL and ECD, 81, 95–99, 123, 163–165, 196, 218, 251
relationship with SO, 2, 7–8, 74, 79–82, 102, 107, 243–244, 318nn1–2
return to Detroit, 93
trip to Santa Fe, 242–244
Ovshinsky, Myrtle (Mashie/Sandra, sister), 21, 96
Ovshinsky, Norma Rifkin (Mrs. Stanford)
anger towards Iris and Stan, 95, 102
background and personality, 38–39
divorce, 99
friction with Bertha, 45
marriage to SO, 39, 54–55
move to New Britain, 58
pregnancy, 46, 310n16
remarriage, 320n21
Ovshinsky, Rosa Young (Mrs. Stanford)
background and education, 343n37
conservatism, 261–262
and the hydrogen car project, 199–200
marriage to SO, 256–258, 260–261
on negotiations with Tatung, 215
ninetieth birthday party for SO, 267, 270
recruitment of engineers from Russia, 201
relationship with SO, 151–152, 253–256, 258, 260–261, 351n2, 353n9
response to Iris's death, 253–254
response to SO's final illness/death, 275–279, 281–282
role in developing phase-change materials, 216
role in promoting electric cars, 343n39
Ovshinsky, Selma (Herb's wife), 50, 58, 71
Ovshinsky, Stanford Robert (Stan), 310n15. *See also* Analogical thinking; Creative process, creativity; Energy Conversion Devices (ECD); Energy Conversion Laboratories (ECL); Intuition; Inventions *and the other members of the Ovshinsky family*
achievements, 4, 7–11, 268, 291–295

Ovshinsky, Stanford Robert (Stan) (cont.)
appearance, 123, 324n4
artistic interests and talents, 24, 225–234
awards and honors, 10, 301–302
boxing, 8, 26–27
cancer diagnosis and metastasis, 275
capacity for multitasking, 5, 151
childhood, 21–25
communication challenges, 6–7, 124, 128, 139, 306n18
creativity, artistic expressions, 27–28, 225–232
curiosity, 25, 86, 225
as a father, 168, 311n28
final wishes and death, 276–278
formal education, 2, 25, 28, 30–31, 84, 306n18
health problems, 38, 213, 242, 259–260, 268, 270, 274–276, 324n58, 349n2
idealism, dedication to social justice, 193, 232–233
integrated view of the sciences, 2, 135
love for toys and models, 232
as "maverick" inventor, 327–328n25
ninetieth birthday party, 267–270, 274
obituaries, memorials and burial, 279–280, 281, 353n12
personality, 3, 8–9, 99, 128, 163–164, 250, 325n18, 336n6
response to Iris's death, 245–247, 253
scientific recognition/rejection, 4, 115–118, 125–128, 152–154, 233–234, 237, 288, 291–294, 306n18, 313n10, 326n18
self-education, love for reading, 2, 5, 24–25, 36–37, 56, 63, 82, 96, 110, 229, 235
self-promotion, 113–114, 126, 203, 205, 293, 325n14
social democratic politics, social vision, 7–8, 10, 95–97, 105–106, 124, 147–148, 163, 292
on the unity of all human spheres of endeavor, 146, 225
visionary predictions, 1, 126
writing skills, 27, 231, 232–233

Ovshinsky effect
definition, 305n11
demonstration of to Fritzsche, 115
and the development of information technologies, 209
discovery and development of, 111–113
electronic vs. thermal mechanisms of action, debates about, 133–134
environment in which discovered, 11
explanation for "cross" pattern, 322n39
mechanisms of action, 133–35, 329n40
and optical phase-change memory, 129–130
Ovshinsky Foundation, 355n16
Ovshinsky Innovation/Ovshinsky Solar (OI/OS)
assets turned over to the Ovshinsky Foundation, 355n16
creation of, 260–261
focus on improving panel production rates, 262
goals, 9
inability to obtain commercial funding, 264
liquidation of following SO's death, 286–287
principals and working team, 262–263
SO's personal financial investment in, 262
Wicker's revival of, 288–289

Pashmakov, Boil, 219–223, 249–250, 262–263, 347n33, 348n41
Patents
the automatic tractor, 61–62
the Benjamin Lathe, 56
electric power steering, Hupp's false claim, 313–314n14
electromagnetic automatic transmission, 68
hydrogen storage alloys for NiMH battery, 341n21
importance of to ECL/ECD success, 116
"Method and Apparatus for Storing and Retrieving of Information," 328n29
nickel metal hydride (NiMH) battery, 190–191

patent litigation, 8, 197, 306n16, 342n32
rechargeable batteries, 340n7, 340n10
roll-to-roll solar panel production machine, 174, 337n9
SO's approach to, 174
"Symmetrical Current Controlling Device" (threshold switch), 116, 323n53
"Thermoelectric Device," 319n16
Pauling, Linus, 155, 305n11, 334n13
Pearson, A. David, 326n16
Pellier, Laurence and René, 132–133
Peterman, Nate, 71
Phase-change memory. *See also* Ovshinsky effect; Threshold switches
  basis for in the Ovshinsky effect, 112
  development of at ECD, 216–218
  discovery of, 4, 118
  slow impact of, 209
  in 3D Xpoint memory chip, 287–288, 354n10
  and use in cognitive computing, 221, 354–355n14
Philips Corporation, 340n7
Phoenix, Arizona, Ovshinsky family in, 40–41
Photovoltaic cell, operation of, 140–141
Physical Review Letters, Ovshinsky paper on chalcogenide switches, 1, 124, 126
Physics Department, ECD, 127–128, 155–156, 328n26
Piori, Emanuel, 334n10
Plasma-enhanced chemical vapor deposition method (PVD), 173, 336n4
Poincaré, Henri, 236
Political activism. *See* Civil rights, political/ union activism
Polsky, Sylvie, 168
Popular Invention magazine, 26
Powell, Max
  employment at ECL, 100–102, 320nn19–20
  friendship with SO, 150
  on Iris, 101
Power steering, 3, 313nn13–14

Power window brake, 68–69, 313n10
"Progress in Understanding the Ovshinsky Effect: Threshold Switching in Chalcogenide Amorphous Semiconductors" (Hudgens), 329n29
Proof of principle, 51, 147, 262, 264, 286, 310n12
Pryor, Roger, 346n24
PVD method. *See* Plasma-enhanced chemical vapor deposition method

Quantrol (threshold switch, original name), 113
Quantum mechanical tunneling, 7, 317n42
Quantum Technologies, Lake Forest, California, 201
Quartet Manufacturing Company, 211
Quartet Ovonics, 211

Rabi, Isidor Isaac, 152–154, 333n6, 334n10
Racism. *See* Civil rights, political/union activism
RCA, discovery of the Staebler-Wronski effect, 176
Read-mostly memory (RMM), 131–132, 139–140
Reger, Arie, 187–189, 191–192, 340n12
Reichman, Benjamin (Benny), 188–189, 191–192, 340n12
Reischauer, Edwin, 162, 335n22
Reischauer, Haru, 335n22
Rewritable CDs/DVDs
  commercial use, 216
  development of, 1, 5, 129
  limited ECD benefits from, 209
Richetti, Mr., 47–48
Rifkin, Abe, 38
Rifkin, Ida Moon, 38
Rifkin, Jerry, 38
Robbins, Lionel, 102, 119, 323n49
Roll-coating, 336n2
Roll-to-roll production approach, 8, 142–143, 171–173, 176–177, 336n3, 337n9

Ross, John, 6, 156–157, 253, 335n17
Rubber, chemistry of, 34, 110–111, 322n37

SAFT (battery company), 342n32
Saito, Freya, 262
Samsung, 213–215, 220, 345n14
Santa Fe trip, 242–244
Sapru, Krishna, 135–137, 157, 187–189, 191, 341n14
Sarbanes-Oxley Act, 2002, 249
Schankler, Sam, 49, 309n8
Schwartz, Brian, 9, 155, 157–159
Science, role in addressing societal problems, 2–3, 7–8, 63, 96, 100–101, 106, 147–149, 294–295, 310n18
Science/scientists, mainstream, recognition/rejection by, 4, 125–128, 152–154, 233–234, 237, 291–294, 306n18, 326n18
Scrap-metal business, father's, 7, 18
Seitz, Frederick, 108
Selenium, 109
Servomechanisms, 84
Sewer socialism, 7, 10, 26, 308n3, 308n16. *See also* Workman's Circle
Shachtman, Max, 27
Shaiken, Harley
  education and career, 329n36
  hiring, 100, 318n5
  and Lagos visit to ECD, 265
  machinist work at ECD, 131–132
  on SO's achievements, 294–295
  at SO's ninetieth birthday party, 268
  on SO's trip to Chile, 265
  Yiddish school, 105
Sharp, Osaka, Japan
  contract with to produce amorphous silicon solar panels, 175
  origins, 337n14
  relationship with ECD, 345n17
  solar-powered calculator, 175, 177
  support for ECD's solar cell production, 175
  thin-film transistors, 175
Shaw, Melvin (Mel), 134, 152, 156, 210, 337n7
Shell Hydrogen, 199
Shnayerson, Michael (*The Car That Could*), 192–193
Sie, Charlie, 148, 328n26
Silicon germanium alloy production, 177
Silicon transistor, 89, 109
Silver, Marvin, 152, 210
Simon, Leon, 87
Siskind, Marvin (Marv)
  on ECD income from patent infringement lawsuits, 197
  on financial blunders by new ECD board, 353n2
  on Momoko's negotiation skills, 160–161
  on OIS relationship with Samsung, 214
  on outcome of patent dispute with Canon, 179
  on problems with Cobasys, 203
  on slow development of optical phase-change memory, 215
  work on solar panel patents, 174
Smart machines. *See* Automation
Smith, Jack, 196
Smith, Richard (Dick), 175
Smith, Robert Holbrook (Dr. Bob), 99
Society for a Sane Nuclear Policy (SANE), 103, 105, 320n22, 320n23
Society of Automotive Engineers, 314n20
Sohio. *See* Standard Oil of Ohio
Solar energy. *See* Thin-film solar panels
Solar-powered calculator, 175, 177
Solid-state physics, 108, 124, 321n28, 321n30, 321n33, 327n22
Sovlux machine, 181–183, 198
Spangenberg, Charles (Chuck), 116
Spear, Walter, 141, 331n61
Spiegel, Anita, 75–76, 78, 82, 95–96, 349–350n5
Staebler-Wronski effect, 176, 264, 332n66
Stalinism, Marxism, 26
Standard Oil of Ohio (Sohio), 175, 178, 336n1, 337n13

Stanford Roberts Machine Company, 48–52, 54–56
Stempel, Robert (Bob)
  on the buyout of Bekaert, 183
  on carbon dioxide exhaust, 341
  and the EV1 car, 194–196, 338n28
  and the hydrogen car, 201–202, 205
  investment in Ovshinsky Innovation, 262
  on loss of support from Canon, 181, 183
  partnership with SO, 342n27
  and SO's office at the Institute for Amorphous Studies, 250
Stevens, William, 124–126
Strand, David (Dave)
  on declining morale at ECD, 250
  on dismantling of OI/OS, 287
  on fun of working at ECD, 152
  leadership of group on physics of phase-change materials, 216–218
  on limitations of the cognitive computer program, 223
  work for Ovshinsky Innovation, 262–263
Stronium 90 fears, 320n24
Structures lab, ECD, 130–131
Superconductivity, 349n87
Swigert, Arthur, 71, 314n18
Symposium on Vitreous Chalcogenide Semiconductors, Leningrad, 127

TA1 and TA2 (Tandems One and Two) machines, 175
Tanaka, Hiroshi, 180
Tann Corporation
  funding for General Automation, 74
  lawsuit related to the Ovitron, 87, 314n24
Tantalum oxide film, 3, 85–86, 108, 317n40
Tatung, Taiwan, 215
Taylor, Frederick Winslow, 308n2
Teller, Edward, 154
Tellurium, 109–111, 138, 322n36
Texaco/Texaco Ovonic Battery System, 199–201
"Thermoelectric Device," 319n16

Thin-film solar panels
  and affordable technology, 319n12
  Canon support for, 179–180
  development work, 1, 4, 139–142, 151
  efficiency improvements, 263–264, 338n21
  gigawatt machine, 185, 260–264, 286, 351n7
  output power ratings, 336n3, 337n15
  resiliency, 338n35
  roll-to-roll production approach, 8, 142–143, 173
  SO's vision for, 142, 171–172, 250–251
  triple junction panels, 181
  underlying technology, 140–141, 173, 331n60, 337n7
Thin-film technologies
  nonsolar applications, 210
  semiconductors, 344n2
Thomas, Norman, 68, 106, 320n23
Threshold switches. *See also* Ovonic thin-film amorphous threshold switch
  durability, 280–281
  electronic vs. thermal mechanisms of action, debates about, 133–134
  invention of, 4
  mechanisms of action and possible applications, 222
  mobility edge, 127
  neural network applications, SO's regaining of control over, 220
  relation to memory switch, 324n55
  three-terminal threshold device, 222–223
  understanding, as early focus of ECD Physics Department, 128
  use in 3D Xpoint memory chip, 287–288, 354n10
Titanium, 40
Toyota Prius
  ECD batteries in, 197
  refitting to run on hydrogen, 201–202
Troy, Michigan, ECD facility, 118–120, 123, 324n1, 324n59
225 alloy, 216–217

The UAW and Walter Reuther (Widick and Howe), 67
Unions. *See* Civil rights, unions, and political activism
United Auto Workers (UAW), 66, 313n6
United Nuclear Corporation (UNC), 145, 333nn73–74
United Solar Systems (USSC), 176, 180–181, 183–185, 338n26, 338n33
United States Advanced Battery Consortium (USABC), 193, 341n20
"The Use of Electro-Mechanical Motion to Replace the Loss of Human Movement" (Ovshinsky), 57

Vanderkirk, Charles, 71
Venkatesan, Srinivasan (Srini)
 development of prototype NiMH battery, 188–189, 341n22
 on ECD's corporate culture, 148
 hydrogen research, 191–192, 200
Vijan, Meera, 198, 342n34, 344n6, 344n9
Visual imagination, Ovshinsky's. *See also* Analogical thinking; Intuition
 as basis for most important work, 6, 134–135, 292–295
 Einstein's, 305n14
 Flasck on, 130
 and SO's struggles with formal math, 307n14

Wagoner, Rick, 342nn30–31
Wall Street Journal, article about Ovshinsky, 124
Watkins, Ed, 71, 87, 314n17
Wayne State University
 neurophysiology studies at, 3, 84
 reaction to the Ovitron at, 89
 Rosa's position at, 262
Welded steel bases, disputes about, 311–312n29
White House roof solar panels, 341n20
Who Killed the Electric Car? (film), 193

Wicker, Guy
 cold fusion experiments, 319n17
 and the development of electrical phase-change devices, 217
 doctoral dissertation, 346–347n30
 on Lowrey, 218
 on OIS components in Samsung displays, 214
 on relationship with Intel, 220, 347n36
 replication of the Ovitron experiment, 317n42, 355n15
 revival of Ovshinsky Innovation, 288–289
Widick, Branko J. (BJ), 27, 66–67, 153, 308n17
Wiener, Norbert, 3, 56, 311nn23-24
Wilhite, Jeff, 206–207, 343n47
Wilkinson, John, 51
Will, Fritz G., 340n7
Williams, J. R. (cartoonist), "Bull of the Woods" character, 30
Wilson, Robert R., 2, 155, 327n25
Wilson, William Griffith (Bill W.), 99, 208
Worcester, Mass., Dibner family move to, 81–82
Worcester Foundation, 316n35
Workmen's Circle
 cemetery, burial of Iris in, 247
 cemetery, burial of SO in, 279–281
 cultural activities, Bund culture, 26
 English language classes, 19
 father's involvement in, 7, 24
 meetings of SO and Iris at, 79
 political activities at, 26
 purpose, 24, 233
 school at, 105, 306n15
World War II, guaranteed profits during, 40, 45
Woz, Roger, 182
Wrinkles in space-time, 235

Yang, Jeff, 150–151, 171–172, 177–180, 258, 337n7
Yang, Moshi, 213

Yaniv, Zvi, 210–215, 344n5, 344n9, 345n19
Yiddish culture, 24–26, 306n15
Yiddish school, 25, 27, 28, 105
Youdina, Irina, 258–259, 267, 270, 279–280, 281, 353n6, 353n9
Young, Rosa. *See* Ovshinsky, Rosa Young (Mrs. Stanford)
Young Peoples' Socialist League, 7, 26
Yukawa, Hideki (*Creativity and Intuition: A Physicist Looks at East and West*), 234, 348n4

Zallen, Richard, 154, 156, 293, 321n29, 327nn22–23
Zero Emission Vehicle mandate (California), 193, 196–197

Mount Laurel Library
100 Walt Whitman Avenue
Mount Laurel, NJ 08054-9539
856-234-7319
www.mountlaurellibrary.org